Springer Handbook
of Crystal Growth

Govindhan Dhanaraj, Kullaiah Byrappa,
Vishwanath Prasad, Michael Dudley (Eds.)

Springer Handbook of Crystal Growth
Organization of the Handbook

Part A Fundamentals of Crystal Growth and Defect Formation
1 Crystal Growth Techniques and Characterization: An Overview
2 Nucleation at Surfaces
3 Morphology of Crystals Grown from Solutions
4 Generation and Propagation of Defects During Crystal Growth
5 Single Crystals Grown Under Unconstrained Conditions
6 Defect Formation During Crystal Growth from the Melt

Part B Crystal Growth from Melt Techniques
7 Indium Phosphide: Crystal Growth and Defect Control by Applying Steady Magnetic Fields
8 Czochralski Silicon Single Crystals for Semiconductor and Solar Cell Applications
9 Czochralski Growth of Oxide Photorefractive Crystals
10 Bulk Crystal Growth of Ternary III–V Semiconductors
11 Growth and Characterization of Antimony-Based Narrow-Bandgap III–V Semiconductor Crystals for Infrared Detector Applications
12 Crystal Growth of Oxides by Optical Floating Zone Technique
13 Laser-Heated Pedestal Growth of Oxide Fibers
14 Synthesis of Refractory Materials by Skull Melting Technique
15 Crystal Growth of Laser Host Fluorides and Oxides
16 Shaped Crystal Growth

Part C Solution Growth of Crystals
17 Bulk Single Crystals Grown from Solution on Earth and in Microgravity
18 Hydrothermal Growth of Polyscale Crystals
19 Hydrothermal and Ammonothermal Growth of ZnO and GaN
20 Stoichiometry and Domain Structure of KTP-Type Nonlinear Optical Crystals
21 High-Temperature Solution Growth: Application to Laser and Nonlinear Optical Crystals
22 Growth and Characterization of KDP and Its Analogs

Part D Crystal Growth from Vapor
23 Growth and Characterization of Silicon Carbide Crystals
24 AlN Bulk Crystal Growth by Physical Vapor Transport
25 Growth of Single-Crystal Organic Semiconductors
26 Growth of III–Nitrides with Halide Vapor Phase Epitaxy (HVPE)
27 Growth of Semiconductor Single Crystals from Vapor Phase

Part E Epitaxial Growth and Thin Films

28 Epitaxial Growth of Silicon Carbide by Chemical Vapor Deposition
29 Liquid-Phase Electroepitaxy of Semiconductors
30 Epitaxial Lateral Overgrowth of Semiconductors
31 Liquid-Phase Epitaxy of Advanced Materials
32 Molecular-Beam Epitaxial Growth of HgCdTe
33 Metalorganic Vapor-Phase Epitaxy of Diluted Nitrides and Arsenide Quantum Dots
34 Formation of SiGe Heterostructures and Their Properties
35 Plasma Energetics in Pulsed Laser and Pulsed Electron Deposition

Part F Modeling in Crystal Growth and Defects

36 Convection and Control in Melt Growth of Bulk Crystals
37 Vapor Growth of III Nitrides
38 Continuum-Scale Quantitative Defect Dynamics in Growing Czochralski Silicon Crystals
39 Models for Stress and Dislocation Generation in Melt Based Compound Crystal Growth
40 Mass and Heat Transport in BS and EFG Systems

Part G Defects Characterization and Techniques

41 Crystalline Layer Structures with X-Ray Diffractometry
42 X-Ray Topography Techniques for Defect Characterization of Crystals
43 Defect-Selective Etching of Semiconductors
44 Transmission Electron Microscopy Characterization of Crystals
45 Electron Paramagnetic Resonance Characterization of Point Defects
46 Defect Characterization in Semiconductors with Positron Annihilation Spectroscopy

Part H Special Topics in Crystal Growth

47 Protein Crystal Growth Methods
48 Crystallization from Gels
49 Crystal Growth and Ion Exchange in Titanium Silicates
50 Single-Crystal Scintillation Materials
51 Silicon Solar Cells: Materials, Devices, and Manufacturing
52 Wafer Manufacturing and Slicing Using Wiresaw

Subject Index

使用说明

1. 《晶体生长手册》原版为一册,分为A~H部分。考虑到使用方便以及内容一致,影印版分为6册:第1册——Part A,第2册——Part B,第3册——Part C,第4册——Part D、E,第5册——Part F、G,第6册——Part H。

2. 各册在页脚重新编排页码,该页码对应中文目录。保留了原书页眉及页码,其页码对应原书目录及主题索引。

3. 各册均给出完整6册书的章目录。

4. 作者及其联系方式、缩略语表各册均完整呈现。

5. 主题索引安排在第6册。

6. 文前介绍基本采用中英文对照形式,方便读者快速浏览。

材料科学与工程图书工作室

联系电话　0451-86412421
　　　　　0451-86414559
邮　　箱　yh_bj@yahoo.com.cn
　　　　　xuyaying81823@gmail.com
　　　　　zhxh6414559@yahoo.com.cn

Springer 手册精选系列

晶体生长手册

晶体生长专题

【第6册】

Springer
Handbook of

Crystal

Growth

〔美〕Govindhan Dhanaraj 等主编

（影印版）

哈尔滨工业大学出版社
HARBIN INSTITUTE OF TECHNOLOGY PRESS

黑版贸审字08-2012-047号

Reprint from English language edition:
Springer Handbook of Crystal Growth
by Govindhan Dhanaraj, Kullaiah Byrappa, Vishwanath Prasad
and Michael Dudley
Copyright © 2010 Springer Berlin Heidelberg
Springer Berlin Heidelberg is a part of Springer Science+Business Media
All Rights Reserved

This reprint has been authorized by Springer Science & Business Media for distribution in China Mainland only and not for export there from.

图书在版编目（CIP）数据

晶体生长手册. 6, 晶体生长专题 = Handbook of Crystal Growth. 6, Special Topics in Crystal Growth：英文 /（美）德哈纳拉等主编. —影印本. —哈尔滨：哈尔滨工业大学出版社, 2013.1

（Springer手册精选系列）
ISBN 978-7-5603-3871-2

Ⅰ.①晶… Ⅱ.①德… Ⅲ.①晶体生长－手册－英文 Ⅳ.①O78-62

中国版本图书馆CIP数据核字(2012)第292357号

责任编辑	杨　桦　许雅莹　张秀华
出版发行	哈尔滨工业大学出版社
社　　址	哈尔滨市南岗区复华四道街10号 邮编 150006
传　　真	0451-86414749
网　　址	http://hitpress.hit.edu.cn
印　　刷	哈尔滨市石桥印务有限公司
开　　本	787mm×960mm 1/16 印张 13.75
版　　次	2013年1月第1版　2013年1月第1次印刷
书　　号	ISBN 978-7-5603-3871-2
定　　价	38.00元

（如因印刷质量问题影响阅读，我社负责调换）

序言

多年以来,有很多探索研究已经成功地描述了晶体生长的生长工艺和科学,有许多文章、专著、会议文集和手册对这一领域的前沿成果做了综合评述。这些出版物反映了人们对体材料晶体和薄膜晶体的兴趣日益增长,这是由于它们的电子、光学、机械、微结构以及不同的科学和技术应用引起的。实际上,大部分半导体和光器件的现代成果,如果没有基本的、二元的、三元的及其他不同特性和大尺寸的化合物晶体的发展则是不可能的。这些文章致力于生长机制的基本理解、缺陷形成、生长工艺和生长系统的设计,因此数量是庞大的。

本手册针对目前备受关注的体材料晶体和薄膜晶体的生长技术水平进行阐述。我们的目的是使读者了解经常使用的生长工艺、材料生产和缺陷产生的基本知识。为完成这一任务,我们精选了50多位顶尖科学家、学者和工程师,他们的合作者来自于22个不同国家。这些作者根据他们的专业所长,编写了关于晶体生长和缺陷形成共计52章内容:从熔体、溶液到气相体材料生长;外延生长;生长工艺和缺陷的模型;缺陷特性的技术以及一些现代的特别课题。

本手册分为七部分。Part A介绍基础理论:生长和表征技术综述,表面成核工艺,溶液生长晶体的形态,生长过程中成核的层错,缺陷形成的形态。

Part B介绍体材料晶体的熔体生长,一种生长大尺寸晶体的关键方法。这一部分阐述了直拉单晶工艺、泡生法、布里兹曼法、浮区熔融等工艺,以及这些方法的最新进展,例如应用磁场的晶体生长、生长轴的取向、增加底基和形状控制。本部分涉及材料从硅和Ⅲ-Ⅴ族化合物到氧化物和氟化物的广泛内容。

第三部分,本书的Part C关注了溶液生长法。在前两章里讨论了水热生长法的不同方面,随后的三章介绍了非线性和激光晶体、KTP和KDP。通过在地球上和微重力环境下生长的比较给出了重力对溶液生长法的影响的知识。

Part D的主题是气相生长。这一部分提供了碳化硅、氮化镓、氮化铝和有机半导体的气相生长的内容。随后的Part E是关于外延生长和薄膜的,主要包括从液相的化学气相淀积到脉冲激光和脉冲电子淀积。

Part F介绍了生长工艺和缺陷形成的模型。这些章节验证了工艺参数和产生晶体质量问题包括缺陷形成的直接相互作用关系。随后的Part G展示了结晶材料特性和分析的发展。Part F和G说明了预测工具和分析技术在帮助高质量的大尺寸晶体生长工艺的设计和控制方面是非常好用的。

最后的Part H致力于精选这一领域的部分现代课题,例如蛋白质晶体生长、凝胶结晶、原位结构、单晶闪烁材料的生长、光电材料和线切割大晶体薄膜。

我们希望这本施普林格手册对那些学习晶体生长的研究生,那些从事或即将从事这一领域研究的来自学术界和工业领域的研究人员、科学家和工程师以及那些制备晶体的人是有帮助的。

我们对施普林格的Dr. Claus Acheron,Dr. Werner Skolaut和le-tex的Ms Anne Strobach的特别努力表示真诚的感谢,没有他们本书将无法呈现。

我们感谢我们的作者编写了详尽的章节内容和在本书出版期间对我们的耐心。一位编者(GD)感谢他的家庭成员和Dr. Kedar Gupta(ARC Energy 的CEO),感谢他们在本书编写期间的大力支持和鼓励。还对Peter Rudolf, David Bliss, Ishwara Bhat和Partha Dutta在A、B、E部分的编写中所给予的帮助表示感谢。

Nashua, New Hampshire, April 2010	G. Dhanaraj
Mysore, India	K. Byrappa
Denton, Texas	V. Prasad
Stony Brook, New York	M. Dudley

Preface

Over the years, many successful attempts have been made to describe the art and science of crystal growth, and many review articles, monographs, symposium volumes, and handbooks have been published to present comprehensive reviews of the advances made in this field. These publications are testament to the growing interest in both bulk and thin-film crystals because of their electronic, optical, mechanical, microstructural, and other properties, and their diverse scientific and technological applications. Indeed, most modern advances in semiconductor and optical devices would not have been possible without the development of many elemental, binary, ternary, and other compound crystals of varying properties and large sizes. The literature devoted to basic understanding of growth mechanisms, defect formation, and growth processes as well as the design of growth systems is therefore vast.

The objective of this Springer Handbook is to present the state of the art of selected topical areas of both bulk and thin-film crystal growth. Our goal is to make readers understand the basics of the commonly employed growth processes, materials produced, and defects generated. To accomplish this, we have selected more than 50 leading scientists, researchers, and engineers, and their many collaborators from 22 different countries, to write chapters on the topics of their expertise. These authors have written 52 chapters on the fundamentals of crystal growth and defect formation; bulk growth from the melt, solution, and vapor; epitaxial growth; modeling of growth processes and defects; and techniques of defect characterization, as well as some contemporary special topics.

This Springer Handbook is divided into seven parts. Part A presents the fundamentals: an overview of the growth and characterization techniques, followed by the state of the art of nucleation at surfaces, morphology of crystals grown from solutions, nucleation of dislocation during growth, and defect formation and morphology.

Part B is devoted to bulk growth from the melt, a method critical to producing large-size crystals. The chapters in this part describe the well-known processes such as Czochralski, Kyropoulos, Bridgman, and floating zone, and focus specifically on recent advances in improving these methodologies such as application of magnetic fields, orientation of the growth axis, introduction of a pedestal, and shaped growth. They also cover a wide range of materials from silicon and III–V compounds to oxides and fluorides.

The third part, Part C of the book, focuses on solution growth. The various aspects of hydrothermal growth are discussed in two chapters, while three other chapters present an overview of the nonlinear and laser crystals, KTP and KDP. The knowledge on the effect of gravity on solution growth is presented through a comparison of growth on Earth versus in a microgravity environment.

The topic of Part D is vapor growth. In addition to presenting an overview of vapor growth, this part also provides details on vapor growth of silicon carbide, gallium nitride, aluminum nitride, and organic semiconductors. This is followed by chapters on epitaxial growth and thin films in Part E. The topics range from chemical vapor deposition to liquid-phase epitaxy to pulsed laser and pulsed electron deposition.

Modeling of both growth processes and defect formation is presented in Part F. These chapters demonstrate the direct correlation between the process parameters and quality of the crystal produced, including the formation of defects. The subsequent Part G presents the techniques that have been developed for crystalline material characterization and analysis. The chapters in Parts F and G demonstrate how well predictive tools and analytical techniques have helped the design and control of growth processes for better-quality crystals of large sizes.

The final Part H is devoted to some selected contemporary topics in this field, such as protein crystal growth, crystallization from gels, in situ structural studies, growth of single-crystal scintillation materials, photovoltaic materials, and wire-saw slicing of large crystals to produce wafers.

We hope this Springer Handbook will be useful to graduate students studying crystal growth and to researchers, scientists, and engineers from academia and industry who are conducting or intend to conduct research in this field as well as those who grow crystals.

We would like to express our sincere thanks to Dr. Claus Acheron and Dr. Werner Skolaut of Springer and Ms Anne Strohbach of le-tex for their extraordinary efforts without which this handbook would not have taken its final shape.

We thank our authors for writing comprehensive chapters and having patience with us during the publication of this Handbook. One of the editors (GD) would like to thank his family members and Dr. Kedar Gupta (CEO of ARC Energy) for their generous support and encouragement during the entire course of editing this handbook. Acknowledgements are also due to Peter Rudolf, David Bliss, Ishwara Bhat, and Partha Dutta for their help in editing Parts A, B, E, and H, respectively.

Nashua, New Hampshire, April 2010	G. Dhanaraj
Mysore, India	K. Byrappa
Denton, Texas	V. Prasad
Stony Brook, New York	M. Dudley

About the Editors

Govindhan Dhanaraj is the Manager of Crystal Growth Technologies at Advanced Renewable Energy Company (ARC Energy) at Nashua, New Hampshire (USA) focusing on the growth of large size sapphire crystals for LED lighting applications, characterization and related crystal growth furnace development. He received his PhD from the Indian Institute of Science, Bangalore and his Master of Science from Anna University (India). Immediately after his doctoral degree, Dr. Dhanaraj joined a National Laboratory, presently known as Rajaramanna Center for Advanced Technology in India, where he established an advanced Crystal Growth Laboratory for the growth of optical and laser crystals. Prior to joining ARC Energy, Dr. Dhanaraj served as a Research Professor at the Department of Materials Science and Engineering, Stony Brook University, NY, and also held a position of Research Assistant Professor at Hampton University, VA. During his 25 years of focused expertise in crystal growth research, he has developed optical, laser and semiconductor bulk crystals and SiC epitaxial films using solution, flux, Czochralski, Bridgeman, gel and vapor methods, and characterized them using x-ray topography, synchrotron topography, chemical etching and optical and atomic force microscopic techniques. He co-organized a symposium on Industrial Crystal Growth under the 17th American Conference on Crystal Growth and Epitaxy in conjunction with the 14th US Biennial Workshop on Organometallic Vapor Phase Epitaxy held at Lake Geneva, WI in 2009. Dr. Dhanaraj has delivered invited lectures and also served as session chairman in many crystal growth and materials science meetings. He has published over 100 papers and his research articles have attracted over 250 rich citations.

Kullaiah Byrappa received his Doctor's degree in Crystal Growth from the Moscow State University, Moscow in 1981. He is Professor of Materials Science, Head of the Crystal Growth Laboratory, and Director of the Internal Quality Assurance Cell of the University of Mysore, India. His current research is in crystal engineering of polyscale materials through novel solution processing routes, particularly covering hydrothermal, solvothermal and supercritical methods. Professor Byrappa has co-authored the Handbook of Hydrothermal Technology, and edited 4 books as well as two special editions of Journal of Materials Science, and published 180 research papers including 26 invited reviews and book chapters on various aspects of novel routes of solution processing. Professor Byrappa has delivered over 60 keynote and invited lectures at International Conferences, and several hundreds of colloquia and seminars at various institutions around the world. He has also served as chair and co-chair for numerous international conferences. He is a Fellow of the World Academy of Ceramics. Professor Byrappa is serving in several international committees and commissions related to crystallography, crystal growth, and materials science. He is the Founder Secretary of the International Solvothermal and Hydrothermal Association. Professor Byrappa is a recipient of several awards such as the Sir C.V. Raman Award, Materials Research Society of India Medal, and the Golden Jubilee Award of the University of Mysore.

Vishwanath "Vish" Prasad is the Vice President for Research and Economic Development and Professor of Mechanical and Energy Engineering at the University of North Texas (UNT), one of the largest university in the state of Texas. He received his PhD from the University of Delaware (USA), his Masters of Technology from the Indian Institute of Technology, Kanpur, and his bachelor's from Patna University in India all in Mechanical Engineering. Prior to joining UNT in 2007, Dr. Prasad served as the Dean at Florida International University (FIU) in Miami, where he also held the position of Distinguished Professor of Engineering. Previously, he has served as a Leading Professor of Mechanical Engineering at Stony Brook University, New York, as an Associate Professor and Assistant Professor at Columbia University. He has received many special recognitions for his contributions to engineering education. Dr. Prasad's research interests include thermo-fluid sciences, energy systems, electronic materials, and computational materials processing. He has published over 200 articles, edited/co-edited several books and organized numerous conferences, symposia, and workshops. He serves as the lead editor of the Annual Review of Heat Transfer. In the past, he has served as an Associate Editor of the ASME Journal of Heat. Dr. Prasad is an elected Fellow of the American Society of Mechanical Engineers (ASME), and has served as a member of the USRA Microgravity Research Council. Dr. Prasad's research has focused on bulk growth of silicon, III-V compounds, and silicon carbide; growth of large diameter Si tube; design of crystal growth systems; and sputtering and chemical vapor deposition of thin films. He is also credited to initiate research on wire saw cutting of large crystals to produce wafers with much reduced material loss. Dr. Prasad's research has been well funded by US National Science Foundation (NSF), US Department of Defense, US Department of Energy, and industry.

Michael Dudley received his Doctoral Degree in Engineering from Warwick University, UK, in 1982. He is Professor and Chair of the Materials Science and Engineering Department at Stony Brook University, New York, USA. He is director of the Stony Brook Synchrotron Topography Facility at the National Synchrotron Light Source at Brookhaven National Laboratory, Upton New York. His current research focuses on crystal growth and characterization of defect structures in single crystals with a view to determining their origins. The primary technique used is synchrotron topography which enables analysis of defects and generalized strain fields in single crystals in general, with particular emphasis on semiconductor, optoelectronic, and optical crystals. Establishing the relationship between crystal growth conditions and resulting defect distributions is a particular thrust area of interest to Dudley, as is the correlation between electronic/optoelectronic device performance and defect distribution. Other techniques routinely used in such analysis include transmission electron microscopy, high resolution triple-axis x-ray diffraction, atomic force microscopy, scanning electron microscopy, Nomarski optical microscopy, conventional optical microscopy, IR microscopy and fluorescent laser scanning confocal microscopy. Dudley's group has played a prominent role in the development of SiC and AlN growth, characterizing crystals grown by many of the academic and commercial entities involved enabling optimization of crystal quality. He has co-authored some 315 refereed articles and 12 book chapters, and has edited 5 books. He is currently a member of the Editorial Board of Journal of Applied Physics and Applied Physics Letters and has served as Chair or Co-Chair for numerous international conferences.

List of Authors

Francesco Abbona
Università degli Studi di Torino
Dipartimento di Scienze Mineralogiche
e Petrologiche
via Valperga Caluso 35
10125 Torino, Italy
e-mail: *francesco.abbona@unito.it*

Mohan D. Aggarwal
Alabama A&M University
Department of Physics
Normal, AL 35762, USA
e-mail: *mohan.aggarwal@aamu.edu*

Marcello R.B. Andreeta
University of São Paulo
Crystal Growth and Ceramic Materials Laboratory,
Institute of Physics of São Carlos
Av. Trabalhador Sãocarlense, 400
São Carlos, SP 13560-970, Brazil
e-mail: *marcello@if.sc.usp.br*

Dino Aquilano
Università degli Studi di Torino
Facoltà di Scienze Matematiche, Fisiche e Naturali
via P. Giuria, 15
Torino, 10126, Italy
e-mail: *dino.aquilano@unito.it*

Roberto Arreguín-Espinosa
Universidad Nacional Autónoma de México
Instituto de Química
Circuito Exterior, C.U. s/n
Mexico City, 04510, Mexico
e-mail: *arrespin@unam.mx*

Jie Bai
Intel Corporation
RA3-402, 5200 NE Elam Young Parkway
Hillsboro, OR 97124-6497, USA
e-mail: *jie.bai@intel.com*

Stefan Balint
West University of Timisoara
Department of Computer Science
Blvd. V. Parvan 4
Timisoara, 300223, Romania
e-mail: *balint@math.uvt.ro*

Ashok K. Batra
Alabama A&M University
Department of Physics
4900 Meridian Street
Normal, AL 35762, USA
e-mail: *ashok.batra@aamu.edu*

Handady L. Bhat
Indian Institute of Science
Department of Physics
CV Raman Avenue
Bangalore, 560012, India
e-mail: *hlbhat@physics.iisc.ernet.in*

Ishwara B. Bhat
Rensselaer Polytechnic Institute
Electrical Computer
and Systems Engineering Department
110 8th Street, JEC 6031
Troy, NY 12180, USA
e-mail: *bhati@rpi.edu*

David F. Bliss
US Air Force Research Laboratory
Sensors Directorate Optoelectronic Technology
Branch
80 Scott Drive
Hanscom AFB, MA 01731, USA
e-mail: *david.bliss@hanscom.af.mil*

Mikhail A. Borik
Russian Academy of Sciences
Laser Materials and Technology Research Center,
A.M. Prokhorov General Physics Institute
Vavilov 38
Moscow, 119991, Russia
e-mail: *borik@lst.gpi.ru*

Liliana Braescu
West University of Timisoara
Department of Computer Science
Blvd. V. Parvan 4
Timisoara, 300223, Romania
e-mail: lilianabraescu@balint1.math.uvt.ro

Kullaiah Byrappa
University of Mysore
Department of Geology
Manasagangotri
Mysore, 570 006, India
e-mail: kbyrappa@gmail.com

Dang Cai
CVD Equipment Corporation
1860 Smithtown Ave.
Ronkonkoma, NY 11779, USA
e-mail: dcai@cvdequipment.com

Michael J. Callahan
GreenTech Solutions
92 Old Pine Drive
Hanson, MA 02341, USA
e-mail: mjcal37@yahoo.com

Joan J. Carvajal
Universitat Rovira i Virgili (URV)
Department of Physics and Crystallography
of Materials and Nanomaterials (FiCMA–FiCNA)
Campus Sescelades, C/ Marcel·lí Domingo, s/n
Tarragona 43007, Spain
e-mail: joanjosep.carvajal@urv.cat

Aaron J. Celestian
Western Kentucky University
Department of Geography and Geology
1906 College Heights Blvd.
Bowling Green, KY 42101, USA
e-mail: aaron.celestian@wku.edu

Qi-Sheng Chen
Chinese Academy of Sciences
Institute of Mechanics
15 Bei Si Huan Xi Road
Beijing, 100190, China
e-mail: qschen@imech.ac.cn

Chunhui Chung
Stony Brook University
Department of Mechanical Engineering
Stony Brook, NY 11794-2300, USA
e-mail: chuchung@ic.sunysb.edu

Ted Ciszek
Geolite/Siliconsultant
31843 Miwok Trl.
Evergreen, CO 80437, USA
e-mail: ted_ciszek@siliconsultant.com

Abraham Clearfield
Texas A&M University
Distinguished Professor of Chemistry
College Station, TX 77843-3255, USA
e-mail: clearfield@chem.tamu.edu

Hanna A. Dabkowska
Brockhouse Institute for Materials Research
Department of Physics and Astronomy
1280 Main Str W.
Hamilton, Ontario L8S 4M1, Canada
e-mail: dabkoh@mcmaster.ca

Antoni B. Dabkowski
McMaster University, BIMR
Brockhouse Institute for Materials Research,
Department of Physics and Astronomy
1280 Main Str W.
Hamilton, Ontario L8S 4M1, Canada
e-mail: dabko@mcmaster.ca

Rafael Dalmau
HexaTech Inc.
991 Aviation Pkwy Ste 800
Morrisville, NC 27560, USA
e-mail: rdalmau@hexatechinc.com

Govindhan Dhanaraj
ARC Energy
18 Celina Avenue, Unit 77
Nashua, NH 03063, USA
e-mail: dhanaraj@arc-energy.com

Ramasamy Dhanasekaran
Anna University Chennai
Crystal Growth Centre
Chennai, 600 025, India
e-mail: rdhanasekaran@annauniv.edu;
rdcgc@yahoo.com

Ernesto Diéguez
Universidad Autónoma de Madrid
Department Física de Materiales
Madrid 28049, Spain
e-mail: *ernesto.dieguez@uam.es*

Vijay K. Dixit
Raja Ramanna Center for Advance Technology
Semiconductor Laser Section,
Solid State Laser Division
Rajendra Nagar, RRCAT.
Indore, 452013, India
e-mail: *dixit@rrcat.gov.in*

Sadik Dost
University of Victoria
Crystal Growth Laboratory
Victoria, BC V8W 3P6, Canada
e-mail: *sdost@me.uvic.ca*

Michael Dudley
Stony Brook University
Department of Materials Science and Engineering
Stony Brook, NY 11794-2275, USA
e-mail: *mdudley@notes.cc.sunysb.edu*

Partha S. Dutta
Rensselaer Polytechnic Institute
Department of Electrical, Computer
and Systems Engineering
110 Eighth Street
Troy, NY 12180, USA
e-mail: *duttap@rpi.edu*

Francesc Díaz
Universitat Rovira i Virgili (URV)
Department of Physics and Crystallography
of Materials and Nanomaterials (FiCMA-FiCNA)
Campus Sescelades, C/ Marcel·lí Domingo, s/n
Tarragona 43007, Spain
e-mail: *f.diaz@urv.cat*

Paul F. Fewster
PANalytical Research Centre,
The Sussex Innovation Centre
Research Department
Falmer
Brighton, BN1 9SB, UK
e-mail: *paul.fewster@panalytical.com*

Donald O. Frazier
NASA Marshall Space Flight Center
Engineering Technology Management Office
Huntsville, AL 35812, USA
e-mail: *donald.o.frazier@nasa.gov*

James W. Garland
EPIR Technologies, Inc.
509 Territorial Drive, Ste. B
Bolingbrook, IL 60440, USA
e-mail: *jgarland@epir.com*

Thomas F. George
University of Missouri-St. Louis
Center for Nanoscience,
Department of Chemistry and Biochemistry,
Department of Physics and Astronomy
One University Boulevard
St. Louis, MO 63121, USA
e-mail: *tfgeorge@umsl.edu*

Andrea E. Gutiérrez-Quezada
Universidad Nacional Autónoma de México
Instituto de Química
Circuito Exterior, C.U. s/n
Mexico City, 04510, Mexico
e-mail: *30111390@escolar.unam.mx*

Carl Hemmingsson
Linköping University
Department of Physics, Chemistry
and Biology (IFM)
581 83 Linköping, Sweden
e-mail: *cah@ifm.liu.se*

Antonio Carlos Hernandes
University of São Paulo
Crystal Growth and Ceramic Materials Laboratory,
Institute of Physics of São Carlos
Av. Trabalhador Sãocarlense
São Carlos, SP 13560-970, Brazil
e-mail: *hernandes@if.sc.usp.br*

Koichi Kakimoto
Kyushu University
Research Institute for Applied Mechanics
6-1 Kasuga-kouen, Kasuga
816-8580 Fukuoka, Japan
e-mail: *kakimoto@riam.kyushu-u.ac.jp*

Imin Kao
State University of New York at Stony Brook
Department of Mechanical Engineering
Stony Brook, NY 11794-2300, USA
e-mail: imin.kao@stonybrook.edu

John J. Kelly
Utrecht University,
Debye Institute for Nanomaterials Science
Department of Chemistry
Princetonplein 5
3584 CC, Utrecht, The Netherlands
e-mail: j.j.kelly@uu.nl

Jeonggoo Kim
Neocera, LLC
10000 Virginia Manor Road #300
Beltsville, MD, USA
e-mail: kim@neocera.com

Helmut Klapper
Institut für Kristallographie
RWTH Aachen University
Aachen, Germany
e-mail: klapper@xtal.rwth-aachen.de;
helmut-klapper@web.de

Christine F. Klemenz Rivenbark
Krystal Engineering LLC
General Manager and Technical Director
1429 Chaffee Drive
Titusville, FL 32780, USA
e-mail: ckr@krystalengineering.com

Christian Kloc
Nanyang Technological University
School of Materials Science and Engineering
50 Nanyang Avenue
639798 Singapore
e-mail: ckloc@ntu.edu.sg

Solomon H. Kolagani
Neocera LLC
10000 Virginia Manor Road
Beltsville, MD 20705, USA
e-mail: harsh@neocera.com

Akinori Koukitu
Tokyo University of Agriculture and Technology
(TUAT)
Department of Applied Chemistry
2-24-16 Naka-cho, Koganei
184-8588 Tokyo, Japan
e-mail: koukitu@cc.tuat.ac.jp

Milind S. Kulkarni
MEMC Electronic Materials
Polysilicon and Quantitative Silicon Research
501 Pearl Drive
St. Peters, MO 63376, USA
e-mail: mkulkarni@memc.com

Yoshinao Kumagai
Tokyo University of Agriculture and Technology
Department of Applied Chemistry
2-24-16 Naka-cho, Koganei
184-8588 Tokyo, Japan
e-mail: 4470kuma@cc.tuat.ac.jp

Valentin V. Laguta
Institute of Physics of the ASCR
Department of Optical Materials
Cukrovarnicka 10
Prague, 162 53, Czech Republic
e-mail: laguta@fzu.cz

Ravindra B. Lal
Alabama Agricultural and Mechanical University
Physics Department
4900 Meridian Street
Normal, AL 35763, USA
e-mail: rblal@comcast.net

Chung-Wen Lan
National Taiwan University
Department of Chemical Engineering
No. 1, Sec. 4, Roosevelt Rd.
Taipei, 106, Taiwan
e-mail: cwlan@ntu.edu.tw

Hongjun Li
Chinese Academy of Sciences
R & D Center of Synthetic Crystals,
Shanghai Institute of Ceramics
215 Chengbei Rd., Jiading District
Shanghai, 201800, China
e-mail: lh_li@mail.sic.ac.cn

Elena E. Lomonova
Russian Academy of Sciences
Laser Materials and Technology Research Center,
A.M. Prokhorov General Physics Institute
Vavilov 38
Moscow, 119991, Russia
e-mail: lomonova@lst.gpi.ru

Ivan V. Markov
Bulgarian Academy of Sciences
Institute of Physical Chemistry
Sofia, 1113, Bulgaria
e-mail: imarkov@ipc.bas.bg

Bo Monemar
Linköping University
Department of Physics, Chemistry and Biology
58183 Linköping, Sweden
e-mail: bom@ifm.liu.se

Abel Moreno
Universidad Nacional Autónoma de México
Instituto de Química
Circuito Exterior, C.U. s/n
Mexico City, 04510, Mexico
e-mail: carcamo@unam.mx

Roosevelt Moreno Rodriguez
State University of New York at Stony Brook
Department of Mechanical Engineering
Stony Brook, NY 11794-2300, USA
e-mail: roosevelt@dove.eng.sunysb.edu

S. Narayana Kalkura
Anna University Chennai
Crystal Growth Centre
Sardar Patel Road
Chennai, 600025, India
e-mail: kalkura@annauniv.edu

Mohan Narayanan
Reliance Industries Limited
1, Rich Branch court
Gaithersburg, MD 20878, USA
e-mail: mohan.narayanan@ril.com

Subramanian Natarajan
Madurai Kamaraj University
School of Physics
Palkalai Nagar
Madurai, India
e-mail: s_natarajan50@yahoo.com

Martin Nikl
Academy of Sciences of the Czech Republic (ASCR)
Department of Optical Crystals, Institute of Physics
Cukrovarnicka 10
Prague, 162 53, Czech Republic
e-mail: nikl@fzu.cz

Vyacheslav V. Osiko
Russian Academy of Sciences
Laser Materials and Technology Research Center,
A.M. Prokhorov General Physics Institute
Vavilov 38
Moscow, 119991, Russia
e-mail: osiko@lst.gpi.ru

John B. Parise
Stony Brook University
Chemistry Department
and Department of Geosciences
ESS Building
Stony Brook, NY 11794-2100, USA
e-mail: john.parise@stonybrook.edu

Srinivas Pendurti
ASE Technologies Inc.
11499, Chester Road
Cincinnati, OH 45246, USA
e-mail: spendurti@asetech.com

Benjamin G. Penn
NASA/George C. Marshall Space Flight Center
ISHM and Sensors Branch
Huntsville, AL 35812, USA
e-mail: benjamin.g.penndr@nasa.gov

Jens Pflaum
Julius-Maximilians Universität Würzburg
Institute of Experimental Physics VI
Am Hubland
97078 Würzburg, Germany
e-mail: jpflaum@physik.uni-wuerzburg.de

Jose Luis Plaza
Universidad Autónoma de Madrid
Facultad de Ciencias,
Departamento de Física de Materiales
Madrid 28049, Spain
e-mail: *joseluis.plaza@uam.es*

Udo W. Pohl
Technische Universität Berlin
Institut für Festkörperphysik EW5-1
Hardenbergstr. 36
10623 Berlin, Germany
e-mail: *pohl@physik.tu-berlin.de*

Vishwanath (Vish) Prasad
University of North Texas
1155 Union Circle
Denton, TX 76203-5017, USA
e-mail: *vish.prasad@unt.edu*

Maria Cinta Pujol
Universitat Rovira i Virgili
Department of Physics and Crystallography
of Materials and Nanomaterials (FiCMA-FiCNA)
Campus Sescelades, C/ Marcel·lí Domingo
Tarragona 43007, Spain
e-mail: *mariacinta.pujol@urv.cat*

Balaji Raghothamachar
Stony Brook University
Department of Materials Science and Engineering
310 Engineering Building
Stony Brook, NY 11794-2275, USA
e-mail: *braghoth@notes.cc.sunysb.edu*

Michael Roth
The Hebrew University of Jerusalem
Department of Applied Physics
Bergman Bld., Rm 206, Givat Ram Campus
Jerusalem 91904, Israel
e-mail: *mroth@vms.huji.ac.il*

Peter Rudolph
Leibniz Institute for Crystal Growth
Technology Development
Max-Born-Str. 2
Berlin, 12489, Germany
e-mail: *rudolph@ikz-berlin.de*

Akira Sakai
Osaka University
Department of Systems Innovation
1-3 Machikaneyama-cho, Toyonaka-shi
560-8531 Osaka, Japan
e-mail: *sakai@ee.es.osaka-u.ac.jp*

Yasuhiro Shiraki
Tokyo City University
Advanced Research Laboratories,
Musashi Institute of Technology
8-15-1 Todoroki, Setagaya-ku
158-0082 Tokyo, Japan
e-mail: *yshiraki@tcu.ac.jp*

Theo Siegrist
Florida State University
Department of Chemical
and Biomedical Engineering
2525 Pottsdamer Street
Tallahassee, FL 32310, USA
e-mail: *siegrist@eng.fsu.edu*

Zlatko Sitar
North Carolina State University
Materials Science and Engineering
1001 Capability Dr.
Raleigh, NC 27695, USA
e-mail: *sitar@ncsu.edu*

Sivalingam Sivananthan
University of Illinois at Chicago
Department of Physics
845 W. Taylor St. M/C 273
Chicago, IL 60607-7059, USA
e-mail: *siva@uic.edu; siva@epir.com*

Mikhail D. Strikovski
Neocera LLC
10000 Virginia Manor Road, suite 300
Beltsville, MD 20705, USA
e-mail: *strikovski@neocera.com*

Xun Sun
Shandong University
Institute of Crystal Materials
Shanda Road
Jinan, 250100, China
e-mail: *sunxun@icm.sdu.edu.cn*

Ichiro Sunagawa
University Tohoku University (Emeritus)
Kashiwa-cho 3-54-2, Tachikawa
Tokyo, 190-0004, Japan
e-mail: i.sunagawa@nifty.com

Xu-Tang Tao
Shandong University
State Key Laboratory of Crystal Materials
Shanda Nanlu 27, 250100
Jinan, China
e-mail: txt@sdu.edu.cn

Vitali A. Tatartchenko
Saint – Gobain, 23 Rue Louis Pouey
92800 Puteaux, France
e-mail: vitali.tatartchenko@orange.fr

Filip Tuomisto
Helsinki University of Technology
Department of Applied Physics
Otakaari 1 M
Espoo TKK 02015, Finland
e-mail: filip.tuomisto@tkk.fi

Anna Vedda
University of Milano-Bicocca
Department of Materials Science
Via Cozzi 53
20125 Milano, Italy
e-mail: anna.vedda@unimib.it

Lu-Min Wang
University of Michigan
Department of Nuclear Engineering
and Radiological Sciences
2355 Bonisteel Blvd.
Ann Arbor, MI 48109-2104, USA
e-mail: lmwang@umich.edu

Sheng-Lai Wang
Shandong University
Institute of Crystal Materials,
State Key Laboratory of Crystal Materials
Shanda Road No. 27
Jinan, Shandong, 250100, China
e-mail: slwang@icm.sdu.edu.cn

Shixin Wang
Micron Technology Inc.
TEM Laboratory
8000 S. Federal Way
Boise, ID 83707, USA
e-mail: shixinwang@micron.com

Jan L. Weyher
Polish Academy of Sciences Warsaw
Institute of High Pressure Physics
ul. Sokolowska 29/37
01/142 Warsaw, Poland
e-mail: weyher@unipress.waw.pl

Jun Xu
Chinese Academy of Sciences
Shanghai Institute of Ceramics
Shanghai, 201800, China
e-mail: xujun@mail.shcnc.ac.cn

Hui Zhang
Tsinghua University
Department of Engineering Physics
Beijing, 100084, China
e-mail: zhhui@tsinghua.edu.cn

Lili Zheng
Tsinghua University
School of Aerospace
Beijing, 100084, China
e-mail: zhenglili@tsinghua.edu.cn

Mary E. Zvanut
University of Alabama at Birmingham
Department of Physics
1530 3rd Ave S
Birmingham, AL 35294-1170, USA
e-mail: mezvanut@uab.edu

Zbigniew R. Zytkiewicz
Polish Academy of Sciences
Institute of Physics
Al. Lotnikow 32/46
02668 Warszawa, Poland
e-mail: zytkie@ifpan.edu.pl

Acknowledgements

H.47 Protein Crystal Growth Methods
by Andrea E. Gutiérrez-Quezada,
Roberto Arreguín-Espinosa, Abel Moreno

One of the authors (Abel Moreno) thanks CONACYT (Mexico) project No. 82888 for sponsorship. Roberto Arreguín-Espinosa acknowledges financial support from the project DGAPA-UNAM, No. IN210007. Andrea E. Gutiérrez-Quezada acknowledges scholarship from SNI-CONACYT.

H.48 Crystallization from Gels
by S. Narayana Kalkura, Subramanian Natarajan

The authors wish to thank Dr. E.K. Girija, S.A. Martin Britto Dhas (Madurai Kamaraj University, India), Dr. George Varghese (St. Berchmans College, Mahatma Gandhi University, India), and Dr. V. Jayanthi (Stanley Medical College, Chennai, India) for all their help in preparing this chapter. S.N.K. thanks AICTE and UGC, India for financial assistance for carrying out majority of the work reported in the review. S.N. thanks the UGC for the SAP funding to his Institution.

H.49 Crystal Growth and Ion Exchange in Titanium Silicates
by Aaron J. Celestian, John B. Parise,
Abraham Clearfield

Support for this work was provided by NSF-CHE-0221934 (CEMS) and DMR-051050 to John B. Parise and NSF-EAR programs. A. Clearfield would like to acknowledge the Department of Energy (DOE) through DE-FG07-01ER63300 and Westinghouse Savannah River Technology Center. We acknowledge the support of the Advanced Photon Source, Argonne National Laboratory, the National Synchrotron Light Source, Brookhaven National Laboratory, the National Institute of Standards and Technology, U.W. Department of Commerce, and the ISIS facility at the Rutherford Appleton Laboratory in providing the neutron research facilities used in this work.

H.50 Single-Crystal Scintillation Materials
by Martin Nikl, Anna Vedda, Valentin V. Laguta

Authors are indebted to K. Nejezchleb, C. W. E. van Eijk and Xue-Jian Liu for providing material for Figs. 50.12, 50.17, 50.29 and 50.30, respectively, M. Dusek for preparation of figures of material structures, P. Bohacek, K. Nejezchleb, N. Senguttuvan, and A. Novoselov for information about crystal growth, E. Jurkova for the help in manuscript preparation, and C. R. Stanek for useful comments and linguistic corrections. Financial support of Czech GACR 202/05/2471, GA AV S100100506, KAN300100802 and Italian Cariplo foundation projects is gratefully acknowledged.

H.52 Wafer Manufacturing and Slicing Using Wiresaw
by Imin Kao, Chunhui Chung,
Roosevelt Moreno Rodriguez

The authors wish to thank Professor Vish Prasad, who has been instrumental in the collaboration of this research on wiresaw slicing and wafer manufacturing. His enthusiasm has always been an inspiration. Several industrial collaborators include Dr. Kedar Gupta, John Talbott, and others. The lead author would also like to thank his previous students who work in wiresaw manufacturing: Drs. Milind Bhagavat, Songbin Wei, Liqun Zhu, and Sumeet Bhagavat, and Mr. Abhiram Govindaraju. The research has been supported by various National Science Foundation (NSF) grants and the United States Department of Energy (DoE) grants.

目 录

缩略语

Part H 晶体生长专题

47 蛋白质晶体生长的方法 ········ 3
    47.1 生物高分子溶液的性质 ········ 4
    47.2 传输现象和形成晶体 ········ 7
    47.3 晶体生长的典型方法 ········ 7
    47.4 扩散-控制方法形成蛋白质晶体 ········ 8
    47.5 晶体生长的新趋势（晶体品质增强） ········ 11
    47.6 原子力显微镜的2-维表征（案例研究） ········ 15
    47.7 X射线衍射的3-维表征和相关方法 ········ 18
    参考文献 ········ 19

48 用凝胶法形成晶体 ········ 27
    48.1 晶体淀积病中的凝胶生长 ········ 28
    48.2 实验方法 ········ 29
    48.3 凝胶系统中的晶格的形成 ········ 30
    48.4 利用凝胶技术的晶体生长 ········ 31
    48.5 晶体淀积病的应用 ········ 34
    48.6 晶体淀积相关的疾病 ········ 36
    48.7 草酸钙 ········ 37
    48.8 磷酸钙 ········ 39
    48.9 羟基磷灰石（HAP） ········ 40
    48.10 二水磷酸氢钙（DCPD） ········ 40
    48.11 硫酸钙 ········ 43
    48.12 尿酸和单钠酸尿 ········ 43
    48.13 L-胱氨酸 ········ 44
    48.14 L-酪氨酸、马尿酸和环丙氟哌酸 ········ 45
    48.15 动脉硬化和胆结石 ········ 45
    48.16 激素的结晶：黄体酮和睾酮 ········ 48
    48.17 胰腺炎 ········ 48

48.18 结 论 ········· 49
参考文献 ········· 50

49 钛硅酸盐中晶体生长和离子交换 ········· 57
49.1 X射线方法 ········· 57
49.2 时间-分辨实验的设备 ········· 62
49.3 检 测 ········· 62
49.4 软 件 ········· 64
49.5 原位细胞的种类 ········· 65
49.6 利用Sitinakite技术对钛硅酸盐（Na-TS）的原位研究 ········· 69
49.7 原位研究的讨论 ········· 78
49.8 总 结 ········· 80
参考文献 ········· 80

50 单晶闪烁材料 ········· 83
50.1 背 景 ········· 83
50.2 闪烁材料 ········· 90
50.3 前景展望 ········· 109
50.4 结 论 ········· 111
参考文献 ········· 111

51 硅太阳能电池：材料、器件和制造 ········· 121
51.1 硅光生伏特 ········· 121
51.2 硅光生伏特的晶体生长技术 ········· 124
51.3 电池制作技术 ········· 131
51.4 总结和讨论 ········· 135
参考文献 ········· 136

52 利用线锯制造和切割晶片 ········· 139
52.1 从晶体锭到基本的晶片 ········· 141
52.2 切割：晶片制造中的第一个后生长工艺 ········· 146
52.3 晶片切割中的现代线据 ········· 150
52.4 总结与展望 ········· 153
参考文献 ········· 153

主题索引 ········· 157

Contents

List of Abbreviations

Part H Special Topics in Crystal Growth

47 Protein Crystal Growth Methods
Andrea E. Gutiérrez-Quezada, Roberto Arreguín-Espinosa, Abel Moreno 1583
- 47.1 Properties of Biomacromolecular Solutions 1584
- 47.2 Transport Phenomena and Crystallization 1587
- 47.3 Classic Methods of Crystal Growth 1587
- 47.4 Protein Crystallization by Diffusion-Controlled Methods 1588
- 47.5 New Trends in Crystal Growth (Crystal Quality Enhancement) 1591
- 47.6 2-D Characterization via Atomic Force Microscopy (Case Study) 1595
- 47.7 3-D Characterization via X-Ray Diffraction and Related Methods 1598
- **References** 1599

48 Crystallization from Gels
S. Narayana Kalkura, Subramanian Natarajan 1607
- 48.1 Gel Growth in Crystal Deposition Diseases 1608
- 48.2 Experimental Methods 1609
- 48.3 Pattern Formation in Gel Systems 1610
- 48.4 Crystals Grown Using Gel Technique 1611
- 48.5 Application in Crystal Deposition Diseases 1614
- 48.6 Crystal-Deposition-Related Diseases 1616
- 48.7 Calcium Oxalate 1617
- 48.8 Calcium Phosphates 1619
- 48.9 Hydroxyapatite (HAP) 1620
- 48.10 Dicalcium Phosphate Dihydrate (DCPD) 1620
- 48.11 Calcium Sulfate 1623
- 48.12 Uric Acid and Monosodium Urate Monohydrate 1623
- 48.13 L-Cystine 1624
- 48.14 L-Tyrosine, Hippuric Acid, and Ciprofloxacin 1625
- 48.15 Atherosclerosis and Gallstones 1625
- 48.16 Crystallization of Hormones: Progesterone and Testosterone 1628
- 48.17 Pancreatitis 1628
- 48.18 Conclusions 1629
- **References** 1630

49 Crystal Growth and Ion Exchange in Titanium Silicates
Aaron J. Celestian, John B. Parise, Abraham Clearfield 1637
- 49.1 X-Ray Methods 1637
- 49.2 Equipment for Time-Resolved Experiments 1642
- 49.3 Detectors 1642
- 49.4 Software 1644
- 49.5 Types of In Situ Cells 1645
- 49.6 In-Situ Studies of Titanium Silicates (Na-TS) with Sitinakite Topology 1649
- 49.7 Discussion of In Situ Studies 1658
- 49.8 Summary 1660
- References 1660

50 Single-Crystal Scintillation Materials
Martin Nikl, Anna Vedda, Valentin V. Laguta 1663
- 50.1 Background 1663
- 50.2 Scintillation Materials 1670
- 50.3 Future Prospects 1689
- 50.4 Conclusions 1691
- References 1691

51 Silicon Solar Cells: Materials, Devices, and Manufacturing
Mohan Narayanan, Ted Ciszek 1701
- 51.1 Silicon Photovoltaics 1701
- 51.2 Crystal Growth Technologies for Silicon Photovoltaics 1704
- 51.3 Cell Fabrication Technologies 1711
- 51.4 Summary and Discussion 1715
- References 1716

52 Wafer Manufacturing and Slicing Using Wiresaw
Imin Kao, Chunhui Chung, Roosevelt Moreno Rodriguez 1719
- 52.1 From Crystal Ingots to Prime Wafers 1721
- 52.2 Slicing: The First Postgrowth Process in Wafer Manufacturing 1726
- 52.3 Modern Wiresaw in Wafer Slicing 1730
- 52.4 Conclusions and Further Reading 1733
- References 1733

Subject Index 1791

List of Abbreviations

μ-PD	micro-pulling-down
1S-ELO	one-step ELO structure
2-D	two-dimensional
2-DNG	two-dimensional nucleation growth
2S-ELO	double layer ELO
3-D	three-dimensional
4T	quaterthiophene
6T	sexithienyl
8MR	eight-membered ring
8T	hexathiophene

A

a-Si	amorphous silicon
A/D	analogue-to-digital
AA	additional absorption
AANP	2-adamantylamino-5-nitropyridine
AAS	atomic absorption spectroscopy
AB	Abrahams and Burocchi
ABES	absorption-edge spectroscopy
AC	alternate current
ACC	annular capillary channel
ACRT	accelerated crucible rotation technique
ADC	analog-to-digital converter
ADC	automatic diameter control
ADF	annular dark field
ADP	ammonium dihydrogen phosphate
AES	Auger electron spectroscopy
AFM	atomic force microscopy
ALE	arbitrary Lagrangian Eulerian
ALE	atomic layer epitaxy
ALUM	aluminum potassium sulfate
ANN	artificial neural network
AO	acoustooptic
AP	atmospheric pressure
APB	antiphase boundaries
APCF	advanced protein crystallization facility
APD	avalanche photodiode
APPLN	aperiodic poled LN
APS	Advanced Photon Source
AR	antireflection
AR	aspect ratio
ART	aspect ratio trapping
ATGSP	alanine doped triglycine sulfo-phosphate
AVT	angular vibration technique

B

BA	Born approximation
BAC	band anticrossing
BBO	BaB_2O_4
BCF	Burton–Cabrera–Frank
BCT	$Ba_{0.77}Ca_{0.23}TiO_3$
BCTi	$Ba_{1-x}Ca_xTiO_3$
BE	bound exciton
BF	bright field
BFDH	Bravais–Friedel–Donnay–Harker
BGO	$Bi_{12}GeO_{20}$
BIBO	BiB_3O_6
BLIP	background-limited performance
BMO	$Bi_{12}MoO_{20}$
BN	boron nitride
BOE	buffered oxide etch
BPD	basal-plane dislocation
BPS	Burton–Prim–Slichter
BPT	bipolar transistor
BS	Bridgman–Stockbarger
BSCCO	Bi–Sr–Ca–Cu–O
BSF	bounding stacking fault
BSO	$Bi_{20}SiO_{20}$
BTO	$Bi_{12}TiO_{20}$
BU	building unit
BaREF	barium rare-earth fluoride
BiSCCO	$Bi_2Sr_2CaCu_2O_n$

C

C–V	capacitance–voltage
CALPHAD	calculation of phase diagram
CBED	convergent-beam electron diffraction
CC	cold crucible
CCC	central capillary channel
CCD	charge-coupled device
CCVT	contactless chemical vapor transport
CD	convection diffusion
CE	counterelectrode
CFD	computational fluid dynamics
CFD	cumulative failure distribution
CFMO	Ca_2FeMoO_6
CFS	continuous filtration system
CGG	calcium gallium germanate
CIS	copper indium diselenide
CL	cathode-ray luminescence
CL	cathodoluminescence
CMM	coordinate measuring machine
CMO	$CaMoO_4$
CMOS	complementary metal–oxide–semiconductor
CMP	chemical–mechanical polishing
CMP	chemomechanical polishing

COD	calcium oxalate dihydrate		DS	directional solidification
COM	calcium oxalate-monohydrate		DSC	differential scanning calorimetry
COP	crystal-originated particle		DSE	defect-selective etching
CP	critical point		DSL	diluted Sirtl with light
CPU	central processing unit		DTA	differential thermal analysis
CRSS	critical-resolved shear stress		DTGS	deuterated triglycine sulfate
CSMO	$Ca_{1-x}Sr_xMoO_3$		DVD	digital versatile disk
CST	capillary shaping technique		DWBA	distorted-wave Born approximation
CST	crystalline silico titanate		DWELL	dot-in-a-well
CT	computer tomography			
CTA	$CsTiOAsO_4$			

E

CTE	coefficient of thermal expansion		EADM	extended atomic distance mismatch
CTF	contrast transfer function		EALFZ	electrical-assisted laser floating zone
CTR	crystal truncation rod		EB	electron beam
CV	Cabrera–Vermilyea		EBIC	electron-beam-induced current
CVD	chemical vapor deposition		ECE	end chain energy
CVT	chemical vapor transport		ECR	electron cyclotron resonance
CW	continuous wave		EDAX	energy-dispersive x-ray analysis
CZ	Czochralski		EDMR	electrically detected magnetic resonance
CZT	Czochralski technique		EDS	energy-dispersive x-ray spectroscopy
			EDT	ethylene dithiotetrathiafulvalene
			EDTA	ethylene diamine tetraacetic acid

D

			EELS	electron energy-loss spectroscopy
D/A	digital to analog		EFG	edge-defined film-fed growth
DBR	distributed Bragg reflector		EFTEM	energy-filtered transmission electron microscopy
DC	direct current			
DCAM	diffusion-controlled crystallization apparatus for microgravity		ELNES	energy-loss near-edge structure
			ELO	epitaxial lateral overgrowth
DCCZ	double crucible CZ		EM	electromagnetic
DCPD	dicalcium-phosphate dihydrate		EMA	effective medium theory
DCT	dichlorotetracene		EMC	electromagnetic casting
DD	dislocation dynamics		EMCZ	electromagnetic Czochralski
DESY	Deutsches Elektronen Synchrotron		EMF	electromotive force
DF	dark field		ENDOR	electron nuclear double resonance
DFT	density function theory		EO	electrooptic
DFW	defect free width		EP	EaglePicher
DGS	diglycine sulfate		EPD	etch pit density
DI	deionized		EPMA	electron microprobe analysis
DIA	diamond growth		EPR	electron paramagnetic resonance
DIC	differential interference contrast		erfc	error function
DICM	differential interference contrast microscopy		ES	equilibrium shape
			ESP	edge-supported pulling
DKDP	deuterated potassium dihydrogen phosphate		ESR	electron spin resonance
			EVA	ethyl vinyl acetate
DLATGS	deuterated L-alanine-doped triglycine sulfate			

F

DLTS	deep-level transient spectroscopy			
DMS	discharge mass spectroscopy		F	flat
DNA	deoxyribonucleic acid		FAM	free abrasive machining
DOE	Department of Energy		FAP	$Ca_5(PO_4)_3F$
DOS	density of states		FCA	free carrier absorption
DPH-BDS	2,6-diphenylbenzo[1,2-*b*:4,5-*b*']diselenophene		fcc	face-centered cubic
			FEC	full encapsulation Czochralski
DPPH	2,2-diphenyl-1-picrylhydrazyl			
DRS	dynamic reflectance spectroscopy			

FEM	finite element method	HIV-AIDS	human immunodeficiency virus–acquired immunodeficiency syndrome
FES	fluid experiment system		
FET	field-effect transistor	HK	high potassium content
FFT	fast Fourier transform	HLA	half-loop array
FIB	focused ion beam	HLW	high-level waste
FOM	figure of merit	HMDS	hexamethyldisilane
FPA	focal-plane array	HMT	hexamethylene tetramine
FPE	Fokker–Planck equation	HNP	high nitrogen pressure
FSLI	femtosecond laser irradiation	HOE	holographic optical element
FT	flux technique	HOLZ	higher-order Laue zone
FTIR	Fourier-transform infrared	HOMO	highest occupied molecular orbital
FWHM	full width at half-maximum	HOPG	highly oriented pyrolytic graphite
FZ	floating zone	HOT	high operating temperature
FZT	floating zone technique	HP	Hartman–Perdok
		HPAT	high-pressure ammonothermal technique
		HPHT	high-pressure high-temperature
		HRTEM	high-resolution transmission electron microscopy

G

GAME	gel acupuncture method
GDMS	glow-discharge mass spectrometry
GE	General Electric
GGG	gadolinium gallium garnet
GNB	geometrically necessary boundary
GPIB	general purpose interface bus
GPMD	geometric partial misfit dislocation
GRI	growth interruption
GRIIRA	green-radiation-induced infrared absorption
GS	growth sector
GSAS	general structure analysis software
GSGG	$Gd_3Sc_2Ga_3O_{12}$
GSMBE	gas-source molecular-beam epitaxy
GSO	Gd_2SiO_5
GU	growth unit

HRXRD	high-resolution x-ray diffraction
HSXPD	hemispherically scanned x-ray photoelectron diffraction
HT	hydrothermal
HTS	high-temperature solution
HTSC	high-temperature superconductor
HVPE	halide vapor-phase epitaxy
HVPE	hydride vapor-phase epitaxy
HWC	hot-wall Czochralski
HZM	horizontal ZM

H

HA	hydroxyapatite
HAADF	high-angle annular dark field
HAADF-STEM	high-angle annular dark field in scanning transmission electron microscope
HAP	hydroxyapatite
HB	horizontal Bridgman
HBM	Hottinger Baldwin Messtechnik GmbH
HBT	heterostructure bipolar transistor
HBT	horizontal Bridgman technique
HDPCG	high-density protein crystal growth
HE	high energy
HEM	heat-exchanger method
HEMT	high-electron-mobility transistor
HF	hydrofluoric acid
HGF	horizontal gradient freezing
HH	heavy-hole
HH-PCAM	handheld protein crystallization apparatus for microgravity
HIV	human immunodeficiency virus

I

IBAD	ion-beam-assisted deposition
IBE	ion beam etching
IC	integrated circuit
IC	ion chamber
ICF	inertial confinement fusion
ID	inner diameter
ID	inversion domain
IDB	incidental dislocation boundary
IDB	inversion domain boundary
IF	identification flat
IG	inert gas
IK	intermediate potassium content
ILHPG	indirect laser-heated pedestal growth
IML-1	International Microgravity Laboratory
IMPATT	impact ionization avalanche transit-time
IP	image plate
IPA	isopropyl alcohol
IR	infrared
IRFPA	infrared focal plane array
IS	interfacial structure
ISS	ion-scattering spectroscopy
ITO	indium-tin oxide
ITTFA	iterative target transform factor analysis
IVPE	iodine vapor-phase epitaxy

J

JDS	joint density of states
JFET	junction FET

K

K	kinked
KAP	potassium hydrogen phthalate
KDP	potassium dihydrogen phosphate
KGW	$KY(WO_4)_2$
KGdP	$KGd(PO_3)_4$
KLYF	$KLiYF_5$
KM	Kubota–Mullin
KMC	kinetic Monte Carlo
KN	$KNbO_3$
KNP	$KNd(PO_3)_4$
KPZ	Kardar–Parisi–Zhang
KREW	$KRE(WO_4)_2$
KTA	potassium titanyl arsenate
KTN	potassium niobium tantalate
KTP	potassium titanyl phosphate
KTa	$KTaO_3$
KTaN	$KTa_{1-x}Nb_xO_3$
KYF	KYF_4
KYW	$KY(WO_4)_2$

L

LACBED	large-angle convergent-beam diffraction
LAFB	L-arginine tetrafluoroborate
LAGB	low-angle grain boundary
LAO	$LiAlO_2$
LAP	L-arginine phosphate
LBIC	light-beam induced current
LBIV	light-beam induced voltage
LBO	LiB_3O_5
LBO	$LiBO_3$
LBS	laser-beam scanning
LBSM	laser-beam scanning microscope
LBT	laser-beam tomography
LCD	liquid-crystal display
LD	laser diode
LDT	laser-induced damage threshold
LEC	liquid encapsulation Czochralski
LED	light-emitting diode
LEEBI	low-energy electron-beam irradiation
LEM	laser emission microanalysis
LEO	lateral epitaxial overgrowth
LES	large-eddy simulation
LG	$LiGaO_2$
LGN	$La_3Ga_{5.5}Nb_{0.5}O_{14}$
LGO	$LaGaO_3$
LGS	$La_3Ga_5SiO_{14}$
LGT	$La_3Ga_{5.5}Ta_{0.5}O_{14}$
LH	light hole
LHFB	L-histidine tetrafluoroborate
LHPG	laser-heated pedestal growth
LID	laser-induced damage
LK	low potassium content
LLNL	Lawrence Livermore National Laboratory
LLO	laser lift-off
LLW	low-level waste
LN	$LiNbO_3$
LP	low pressure
LPD	liquid-phase diffusion
LPE	liquid-phase epitaxy
LPEE	liquid-phase electroepitaxy
LPS	$Lu_2Si_2O_7$
LSO	Lu_2SiO_5
LST	laser scattering tomography
LST	local shaping technique
LT	low-temperature
LTa	$LiTaO_3$
LUMO	lowest unoccupied molecular orbital
LVM	local vibrational mode
LWIR	long-wavelength IR
LY	light yield
LiCAF	$LiCaAlF_6$
LiSAF	lithium strontium aluminum fluoride

M

M–S	melt–solid
MAP	magnesium ammonium phosphate
MASTRAPP	multizone adaptive scheme for transport and phase change processes
MBE	molecular-beam epitaxy
MBI	multiple-beam interferometry
MC	multicrystalline
MCD	magnetic circular dichroism
MCT	HgCdTe
MCZ	magnetic Czochralski
MD	misfit dislocation
MD	molecular dynamics
ME	melt epitaxy
ME	microelectronics
MEMS	microelectromechanical system
MESFET	metal-semiconductor field effect transistor
MHP	magnesium hydrogen phosphate-trihydrate
MI	morphological importance
MIT	Massachusetts Institute of Technology
ML	monolayer
MLEC	magnetic liquid-encapsulated Czochralski

MLEK	magnetically stabilized liquid-encapsulated Kyropoulos	NTRS	National Technology Roadmap for Semiconductors
MMIC	monolithic microwave integrated circuit	NdBCO	$NdBa_2Cu_3O_{7-x}$

O

MNA	2-methyl-4-nitroaniline
MNSM	modified nonstationary model
MOCVD	metalorganic chemical vapor deposition
MOCVD	molecular chemical vapor deposition
MODFET	modulation-doped field-effect transistor
MOMBE	metalorganic MBE
MOS	metal–oxide–semiconductor
MOSFET	metal–oxide–semiconductor field-effect transistor
MOVPE	metalorganic vapor-phase epitaxy
mp	melting point
MPMS	mold-pushing melt-supplying
MQSSM	modified quasi-steady-state model
MQW	multiple quantum well
MR	melt replenishment
MRAM	magnetoresistive random-access memory
MRM	melt replenishment model
MSUM	monosodium urate monohydrate
MTDATA	metallurgical thermochemistry database
MTS	methyltrichlorosilane
MUX	multiplexor
MWIR	mid-wavelength infrared
MWRM	melt without replenishment model
MXRF	micro-area x-ray fluorescence

OCP	octacalcium phosphate
ODE	ordinary differential equation
ODLN	opposite domain LN
ODMR	optically detected magnetic resonance
OEIC	optoelectronic integrated circuit
OF	orientation flat
OFZ	optical floating zone
OLED	organic light-emitting diode
OMVPE	organometallic vapor-phase epitaxy
OPO	optical parametric oscillation
OSF	oxidation-induced stacking fault

P

PAMBE	photo-assisted MBE
PB	proportional band
PBC	periodic bond chain
pBN	pyrolytic boron nitride
PC	photoconductivity
PCAM	protein crystallization apparatus for microgravity
PCF	primary crystallization field
PCF	protein crystal growth facility
PCM	phase-contrast microscopy
PD	Peltier interface demarcation
PD	photodiode
PDE	partial differential equation
PDP	programmed data processor
PDS	periodic domain structure
PE	pendeo-epitaxy
PEBS	pulsed electron beam source
PEC	polyimide environmental cell
PECVD	plasma-enhanced chemical vapor deposition
PED	pulsed electron deposition
PEO	polyethylene oxide
PET	positron emission tomography
PID	proportional–integral–differential
PIN	positive intrinsic negative diode
PL	photoluminescence
PLD	pulsed laser deposition
PMNT	$Pb(Mg, Nb)_{1-x}Ti_xO_3$
PPKTP	periodically poled KTP
PPLN	periodic poled LN
PPLN	periodic poling lithium niobate
ppy	polypyrrole
PR	photorefractive
PSD	position-sensitive detector
PSF	prismatic stacking fault

N

N	nucleus
N	nutrient
NASA	National Aeronautics and Space Administration
NBE	near-band-edge
NBE	near-bandgap emission
NCPM	noncritically phase matched
NCS	neighboring confinement structure
NGO	$NdGaO_3$
NIF	National Ignition Facility
NIR	near-infrared
NIST	National Institute of Standards and Technology
NLO	nonlinear optic
NMR	nuclear magnetic resonance
NP	no-phonon
NPL	National Physical Laboratory
NREL	National Renewable Energy Laboratory
NS	Navier–Stokes
NSF	National Science Foundation
nSLN	nearly stoichiometric lithium niobate
NSLS	National Synchrotron Light Source
NSM	nonstationary model

PSI	phase-shifting interferometry	RTV	room temperature vulcanizing
PSM	phase-shifting microscopy	R&D	research and development
PSP	pancreatic stone protein		
PSSM	pseudo-steady-state model		

S

PSZ	partly stabilized zirconium dioxide	S	stepped
PT	pressure–temperature	SAD	selected area diffraction
PV	photovoltaic	SAM	scanning Auger microprobe
PVA	polyvinyl alcohol	SAW	surface acoustical wave
PVD	physical vapor deposition	SBN	strontium barium niobate
PVE	photovoltaic efficiency	SC	slow cooling
PVT	physical vapor transport	SCBG	slow-cooling bottom growth
PWO	$PbWO_4$	SCC	source-current-controlled
PZNT	$Pb(Zn, Nb)_{1-x}Ti_xO_3$	SCF	single-crystal fiber
PZT	lead zirconium titanate	SCF	supercritical fluid technology
		SCN	succinonitrile

Q

		SCW	supercritical water
QD	quantum dot	SD	screw dislocation
QDT	quantum dielectric theory	SE	spectroscopic ellipsometry
QE	quantum efficiency	SECeRTS	small environmental cell for real-time studies
QPM	quasi-phase-matched		
QPMSHG	quasi-phase-matched second-harmonic generation	SEG	selective epitaxial growth
		SEM	scanning electron microscope
QSSM	quasi-steady-state model	SEM	scanning electron microscopy
QW	quantum well	SEMATECH	Semiconductor Manufacturing Technology
QWIP	quantum-well infrared photodetector		

R

		SF	stacking fault
		SFM	scanning force microscopy
RAE	rotating analyzer ellipsometer	SGOI	SiGe-on-insulator
RBM	rotatory Bridgman method	SH	second harmonic
RC	reverse current	SHG	second-harmonic generation
RCE	rotating compensator ellipsometer	SHM	submerged heater method
RE	rare earth	SI	semi-insulating
RE	reference electrode	SIA	Semiconductor Industry Association
REDG	recombination enhanced dislocation glide	SIMS	secondary-ion mass spectrometry
RELF	rare-earth lithium fluoride	SIOM	Shanghai Institute of Optics and Fine Mechanics
RF	radiofrequency		
RGS	ribbon growth on substrate	SL	superlattice
RHEED	reflection high-energy electron diffraction	SL-3	Spacelab-3
		SLI	solid–liquid interface
RI	refractive index	SLN	stoichiometric LN
RIE	reactive ion etching	SM	skull melting
RMS	root-mean-square	SMB	stacking mismatch boundary
RNA	ribonucleic acid	SMG	surfactant-mediated growth
ROIC	readout integrated circuit	SMT	surface-mount technology
RP	reduced pressure	SNR	signal-to-noise ratio
RPI	Rensselaer Polytechnic Institute	SNT	sodium nonatitanate
RSM	reciprocal space map	SOI	silicon-on-insulator
RSS	resolved shear stress	SP	sputtering
RT	room temperature	sPC	scanning photocurrent
RTA	$RbTiOAsO_4$	SPC	Scientific Production Company
RTA	rapid thermal annealing	SPC	statistical process control
RTCVD	rapid-thermal chemical vapor deposition	SR	spreading resistance
RTP	$RbTiOPO_4$	SRH	Shockley–Read–Hall
RTPL	room-temperature photoluminescence	SRL	strain-reducing layer
RTR	ribbon-to-ribbon	SRS	stimulated Raman scattering

SRXRD	spatially resolved XRD		TTV	total thickness variation
SS	solution-stirring		TV	television
SSL	solid-state laser		TVM	three-vessel solution circulating method
SSM	sublimation sandwich method		TVTP	time-varying temperature profile
ST	synchrotron topography		TWF	transmitted wavefront
STC	standard testing condition		TZM	titanium zirconium molybdenum
STE	self-trapped exciton		TZP	tetragonal phase
STEM	scanning transmission electron microscopy			

U

STM	scanning tunneling microscopy
STOS	sodium titanium oxide silicate
STP	stationary temperature profile
STS	space transportation system
SWBXT	synchrotron white beam x-ray topography
SWIR	short-wavelength IR
SXRT	synchrotron x-ray topography

UC	universal compliant
UDLM	uniform-diffusion-layer model
UHPHT	ultrahigh-pressure high-temperature
UHV	ultrahigh-vacuum
ULSI	ultralarge-scale integrated circuit
UV	ultraviolet
UV-vis	ultraviolet–visible
UVB	ultraviolet B

T

TCE	trichloroethylene
TCNQ	tetracyanoquinodimethane
TCO	thin-film conducting oxide
TCP	tricalcium phosphate
TD	Tokyo Denpa
TD	threading dislocation
TDD	threading dislocation density
TDH	temperature-dependent Hall
TDMA	tridiagonal matrix algorithm
TED	threading edge dislocation
TEM	transmission electron microscopy
TFT-LCD	thin-film transistor liquid-crystal display
TGS	triglycine sulfate
TGT	temperature gradient technique
TGW	Thomson–Gibbs–Wulff
TGZM	temperature gradient zone melting
THM	traveling heater method
TMCZ	transverse magnetic-field-applied Czochralski
TMOS	tetramethoxysilane
TO	transverse optic
TPB	three-phase boundary
TPRE	twin-plane reentrant-edge effect
TPS	technique of pulling from shaper
TQM	total quality management
TRAPATT	trapped plasma avalanche-triggered transit
TRM	temperature-reduction method
TS	titanium silicate
TSC	thermally stimulated conductivity
TSD	threading screw dislocation
TSET	two shaping elements technique
TSFZ	traveling solvent floating zone
TSL	thermally stimulated luminescence
TSSG	top-seeded solution growth
TSSM	Tatarchenko steady-state model
TSZ	traveling solvent zone

V

VAS	void-assisted separation
VB	valence band
VB	vertical Bridgman
VBT	vertical Bridgman technique
VCA	virtual-crystal approximation
VCSEL	vertical-cavity surface-emitting laser
VCZ	vapor pressure controlled Czochralski
VDA	vapor diffusion apparatus
VGF	vertical gradient freeze
VLS	vapor–liquid–solid
VLSI	very large-scale integrated circuit
VLWIR	very long-wavelength infrared
VMCZ	vertical magnetic-field-applied Czochralski
VP	vapor phase
VPE	vapor-phase epitaxy
VST	variable shaping technique
VT	Verneuil technique
VTGT	vertical temperature gradient technique
VUV	vacuum ultraviolet

W

WBDF	weak-beam dark-field
WE	working electrode

X

XP	x-ray photoemission
XPS	x-ray photoelectron spectroscopy
XPS	x-ray photoemission spectroscopy
XRD	x-ray diffraction
XRPD	x-ray powder diffraction
XRT	x-ray topography

Y

YAB	$YAl_3(BO_3)_4$
YAG	yttrium aluminum garnet
YAP	yttrium aluminum perovskite
YBCO	$YBa_2Cu_3O_{7-x}$
YIG	yttrium iron garnet
YL	yellow luminescence
YLF	$LiYF_4$
YOF	yttrium oxyfluoride
YPS	$(Y_2)Si_2O_7$
YSO	Y_2SiO_5

Z

ZA	$Al_2O_3\text{-}ZrO_2(Y_2O_3)$
ZLP	zero-loss peak
ZM	zone-melting
ZNT	ZN-Technologies
ZOLZ	zero-order Laue zone

Part H Special Topics in Crystal Growth

47 Protein Crystal Growth Methods
Andrea E. Gutiérrez-Quezada, Mexico City, Mexico
Roberto Arreguín-Espinosa, Mexico City, Mexico
Abel Moreno, Mexico City, Mexico

48 Crystallization from Gels
S. Narayana Kalkura, Chennai, India
Subramanian Natarajan, Madurai, India

49 Crystal Growth and Ion Exchange in Titanium Silicates
Aaron J. Celestian, Bowling Green, USA
John B. Parise, Stony Brook, USA
Abraham Clearfield, College Station, USA

50 Single-Crystal Scintillation Materials
Martin Nikl, Prague, Czech Republic
Anna Vedda, Milano, Italy
Valentin V. Laguta, Prague, Czech Republic

51 Silicon Solar Cells: Materials, Devices, and Manufacturing
Mohan Narayanan, Gaithersburg, USA
Ted Ciszek, Evergreen, USA

52 Wafer Manufacturing and Slicing Using Wiresaw
Imin Kao, Stony Brook, USA
Chunhui Chung, Stony Brook, USA
Roosevelt Moreno Rodriguez, Stony Brook, USA

1582

47. Protein Crystal Growth Methods

Andrea E. Gutiérrez-Quezada, Roberto Arreguín-Espinosa, Abel Moreno

Nowadays, advances in genomics as well as in proteomics have produced thousands of new biological macromolecules for study in structural biology, biomedicine research, and drug design projects.

Novel and classical methods of protein crystallization as well as modern techniques for two-dimensional (2-D) and three-dimensional (3-D) characterization of different biomolecules are reviewed in this chapter. Production of high-quality single crystals will be analyzed in detail from classical approaches to modern, high-throughput crystal growth methods for x-ray diffraction, as will new strategies for reducing the amount of raw materials used, accelerating the work, and increasing success rates. It will be pointed out that this work on crystallization as well as characterization is multidisciplinary. These scientific efforts are also interrelated and require close collaboration between biochemists, biophysicists, microbiologists, and molecular biologists, as well as physicists and engineers to develop new strategies and equipment for structural purposes. Finally, some of the problems faced and plans for solving them by using x-ray diffraction, neutron diffraction, and electron microscopy will be revised.

- 47.1 Properties of Biomacromolecular Solutions 1584
- 47.2 Transport Phenomena and Crystallization 1587
- 47.3 Classic Methods of Crystal Growth 1587
- 47.4 Protein Crystallization by Diffusion-Controlled Methods 1588
  - 47.4.1 Crystallization in Microgravity Environments 1588
  - 47.4.2 Crystallization in Gels 1589
  - 47.4.3 Crystallization in Capillary Tubes .. 1590
- 47.5 New Trends in Crystal Growth (Crystal Quality Enhancement) 1591
  - 47.5.1 Crystallization Under Electric Fields 1591
  - 47.5.2 Crystallization Under Magnetic Fields 1592
  - 47.5.3 Combining Electric and Magnetic Fields 1593
  - 47.5.4 Robotics and High-Throughput Protein Crystallization 1593
- 47.6 2-D Characterization via Atomic Force Microscopy (Case Study) 1595
  - 47.6.1 General Overview 1595
  - 47.6.2 Coupling AFM and Electrochemistry for Protein Crystal Growth 1596
  - 47.6.3 AFM Characterization by Protein Immobilization by Means of Polypyrrole Films Deposited on Different Electrodes (HOPG and ITO) 1597
- 47.7 3-D Characterization via X-Ray Diffraction and Related Methods 1598
- References 1599

Progress made in the biological sciences (biology, biochemistry, and biomedicine) during the last 25 years has been deeply dependent on the structural knowledge of atomic or molecular resolution of different types of biological macromolecules: proteins, nucleic acids (DNA more than RNA), and a small number of polysaccharides. Much effort has been made worldwide to stimulate the structural study of proteins and of the different conformations they adopt in nature. The final aim is to understand the diversity of protein structural families, their folding, and the relation that exists between their composition and structure/function. Up to now, the redundancy in the motifs and structural elements found in nature suggests that the number of different conformations is finite and manageable. Once most of them are known, it will be possible to predict the function of new, unknown protein domains. In this way, the modeling of these biological macromolecules will become one of the most promising tools for protein structure predictions in the near future.

On the other hand, structural biology has also famously influenced the field of protein engineering. While recombinant DNA techniques are used as synthetic tools, structure elucidation is used as an analytical tool. Advances in genomics have allowed the expression of proteins in live systems, the study of their activities, and based on their structure, their genetic modification for any practical purpose, such as increasing their stability or affinity for a substrate, inhibitor, etc. All this will revolutionize human life in different aspects: economics, health sciences, food sciences (nutrition), as well as in the development of new biomaterials with tailored properties based on the structure of biological systems.

Beyond the impact that structural biology has had on biochemistry, the three-dimensional structures of macromolecules have been demonstrated to be of formidable value in biotechnology. Nowadays, structural biology promotes the pharmacology field, through rational drug design based on the high-resolution structure of target macromolecules. This will have great impact on such diverse problems as curing human diseases, solving veterinary problems, and attacking crop damages [47.1].

Advances in genomics as well as in proteomics have produced thousands of new proteins for study in structural biology and drug design projects. The complete sequencing of vertebrate and invertebrate genomes [47.2] has accelerated international efforts to develop high-throughput methods and technologies that allow fast, three-dimensional protein structure determination [47.3]. Since the number of new proteins will continue to increase, as well as the number of scientists who study them, the necessity for new, efficient, and effective methods of structure determination has emerged [47.4]. Up to now, and in the near future, x-ray diffraction of single crystals of specific macromolecules has been the only technique that can provide structural data at atomic resolution for these purposes. Other techniques that generate structural and molecular dynamic data do exist, but they are not used for the purposes expressed previously [47.1].

Some public and private projects have emerged under the names of structural genomics and structural proteomics. More recently, new terms have been created such as crystallomics, crystallogenesis, and chemotronics (a new branch between science and technology, related to the creation of liquid electrochemical converters in which ions in solution instead of electrons perform the role of current carriers). These efforts need fast and efficient techniques for three-dimensional structure elucidation. They are focused on high-throughput crystal growth for x-ray diffraction, considering new strategies for reducing the amount of raw material used, accelerating the work, and increasing success rates. These efforts are multidisciplinary and interrelated, and need close collaboration between biochemists, biophysicists, microbiologists, and molecular biologists, as well as physicists and engineers to develop new strategies and equipment. Herein, some of the problems faced and plans for solving them will be reviewed.

For x-ray crystallography to be applied to crystals of adequate size and quality, precise data collection is required. This converts crystals into the key to the whole process, and their production into the bottleneck. The problem of growing adequate crystals involves diverse aspects; in this chapter some of them will be commented on, such as model biomolecules, in addition to some novel and ingenious approaches to solving them, as well as their growth in a high-quality crystalline state.

One of the difficulties in obtaining high-quality crystals is the natural convection that exists in every experiment performed under normal Earth gravity conditions. In addition, problems involved in protein crystallization from solution, transport phenomena, and methods of crystal growth will be carefully reviewed. This is the reason why protein crystallization is so difficult: because of the lack of understanding of many of their physicochemical properties.

47.1 Properties of Biomacromolecular Solutions

Since all chemical reactions in biological systems take place in aqueous solutions, the question is: What is a solution? By solution we understand a single phase, or a homogeneous system of variable composition formed by at least two independent components. A homogeneous system is characterized by the absence of an interphase between the component parts of the solution, and by uniform composition and properties throughout the entire volume.

The components of a solution are the individual chemical substances which can be isolated from the system and which can exist in an isolated state. For example, an aqueous solution of lysozyme (model protein) consists of water (or buffer solution) and molecules of lysozyme. The solution components are the solute (protein) and the solvent (in general, water, for biological systems). The solute is the component of the solution in which the state of aggregation in normal conditions differs from that of the solution. The other components of the solution are known as solvents. In the case of liquid solutions, solutes are substances that under normal conditions are solid or gaseous, while the solvents are liquids.

The composition of a solution, in contrast to the composition of definite chemical compounds, can vary continually within wide limits. In this respect solutions are similar to chemical mixtures, though differing from them in their homogeneity and the change of many properties on mixing. The properties of a solution depend on the interactions between the particles of the solute, on the interaction between the particles of the solvent, and on the interaction between the particles of the solute and the solvent [47.5].

In the particular case of biological macromolecules, solubility is defined according to the solvent properties or solution–crystal equilibrium. In general the solvent is water or a buffer solution, so that the protein will have a proper chemical composition to bond to the solution at a specific pH value, or ionic strength (given by salts), and finally some additives (detergents, divalent cations, etc.). All these factors will affect the properties of the protein solubility in a different way. For instance, all soluble proteins in water or buffer solutions will keep hydrophobic amino acids inside the internal structure. On the other hand the hydrophilic amino acids will be distributed around the protein, exposed to the solvent. This process gives stability to the protein molecule and keeps the biological system in equilibrium with the solvent. As a consequence, the stability of any biomolecule in a solution will depend on the interaction between solvent and solute. The balance of these interactions could be modified by chemical and physical parameters such as temperature, pH, salts, additives, organic solvents, etc. The appropriate way to look at the physical or chemical behavior of macromolecular solutions must

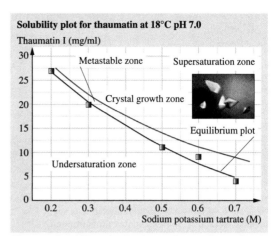

Fig. 47.1 Solubility plot for thaumatin experimentally obtained at 18 °C

be by means of a solubility plot showing the variation of the solubility of the protein versus the variation of the crystallization factor (the concentration of a precipitating agent, pH or temperature). A typical solubility curve shows different areas of the phase diagram where different types of phase separation take place. Figure 47.1 shows an example solubility curve for thaumatin, obtained at 18 °C.

This plot is divided into different zones where the crystal growth process takes place. In the area located under the equilibrium curve, the molecules of the protein are freely distributed throughout the solvent. This means that many biomolecules are needed to saturate the system. In the upper part of this equilibrium curve, there is a parallel line where a quasisteady state is obtained. This area is called the metastable zone, where nuclei are forming and dissolving at the same time and two types of forces are participating in the process: surface and volume forces. In the nucleation zone, the crystal growth process takes place, and larger crystals will grow at lower supersaturation values. However, at very high supersaturation, the system is highly supersaturated and protein–protein interactions will occur at high velocity, yielding amorphous precipitation. It is worth mentioning that everything is controlled by the driving force known as supersaturation. This is related to the differences in the chemical potential of the sample and connected to the Gibbs free energy. There is another way of investigating the solubility behavior of biomacromolecular solutions: by evaluating the solubility of protein molecules as a function of temperature. This physicochemical parameter will give us additional

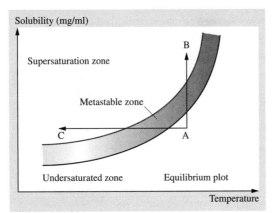

Fig. 47.2 Solubility plot as a function of temperature, the pathway from A to B is called isothermal growth and from A to C is called isotonic growth

information related to the stability of the protein molecule at a specific pH value. For instance, Fig. 47.2 shows a theoretical plot of solubility behavior versus temperature. This plot contains the same areas as those shown in Fig. 47.1.

From this curve it is possible to see two pathways which permit the solubility to be reduced: (1) traveling from the undersaturated region A to region B at constant temperature, (2) traveling from region A to region C while reducing the temperature. Both sides of this plot will permit the reduction of the solubility to obtain phase separation, and depending on the velocity, this solid phase could be a crystalline phase. We must take into account that there are several parameters that participate simultaneously in the crystallization process of a biological sample, and we must also take into account that understanding the nucleation step based on a solubility plot will give us the possibility of growing high-quality single crystals.

Nucleation is a major step in the crystallization process. It is primarily defined in terms of nuclei formation and size distribution. As soon as crystallites are detectable, the phenomenon is called crystal growth. During nucleation several events occur simultaneously on various time scales, namely: molecular conformation changes that take place in ≈ 0.01 ns, surface structure and defect displacements occurring within 1 ns, surface step displacement in 1 μs, growth of one atomic layer in 1 ms, hydrodynamic transport in about 1 s, and finally homogenous nucleation, which needs no more than a few minutes [47.6]. Chemical and physical interactions between different molecules can be monitored, but only some methods provide sufficient resolution in terms of particle size and time scale. Static and dynamic light-scattering methods have been employed to verify protein homogeneity and measure protein–protein interactions under precrystallization conditions. They have also been applied without a sophisticated data reduction scheme to predict protein solubility and crystallizability [47.7, 8]. As a consequence, the combination of spectroscopic and crystallographic data may provide an insight into the energetics of nucleation. The free-energy barrier of this process is controlled by the supersaturation value, which is normally the driving force and is related to the spontaneity of the system by (47.1).

$$\Delta G = -\left[\frac{\left(\frac{4}{3}\pi r^3\right)}{\Omega}\right] k_\mathrm{B} T \ln \beta + 4\pi r^2 \gamma , \qquad (47.1)$$

where Ω is the molar volume occupied by a unit in the crystal, r is the hydrodynamic radius of the macromolecule or the cluster, k_B and T are the Boltzmann constant and absolute temperature, respectively, and γ is the surface energy (expressed in units of $\mathrm{J/cm^2}$).

Supersaturation can be defined by its absolute value $\beta = C/C_\mathrm{e}$, or its relative value, $\alpha = (C - C_\mathrm{e})/C_\mathrm{e}$, where

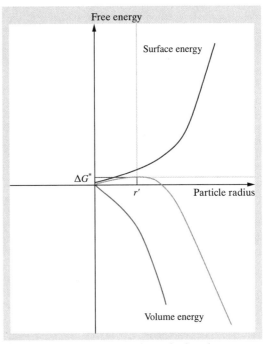

Fig. 47.3 Theoretical plot for the nucleation phenomena

C_e represents the actual value of solubility at certain concentrations of precipitant and protein, i.e., one of the (x, y) values of the plot shown in Fig. 47.1. Nucleation behavior can be easily computed if we have the values of Ω at which the molar volume is occupied by a unit in the crystal from x-ray diffraction, and the hydrodynamic radius (r) from the hydrodynamic properties of the solution as well as the surface energy (γ), both of which are usually obtained by dynamic light-scattering methods. Figure 47.3 shows the theoretical behavior of the nucleation process, where surface and volume forces are competing. The result of the summation of both plots will give us the variation of the Gibbs free energy versus the size in units of the monomer radius [47.9].

47.2 Transport Phenomena and Crystallization

Transport processes, and in particular mass transport, are very important for crystal growth from aqueous solutions [47.10–12]. Mass and heat transport processes are critical to the final quality and characteristics of the crystals [47.13]. Many crystallogenesis techniques have been explicitly developed for controlling the relative contributions of convective and diffusive transport in crystal growth [47.14]. During the active incorporation of ions or molecules into the three-dimensional lattice, density differences are generated in the proximal area of the developing faces, leading to convective flux in the surroundings of the crystal [47.15–17]. Convective transport of molecules competes with pure diffusive transport, and the interaction between them will determine the way and the kinetics of nutrient presentation of the growing crystal.

Transport phenomena not only affect the nutrients of the crystals but also the rate of adsorption and incorporation of impurities, which affects the size, morphology development, and perfection of the crystal [47.18]. On the other hand, convective transport only occurs in the presence of gravity. Only then can heavier fluids fall and lighter fluids rise, letting convective currents emerge in the bulk of the solution. Other types of convection do exist, such as convection due to surface tension [47.19], but they are not significant in the crystallization process of solutions.

47.3 Classic Methods of Crystal Growth

Medium-sized single crystals with near-to-perfect habits made of molecular arrangements with no defects that produce well-resolved and intense diffraction patterns are the dream of every protein crystallographer. Four basic crystallization methods are presently at the disposal of crystal growers to grow such crystals. Like half a century ago in chemistry laboratories, crystallization assays can be set up either (1) in batch mode, (2) by vapor diffusion between hanging (or sitting) droplets and a reservoir of precipitant, (3) by dialysis across a semipermeable membrane or (4) based on free-interface diffusion at the interface of two liquids [47.20–23].

In batch crystallization, nuclei can form and grow immediately in a constant volume once all ingredients, i.e., protein, buffer, precipitant, and additives, have been mixed. An automated version successfully employs microliter droplets that are deposited under a layer of oil [47.24]. In the vapor diffusion method, a small volume of macromolecular solution is equilibrated against a larger volume of precipitant. Equilibration occurs with the transfer (at constant temperature) of a volatile compound (usually water) until isopiestic pressure is reached [47.25]. Its rate is viscosity and vapor pressure dependent, and the volume of the initial protein solution decreases proportionally to the precipitant concentration difference. Most commercial crystallization robots employed for high-throughput applications dispense a great number of submicroliter volume droplets onto dedicated crystallization plates [47.26]. On the other hand, dialysis permits us to approach the equilibrium differently, and can reach the supersaturation zone more smoothly and in a better controlled manner [47.27, 28]. Also samples can be reused as long as they are not irreversibly denatured. This technique is not so popular, probably because samples smaller than a few microliters are difficult to handle and because it has not yet been automated. Finally, in Earth-based laboratories, convectional flow triggered by density differences restricts the use of diffusion at the interface between macromolecular and precipitant solutions to small sample volumes contained in capillary tubes [47.29]. Free-interface diffusion has been successfully applied in quasiweightlessness, where solutal convection is naturally weak [47.30].

47.4 Protein Crystallization by Diffusion-Controlled Methods

47.4.1 Crystallization in Microgravity Environments

Many years of experimentation with diverse crystals has confirmed the notion that, by minimizing convective mass transport, better quality crystals can be obtained, with improved mechanical and optical properties, reduced density of defects, and larger size.

How is it possible to suppress the natural convection in crystallogenesis? Nowadays, different approaches have arisen for removing or at least reducing it. One of them is crystallization of macromolecules in space, where in the absence of gravity, convection disappears. In the last decade, a new approach that involves the use of magnetic fields has appeared. Magnetic forces opposed to gravity can reduce natural convection inside solutions [47.31, 32]. Also, methods for crystallizing macromolecules in gels are good and well-accepted alternatives for eliminating natural convection [47.33].

It is natural to think that, at zero or reduced gravity, crystals with superior properties can be grown [47.18]. Is this possible? Observations and experimental data support the hypothesis that convective flows can be related to the introduction of statistical disorder, defects, and dislocations on growing crystal surfaces [47.34–36]. Convective transport tends to be variable and random, producing variations in the supersaturation levels in the environment of the developing faces, exposing them permanently to high levels of nutrients, similar to those of bulk crystallization. On the contrary, under microgravity, convection is suppressed and the concentration of nutrients at the interface of the crystal is reduced. Mass transport is purely diffusive, which for proteins is very slow, and a region of depletion of nutrients is established around the nucleus. Thanks to the absence of gravity, this zone is quasistable. We can imagine one crystal in the center and the gradient of nutrients on the right-hand side and on the left-hand side aggregates and large impurities. In the right-hand part, nutrient molecules diffuse very slowly because of their size, lengthening the effect. On the left, impure molecules diffuse more slowly than monomers do. As a consequence, the depletion zone acts as a *diffusive filter*, avoiding the incorporation of impurities into the growing crystal. Apparently, this is the principal mechanism for improvement in crystal quality under microgravity. This hypothesis is not only supported by the experimental data but also by mathematical models that explain the mass transport process involved. It has also been shown (in a particular case) that up to a 40% reduction of nutrient molecules near the nucleus can be achieved in the absence of gravity, relative to the bulk concentration [47.37].

In the past, the aim of convection suppression was the inspiration for many scientists who devoted their work to developing new techniques and devices. The first serious experiment for crystallizing macromolecules under microgravity was done by a German team under the direction of Prof. Littke in 1978. On that occasion, lysozyme and β-galactosidase were successfully crystallized by the liquid–liquid diffusion method in a series of reactors [47.38–41].

The progression made in this crystal growth area was hard and slow, owing to the sparse literature on experiments performed in microgravity during the 1970s and 1980s. Moreover, many results were unavailable to the research community because they belonged to private companies or because of the lack of communication between Oriental and Western scientists [47.18]. It was not until 1989 that the first formal scientific paper was published in *Science*, in which x-ray diffraction analysis of many crystals grown in microgravity was reported. They presented higher intensity/estimated error $(I/\sigma(I))$ rates through the whole range of resolution, and higher resolution values [47.42]. This article provided tangible proof that crystals formed in microgravity environments generate more data of higher quality, producing more precise structures.

How are experiments performed in microgravity? Nowadays many devices exist, based primarily on two techniques: vapor diffusion and liquid–liquid diffusion [47.43], and the thermal-induced batch technique has also been successful in growing some large crystals that gave very high resolution [47.44]. The experiments are performed by governmental space agencies or private consortia, such as Payload Systems (USA), Intospace (a European consortium, recently dissolved), and Bioserved (a center for commercial development sponsored by NASA). Each consortium has developed its own devices for its experiments [47.43]. Among the vapor diffusion devices, the vapor diffusion apparatus (VDA), designed and built by *Bugg* and collaborators from Alabama University at Birmingham, USA [47.45, 46], is the most commonly used device, with more than 25 missions. Although originally the device was very simple, it was possible to crystallize many diverse proteins, such as lysozyme, canavaline, bovine serum albumin, and others. Presently, newer and more complex versions of microgravity devices exist, offering major advantages and possibilities for controlled experimentation.

As mentioned above, experiments done in space do not evolve in the same manner as those performed on Earth. For example, for vapor diffusion experiments done in both environments, the equilibrium kinetics is different, and this difference is stronger in liquid–liquid diffusion experiments. In spite of this, the superiority of crystals grown in space compared with those grown under Earth's gravity has been established by comparison according to four well-chosen criteria. First, a *visual examination* is fulfilled (this is a subjective analysis based on observation of the crystal under the microscope). Then, the *sizes and the distribution* of those values in the experiment are analyzed. *Morphology* is another evaluation criterion, as many protein crystals grow with different crystalline habits depending on whether the process was done on Earth or in space [47.42]. Finally, the properties of the x-ray diffraction pattern generated by a crystal must be considered (as the internal order of a crystal relies on the growth kinetics).

Following these criteria, a number of advantages found in crystals grown in microgravity can be summarized [47.42, 47, 48]:

- Visual superiority.
- Larger crystals, many orders of magnitude larger than the biggest crystals grown on Earth [47.18].
- Higher resolutions achieved in x-ray diffraction patterns.
- Better $I/\sigma(I)$ signals (Wilson plot); in the entire range of the resolution, higher values of $I/\sigma(I)$ for microgravity crystals were found. However, the physical properties responsible for this observation are as yet unknown. Apparently, the reason is a lower defect density.
- Sharper x-ray diffraction intensity peaks, showing quantitatively the better internal order and perfection of protein crystals grown in space.
- Not only proteins can be crystallized in the absence of gravity, but also other macromolecules such as viruses (satellite tobacco mosaic virus), DNA, pharmaceutical targets (HIV reverse transcriptase) or membrane proteins (bacteriorhodopsin).
- Under microgravity, the sedimentation effect disappears. In space, crystals keep their defined and stable positions over long periods of time. It is therefore a favorable environment for multiple crystal growth, minimizing the superposition of diffusive fields and assuring more or less uniform access to nutrients by all faces [47.49]. Besides, the incorporation of microcrystals or three-dimensional nuclei by sedimentation into the growing faces is avoided.

On the other hand, among its main disadvantages are the cost and the waiting time. It is an expensive method and requires a lot of time since the missions are not daily and they last for many days, especially when consecutive experiments are desired. Therefore, the disadvantages of microgravity protein crystallization are: high cost, low reproducibility, low reliability, lack of correspondence between Earth- and space-based observations, and lack of understanding of microgravity observations. Finally, based on these disadvantages, we could say that nowadays most protein crystallization programs have been abandoned by the respective space agencies.

47.4.2 Crystallization in Gels

As early as the end of the 19th century, inorganic compounds were crystallized in gelatin or gels made of siloxane [47.50, 51]. Once the actual role of the gel was recognized, it was added to crystallization media to reduce convectional flow [47.52, 53], favor mass transfer by diffusion [47.54, 55], and immobilize crystal nuclei in the network and suppress gravity-driven sedimentation [47.50, 51, 56]. After the first report by *Low* and *Richards* in 1954 of the use of gelatin in the crystallization of albumin [47.57], it took a while before inorganic and organic gels were reintroduced for protein crystallization [47.55]. This led the way to the production of crystals having many fewer defects. Indeed, protein and virus crystals grown in such gels have enhanced diffraction properties, including sharper Bragg reflections, higher diffraction intensities compared with background noise, or a higher diffraction limit compared with crystals grown in solution [47.58, 59]. Despite the discovery of these benefits, gels are largely underexploited by protein crystal growers. The same holds for counterdiffusion [47.60, 61], a crystallization method in which capillary forces exerted in cylindrical tubes of small diameter strongly reduce convection and stabilize the concentration gradients that exist around growing crystals [47.62, 63].

47.4.3 Crystallization in Capillary Tubes

The aim of this section is to summarize technically simple and efficient crystallization methods for optimizing the quality of crystals of a variety of biological particles, including proteins, nucleic acids, multimolecular assemblies, and viruses. Crystallization assays in microporous, chemical or physical hydrogels (as mentioned above) or inside capillary tubes (as described here) only require inexpensive chemicals and simple laboratory equipment. Conditions identified by a preliminary sparse-matrix approach in solution can be readily transferred to gels or capillaries in an attempt to enhance the diffraction properties of crystals. Gels and capillaries are also convenient for crystal storage and transport. In the course of this study, novel gel-forming compounds have been tested and a novel crystallization method for reducing convectional flow has been implemented. Furthermore, crystal content analyses by mass spectrometry have been performed in order to determine the limits of the impurity-sieving effect of agarose on the incorporation of isoforms in thaumatin crystals. Crystallization and crystallographic results obtained with small model proteins and large enzymes involved in gene expression are presented. The contributions of diffusive media and of counterdiffusion methods are discussed and practical recommendations given. Another way to reduce the natural convection under Earth's gravity is to incorporate gelled media into the solutions. In 1968, *Zeppezauer* and coworkers described the use of microdialysis cells, made of capillary tubes sealed with gel stoppers (polyacrylamide), to reduce convection in crystallization solutions to obtain better crystals [47.64]. Then, in 1972, *Salemme* also succeeded in the crystallization of proteins inside a glass capillary tube [47.29]. He put a protein solution into contact with a precipitant agent and let the system reach equilibrium by counterdiffusion [47.38]. Some years later, ribosomal subunits were crystallized successfully with the same setup [47.65].

After many years of investigation, Prof. García-Ruiz proposed the use of gelled media for crystallizing macromolecules by counterdiffusion. This technique combined the principle of reduced convection and the advantage of having a wide range of conditions in a single experiment [47.66]. All this progress allowed *García-Ruiz* and colleagues in 1993 to develop of a new technique called the gel acupuncture method (GAME) [47.67, 68]. This novel technique consists of the permeation of the precipitating agent solution through the gel and the penetration by capillary force into the capillary tube, filled with a protein solution, allowing for crystallization [47.68]. This technique is well known today, and various types of gels, capillary tubes, additives, and precipitating agents have been evaluated for its use [47.68, 69]. A difference from other methods is that inside a capillary tube there is not only one supersaturation level, thus precipitation zones will be found in regions of very high supersaturation, nucleation will happen at high supersaturation levels, and the growth of those nuclei will be found in regions of lower supersaturation levels. This increases the probability of finding the right conditions for crystallization [47.70]. Other advantages are the possibility of crystallizing proteins inside capillary tubes ready for direct x-ray diffraction data collection, avoiding the usual physical manipulation of the crystals and thus reducing the risk of breakage at the moment of mounting them or transporting them to the synchrotron [47.71], or the use of cryoprotectors and/or heavy metals inside the crystallization solutions by using a counterdiffusion method in capillary tubes [47.72].

Through the counterdiffusion methods in capillary tubes it was possible to crystallize diverse proteins of different molecular weight and a wide range of isoelectric point, viruses, and protein–DNA complexes [47.73]. In addition, thanks to the advances in structural genomics, a new device was developed for executing multiple and independent experiments, which is appropriate for effective screening of crystallization conditions of biological macromolecules [47.71]. It combines the benefits of multiple conditions in one capillary, thereby increasing the chances of finding the optimal ones, with the possibility of direct x-ray diffraction analysis in the device. All these advantages have converted this device into the first totally all-inclusive system since the initial steps toward data collection for structural analysis [47.71].

47.5 New Trends in Crystal Growth (Crystal Quality Enhancement)

At the beginning of the new millennium, several structural projects were devoted to finding a cure for diseases based on knowledge of the three-dimensional structure of specific biological targets; the problem has been to obtain high-quality single biocrystals to be investigated by x-ray diffraction. There are several novel approaches to overcoming the poor quality usually obtained in biological crystals for high-resolution x-ray crystallography. Some of these are the application of an in-situ internal electric field in the crystal growth process [47.74–77], the use of external electric fields in protein crystallization [47.78–81], the application of strong magnetic fields and high pressure [47.82–84], the combination of electric and magnetic fields [47.80], the use of ultrasonic fields [47.77], the use of femtosecond laser irradiation (FSLI), and the solution-stirring (SS) method [47.85], as well as addition of nucleants to crystallization droplets [47.86–91]. Basically, the idea behind all of these approaches is related to placing the system in the nucleation regime in the solubility plot and providing the system energy for spontaneous first nucleation. In order to separate the nucleation phenomena and the crystal growth process, it is necessary to look for a precise technique to investigate the limits of these two processes. In this regard, dynamic light-scattering methods usually help to define those areas where nucleation is happening [47.8, 92–94] while atomic force microscopy or video microscopy is appropriate to investigate mechanisms of crystal growth (reviewed later).

47.5.1 Crystallization Under Electric Fields

The study of the effect of electric fields on protein crystallization had not been explored until the pioneering work on estradiol 17β-dehydrogenase electrocrystallization [47.95]. Recent studies performed by Aubry and coworkers, with electric fields external to the crystallization solution, showed that it is possible to reduce lysozyme nucleation and increase the crystal growth rate. Then, the same group evaluated the crystal growth kinetics and found an increase in the protein concentration near drops that were close to the cathode [47.75]. *Nanev* and *Penkova* [47.76] came up with similar results when crystallizing lysozyme by batch method in the presence of an external electric field. They reported that lysozyme crystals grew with a definite orientation towards the cathode. Recently a full review of the effect of electric fields on protein crystallization has been published [47.77].

Biological macromolecule crystallization, in the presence of an internal electric field, uses a similar setup to that used by the gel acupuncture method, except that an inert electrode (Pt) is introduced into the capillary tube and comes into contact with the protein solution, and another electrode is set collinear to it, in the gel [47.74]. Protein molecules are charged since the pH solution is far from the protein's isoelectric point. When a small direct constant current is imposed on the system, a potential difference is established between the electrodes, provoking an orientation effect over the macroions (protein molecules). Lysozyme and thaumatin crystals were found to be firmly attached to the anode during the crystallization process.

To understand what is going on when the potential difference is established, it is important to comprehend how the solution is structured and the effects the electric field has on it. The proposed hypothesis for nucleation inside the capillary tube, in contact with an anode, considers the presence of an electric double layer in the surroundings of the electrode. When an electrode comes into contact with an electrolyte solution, the ions feel asymmetric forces and order themselves, forming an electric double layer. The first layer is composed of water molecules with their dipoles oriented, whereas the second layer is composed of counterions. Positively charged protein molecules need a negative ion for favorable protein–protein interactions, for example, CL^- for lysozyme [47.96] or potassium sodium tartrate for thaumatin [47.97]. The potential difference established in the cell allows the migration of anions of the precipitating agent into the capillary tube, encouraging interaction with the protein molecules in the solution, in the first instance. Then the counterions of the electric double layer will act as supports for positively charged protein molecules or nuclei, allowing the crystal grow over the anode. This behavior was not found when a cathode was placed inside the capillary tube [47.74].

This technique is very new and has only been evaluated with model proteins, such as lysozyme and thaumatin. However, it seems very promising as crystals with similar quality to those grown with the gel acupuncture method were obtained, but with shorter nucleation-induction times and without affecting three-dimensional growth rates [47.74]. This is very interesting from a biotechnology point of view, since shortening the crystalline production time is desirable.

The reduction of the induction time in crystallization had also been observed by *Moreno* and *Sazaki* [47.79]. Lately, these scientists have studied the effect of an electric field with a different setup, using the batch method and with parallel electrodes. Despite these variations, they noticed an induction time three times shorter in the presence of an internal electric field, corroborating the results found with the gel acupuncture method plus an internal electric field [47.74]. Besides, for the same work, the benefit of nucleation control is remarkable, since they obtained fewer lysozyme crystals with homogeneous size distribution.

47.5.2 Crystallization Under Magnetic Fields

It is possible to reduce natural convection on the Earth with the help of magnetic fields. Depending on whether they are homogenous or inhomogeneous, the fields act upon a sample in different ways. Inhomogeneous magnetic fields are responsible for reducing the effective gravity that a solution feels through the action of a magnetization force [47.32]. If a magnetic field gradient is applied vertically, a magnetization force will be generated. When this force opposes gravitational force, a reduction of vertical acceleration (effective gravity) is obtained. Hence, a decrease in natural convection is accomplished.

With a mathematical model of a crystallization system under a magnetic field, the concentrations of macromolecules in the surroundings of a growing crystal were estimated [47.98, Fig. 2], verifying that a magnetic gradient of $-685\,\mathrm{T^2/m}$ reduces convection by 50%, while a magnetic gradient of $-1370\,\mathrm{T^2/m}$ practically eliminates convection, producing similar conditions to that of microgravity [47.98]. Experimentally, high-quality high-resolution crystals were obtained, in agreement with the mathematical model [47.99]. Moreover, *Wakayama* and colleagues found that, in the presence of a magnetizing force opposite to g, fewer lysozyme crystals were obtained than in the absence of the magnetic force [47.32].

When a homogeneous magnetic field is applied, high-quality crystals are also observed [47.100], even though the mechanism involved is different. An increase in viscosity near the growing crystal was observed when a magnetic field of 10 T was applied [47.101, 102]. This increase in viscosity means that there is a reduction of natural convection inside the solution. Furthermore, an orientation effect was observed upon the crystals formed under high magnetic fields [47.31, 32].

More recently, in another study, the decrease of the diffusion coefficient of lysozyme inside a crystallization solution under a homogeneous magnetic field of 6 and 10 T was evaluated [47.103]. All these observations are interrelated and are the consequence of the orientation effect by the magnetic field at a microscopic level. In a supersaturated solution, proteinaceous nuclei are suspended in the solution bulk, and sediment upon reaching an adequate size, which depends on the magnitude of the magnetic field applied. These nuclei act as blocks, avoiding the free diffusion of monomers, turning the solution more viscous and as a result reducing convection [47.103].

The research field of crystal growth under magnetic fields is relatively new and needs more study. A lot of things still remain to be understood about the effects of magnetic fields (both homogeneous and inhomogeneous) on macromolecular solutions. Evidently, a strong external magnetic field induces a magnetizing force, increases the viscosity of the protein solution, orients the growing crystals, and affects the growth process in a complex manner. All these phenomena seem to favor the resulting crystal quality, although a more complete investigation is needed to understand the mechanism better [47.102].

Finally, it is worth mentioning that the use of strong magnetic fields is still very expensive to be performed only in crystallization experiments. Maintenance of the superconductive magnets and the magnet itself are limiting factors for economic reasons. However, there are recent publications that have demonstrated that applying only a strong magnetic field, coupled with growth in a gel, improves the resolution limit as well as the crystal quality [47.104]. The combined effects of a magnetic field and magnetic field gradients on convection in crystal growth were published by Wakayama and colleagues [47.103, 104].

More recently a novel experiment using a popular magnet usually used for nuclear magnetic resonance (NMR) commonly used in chemistry laboratories was published [47.105]. This combined the batch method in gels under the presence of strong magnetic fields of 7 and 10 T in capillary tubes inserted in silicon hydrogels. A significant effect on the orientation of the crystals was noticed after 48 h in the presence of a 10 T magnetic field. The size of the crystals of lysozyme, thaumatin, and ferritin was improved while the number of crystals was decreased compared with the control. For instance, the crystal quality of thaumatin reached 1.15 Å resolution, compared with the control which produced crystals of the same thaumatin at 1.7–1.8 Å resolution.

47.5.3 Combining Electric and Magnetic Fields

The simultaneous effect of magnetic and electric fields is a new field of research. Depending on the configuration of the system, great advantages can be acquired, such as homogeneity in crystal size, thanks to apparent suppression of secondary nucleation events, permitting the continuous growth of previously formed nuclei [47.103, 106]. Also, an orientation effect is noticed when the magnetic field is parallel to the electrodes, having the same effect as described in previous works [47.79, 80].

47.5.4 Robotics and High-Throughput Protein Crystallization

In the last few years, genomic advances have encouraged high-throughput structural biology studies. So much so that it has received the name of structural genomics and many huge public grants have been given to academic laboratories and private enterprises, such as pharmaceutical industries, around the world. The projects are diverse and extend from the study of the structure–function relation of proteins, through the mechanisms involved in protein folding [47.107], to the more pragmatic approach of rational drug design based on the structure of target molecules [47.108]. X-ray diffraction crystallography is critical in these studies, being the battle-horse in such initiatives. In this way, the crystallogenesis of biological macromolecules arises as a vital step in the whole process, although it is the most complicated and last understood step in structural biology. In this section some aspects involved in high-throughput crystal growth of biological macromolecules will be reviewed.

The growing of crystals of biological macromolecules in large quantities takes place in several steps: protein production in large amounts (by heterologous expression or from its natural source), purification, crystallization trials, and their corresponding inspections. Since many proteins are evaluated at the same time it is mandatory to have automated systems that accelerate the work, while at the same time they should be trustworthy since the efficiency of each step affects the next one.

Heterologous protein production is almost completely automated for massive aims. This includes cloning, transformation, and gene expression. These stages involved DNA molecules and, thanks to their high stability, they can be automated easily [47.109, 110]. Nevertheless, there is not yet a totally automated system for protein purification on a large scale with applications in structural biology. Proteins differ in their expression levels, solubility, and physicochemical properties, so that most scientists prefer to adopt a combination of manual techniques to obtain pure proteins. Despite this, many biotechnological companies have developed more integral solutions for large-scale purification. For example, Syrrx (San Diego, USA) uses a purification system developed at the Genomics Institute of Novartis Research Foundation, which combines centrifugation and robotized sonication with a system of column chromatography arranged in parallel, which is able to purify 96–162 proteins per day [47.110]. Affinium Pharmaceuticals (Toronto) has also developed an integral purification system. Proteomax covers all the steps, from the cellular extract processing to the pure concentrated sample, ready for analysis. This equipment can clarify the lysate, perform column chromatography, and desalt and concentrate samples, giving in the best cases a pure protein ready for structural studies. So many automated steps results in very useful purification equipment for large-scale studies [47.108].

Structural biology laboratories are capable of handling more than 1000 different proteins in a month. As a consequence they require the maximum possible automation of every stage, including crystallogenesis. This is not such a big problem, particularly considering that the vapor diffusion and microbatch crystallization techniques are the most frequently used. Therefore, diverse robots that can perform these functions exist on the market. Decode Biostructures produces ROBO-HTC, composed of a robot that prepares all different conditions (Matrix Maker) and another robot that dispenses the drops. Douglas Instruments is responsible for ORIX 6, which performs vapor diffusion with sitting drops or microbatch assays. This robot can process about 240 cells per hour. Another commonly used robot is Mosquito, from Molecular Dimensions. Mosquito is built by TTP LabTech, of TTP Group plc, one of the most successful technology companies in the world. This robot contains a set of precision micropipettes mounted on a continuous band, which deposit small volumes from 50 nl to 1.2 µl. Moreover, the micropipettes are disposable, avoiding cross-contamination problems and exhaustive washing. The robot dispenses drops for microbatch or vapor diffusion, and hanging- or sitting-drop crystallization experiments. Also, it can be used with 96-, 384-, and 1,536-well plates, and comes with an easy-to-use software system that can be programmed.

Once the assays are set for incubation, regular inspection is required in order to find the adequate conditions for obtaining high-quality crystals. This is the most arduous part of high-throughput crystallization. For any one protein 1000 experiments are needed, on average, to obtain an appropriate crystal for x-ray diffraction analysis [47.108]. Many companies use human inspection, a very tedious and laborious step, so the idea of designing an automated inspection system is very tempting. An ideal inspection system should have the following characteristics:

1. The identification and elimination of clear drops, with 0% error (without risk of losing any crystals)
2. The capture of kinetic data (of the growing process or precipitation versus time)
3. The ability to distinguish crystals from precipitates and to find crystals inside precipitates in most cases
4. The determination of size and form of the crystal
5. The ability to improve points 3 and 4 based on *internal learning* (databases)

Some promising advances in image technology have been made by Decode BioStructures, which offers the *Crystal Monitor Workstation*. This equipment has a stereoscopic microscope, digital camera, voice control, and a database interface and can be coupled with ROBOHTC (from the same company). There also exists *Crystal Score* by Diversified Scientific, which provides a microscope over a motorized plate. It comes with a device that counts and sizes the crystals. RoboDesign has two options on the market: RoboMicroscope II, which can localize the drops, focus them, capture a color image, and store them automatically, while CPXO can classify drops into clear, precipitated, with crystals, and other categories.

Some of the advantages of automated image capturing are the high frequency with which images are registered, at precise times, and the possibility of evaluating them with diverse computer software, and applying artificial intelligence. Although the equipment does not have human experience it can be trained to develop its own database. However, in the end, there is no system that can totally ignore human inspection. Crystals are very difficult to obtain and, as long as an automated system with 0% false negatives does not exist, the human eye will be indispensable [47.108].

So far the advances and problems of high-throughput structural biology have been mentioned in a general way. However, most of those studies are on soluble proteins. Membrane proteins entail other kinds of problems, as previously mentioned (in the membrane crystallization section).

What are the specific challenges of membrane protein structural biology? Diverse stages can be mentioned, such as protein production, purification, and crystallogenesis. Unlike soluble proteins, where more than 90% of the new protein structures come from recombinant samples, membrane proteins obtained from molecular biology techniques make up less amount than 50% [47.111]. This is in part because the strategies developed for overexpressing proteins are designed for soluble ones and do not favor integral membrane proteins [47.109]. The synthesis of membrane proteins makes use of the cell's secretory system, certain directionality, and their insertion inside membranes [47.110, 112]. Besides, many cells are not equipped to withstand such a flux in new membrane proteins, which saturate their secretory pathways and generate inclusion bodies in the cytosol or toxic intermediates for those cells. So, the choice of the expression system is an important issue in the production of membrane proteins.

Prokaryotic integral proteins can be expressed in prokaryotic systems with promising results. Some examples are structures of ionic channels [47.113] and certain proteins from the outer membrane [47.114], among others. On the contrary, eukaryotic membrane proteins overexpressed in prokaryotic organisms have been harder to achieve [47.115], since prokaryotic membranes have a different lipidic composition and can have a hostile environment for heterologous proteins. Besides, posttranslational modifications are necessary for correct folding of eukaryotic proteins or their insertion into membranes, but these are absent in prokaryotic cells [47.116]. Despite this, some successful isolated examples do exist, such as the in case of an enzyme bound to mammal membranes overexpressed in *E. coli*, when it was crystallized [47.117] and its structure solved [47.118], or the overexpression of a eukaryotic receptor coupled to G-protein in *Halobacterium salinarum* [47.119]. On the other hand, yeasts are good overexpression systems for eukaryotic proteins since they are easy to handle and powerful genetics tools [47.120, 121]. Mammal cells are the best choice for preserving the structural and functional integrity of mammal membrane proteins. However they are expensive and very complex to use, and as a result they are the last choice.

With regard to purification, detergents are usually used to dissolve membrane proteins. The choice of detergent is essential, especially when designing studies on a large scale. Ideally a detergent should solubilize the membrane protein without forming aggregates [47.122]. An appropriate detergent is one that can selectively stabilize the native structure of proteins [47.123]. Every protein behaves in a particular way, which means that a specific purification protocol is needed for each one. This concept is in opposition to *a unique measure for all*, the motto of en masse experiments. Diverse strategies have been thought up to resolve this, like fusing a protein which has a certain affinity for a ligand to the end of a membrane protein that one wants to overexpress in order to facilitate its purification [47.124, 125].

Once pure, integral proteins are ready for crystallizing. Two alternative paths can be applied:

1) Crystallize the protein–detergent complex directly
2) Incorporate the protein once again into a lipidic bilayer environment, previous to its crystallization

Most of the structures solved by x-ray diffraction analysis come from crystals formed by the first path, by vapor diffusion or microbatch techniques. The method is similar to that used on soluble proteins but in this case the solute is the detergent–protein complex [47.122]. The use of robots simplifies the handling of these mixtures. The other path consists of restoring the membrane proteins to a lipidic bilayer environment, before setting the crystallization experiments. This approach has its major model in the lipidic cubic phases method, as explained previously [47.126]. Automated equipment that can use this criterion for high-throughput membrane protein crystallization is under construction [47.127].

Finally, we can understand how the advances in the processes involved and the automation achieved in recent years have influenced the development of structural biology throughout the world. Many laboratories can successfully clone, express, purify, and crystallize soluble proteins on a scale that was unthinkable years ago. Nevertheless, there is still much to control and predict in many different stages of the general process. In relation to soluble proteins, the process for integral membrane proteins has not been achieved. In the near future, the problem of the amount of protein available for crystallization must be overcome by increasing protein production levels and improving the system of purification, especially the detergent choice. All this will help to obtain more successful crystallization trials. Besides, studies of the mechanisms that rule the process will aid in its scaling up and automation. Moreover, so many crystallization experiments will enrich databases and in consequence it will be possible to extract the tendencies and crystallization patterns. This will be a valuable ability for structural genomics, particularly for the aim of crystallizing new proteins.

47.6 2-D Characterization via Atomic Force Microscopy (Case Study)

47.6.1 General Overview

Understanding the nucleation and crystallization processes and the control of size and quality of macromolecular crystals of different proteins for structural investigations in biology and biomedical sciences is still a challenge in many laboratories worldwide. Particularly, the mechanisms of crystal growth are important in order to understand the history of the crystallization process. The pioneering efforts devoted to this investigation of those mechanisms of crystal growth were done by using scanning electron microscopy methods, examining the surfaces of lysozyme crystals as well as model proteins and virus particles, and deducing recent mechanisms of crystal growth [47.128–141], as reviewed by *McPherson* et al. in 2000 [47.142]. These results showed that crystal growth occurred by a lattice defect mechanism at low supersaturation and by two-dimensional nucleation at high supersaturation. Step velocities and two-dimensional nucleation rates were obtained, and their dependence on supersaturation was compared with theory. Preliminary results on the early stages of nucleation and the phenomenon of cessation of growth have been presented. More recently, atomic force microscopy (AFM) has become a common tool in biophysical studies of proteins (mainly due to its ability to perform characterizations near to physiological conditions) [47.143–151]. Tertiary and quaternary structures, forces driving folding–unfolding processes, and secondary structure elements can be studied in their native environments, allowing a high resolution level associated with small distortions. It is important to remark that surface characterization techniques are not limited to AFM. Several groups have carried out insightful electron microscopy characterizations. Another prolific method has been interferometry, in particular two-beam

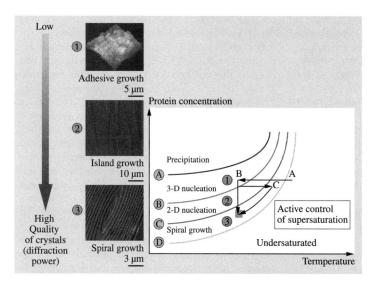

Fig. 47.4 Conceptual plot for protein crystal growth mechanisms observed by atomic force microscopy (designed by Dr. Gen Sazaki, Tohoku University Japan)

and phase-shifting methods, with several recent improvements. However, as a case study and reflecting the expertise of one of the authors, AFM methods coupled to electrochemical techniques will be revised in this chapter.

From the crystal growth viewpoint, we can use atomic force microscopy to show different areas where these crystal growth mechanisms are obtained on the solubility plot. Figure 47.4 shows those areas where high-quality single crystals can be obtained and their mechanisms of crystal growth can be studied by these atomic force microscopy methods. Knowledge of the limiting areas A, B, C, and D permits us to explain why a crystal that grows at high supersaturation will diffract the x-ray radiation poorly (due to adhesive crystal growth mechanism). On the contrary, crystals that grow at intermediate and low supersaturation values will diffract the x-ray radiation properly. The reason for this enhancement of crystal quality is that these crystals will grow by the following mechanisms of crystal growth: island growth or spiral growth. This plot also shows the best way to produce crystals of high quality by varying the temperature as a crystallizing parameter (as shown in Fig. 47.2). This overview image of the crystal growth mechanisms shown in Fig. 47.4 also explains why we can produce high-quality single crystals by microseeding methods. For instance, if one crystal nucleates at high supersaturation, the crystal growth cell will be filled with tiny, poorly shaped crystals. Most of the time these crystals will be poor scatterers of the x-ray diffraction due to the adhesive crystal growth

mechanism, which controls the crystal growth process at the beginning of the nucleation phenomena.

We can infer from Fig. 47.4 that the crystal will grow at the beginning by adhesive crystal growth at high supersaturation, and then the supersaturation will be reduced by the system itself (due to mass consumption). Then the crystal will continue to grow by island growth, finishing up with spiral growth at the end. This type of crystal is not available to diffract x-ray radiation (due to its poor internal crystalline order). However, since they finish growing by island or spiral growth mechanism, these crystalline species can be used as a source of microcrystals to be added to preequilibrated droplets by using the microseeding method, as pointed out by *Stura* and *Wilson* in 1991 [47.152].

47.6.2 Coupling AFM and Electrochemistry for Protein Crystal Growth

Nowadays, most atomic force microscopes are coupled to electrochemical devices (potentiostats and galvanostats). So, we can use these techniques from electrochemistry to produce electrocrystallization in different ways. The first is to produce compatible nuclei on the surface of various electrodes (mostly graphite). Then a potential difference can be applied to reduce the solubility of the solution of the biomolecule. Unfortunately, not all proteins are suitable for electrocrystallization. The most promising biological macromolecules are those which contain some metals as cofactors or that are covalently bonded to porphyrin groups, some amino acids or certain disulfide bridges. There are only a few cases where the idea of electrocrystallization has been applied to some biological macromolecules. The pioneering work was performed on the crystallization of human placental estradiol 17β-dehydrogenase, published by *Chin* et al. in 1976 [47.95]. There was another group who investigated the codeposition of electrocrystallized calcium phosphate and bovine serum albumin on the surfaces of titanium alloy (Ti-6Al-4V) under different conditions. Infrared (IR) and ultraviolet (UV) spectra showed that: (1) the content of the protein formed by electrochemical coprecipitation in the solution of calcium phosphate was higher than that formed by simple absorption; and (2) the protein formed at high direct current was more than that at low direct current. These results provided useful information about biocoating techniques of prosthetic implant materials [47.153]. The second approach focused on real electrocrystallization and was done by *Moreno* and *Rivera* [47.154]. They investigated the role of electrochemical processes on iron and $CdSO_4$ in the crystallization of horse spleen ferritin by using the cyclic voltammetry technique. It was found that, although both species exhibited important redox properties in the presence of an external applied potential, $CdSO_4$ played a leading role not only in the nucleation process but also in the growth behavior and morphology control of ferritin crystals.

47.6.3 AFM Characterization by Protein Immobilization by Means of Polypyrrole Films Deposited on Different Electrodes (HOPG and ITO)

Enzyme immobilization on electrode surfaces has been limited to soluble enzymes [47.155, 156]. However, recent developments in protein crystallization have created an important interest in the study of solid-state electrochemistry of protein single crystals in order to understand the mechanisms of crystal growth. Unfortunately, the fixation of these monocrystals to an electrode surface is difficult since the monocrystals break easily under mechanical pressure, and therefore they cannot be immobilized as other inorganic crystals [47.157, 158]. From this point of view, it is feasible to grow ex situ redox metalloprotein single crystals (such as catalases, ferritins, cytochromes, etc.) so that they can be introduced into the fluid cell of the atomic force microscope (AFM). In order to immobilize these crystals we can use polypyrrole (ppy) films deposited on highly oriented pyrolytic graphite electrodes (HOPG) or indium-tin oxide electrodes (ITO) for structural investigations by AFM techniques. It is worth mentioning that ppy films are conductive and therefore their application as chemical cements will expand the number of future structural investigations into soft biological single crystals.

In general, we can fix any type of biocrystals by using the method of *Hernández-Pérez* et al. (2002) in which the polypyrrole film had been used as chemical glue for AFM investigations [47.159]. Figure 47.5 shows a case study in which we can see a ferritin monocrystal chemically fixed to a HOPG electrode by means of these ppy films. Recently, cytochrome c has been one of the most studied proteins in electrochemistry due to its electron-transfer properties and capability of being used as a solid-state electron-transfer device [47.160]. Even though only bovine cytochrome c is commercially available, its three-dimensional structure has recently been published at 1.5 Å resolution [47.161]. Therefore, in the near future this promising, natural electron-transfer protein can either be used for electrocrystallization investigations or as an electric biosensor component [47.160]. Pyrrole molecules can be polymerized in aqueous solutions by mixing 0.077 M pyrrole solution and 0.34 M $LiClO_4$ and applying a current of 50 mV/s in an electrochemistry AFM fluid cell (Veeco Co., Santa Barbara, USA). Figure 47.5a shows the surface of the HOPG electrode

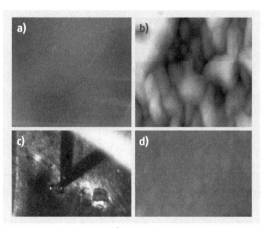

Fig. 47.5 (a) $12 \times 12\,\mu m^2$ scanning AFM image of the HOPG surface. (b) $12 \times 12\,\mu m^2$ AFM image of the HOPG surface after polymerization of the pyrrole. (c) Full optical image of the ferritin crystal when scanning and after polymerization of pyrrole. (d) $12 \times 12\,\mu m^2$ scanning AFM image of the ferritin crystal surface

without polypyrrole films. On applying a potential difference, the pyrrole polymerizes and the morphology of the HOPG surface is changed (Fig. 47.5b). Figure 47.5c shows an overall image of ferritin crystals obtained by batch mixing method with 1 : 1 : 1 ratio 0.1 M cadmium sulfate/0.5 M ammonium sulfate, ferritin solution of 20 mg/ml, and pyrrole/LiClO$_4$ mixture as mentioned above for polymerization. This polymerization of pyrrole was performed by means of a fluid cell in an EC-AFM Nanoscope IIIa. Finally, the AFM image of the surface of this ferritin crystal is shown in Fig. 47.5d. This image is different when compared with the one containing the polypyrrole. This experiment is particularly important because we can see from that image that polypyrrole film will surround the crystal so as to fix it properly onto the surface of the electrode. Later, after scanning the surface by AFM (Fig. 47.5d), we can see that there is no pyrrole on the surface of the biocrystal (ferritin), which means that the protein crystal has been fixed to the electrode by a chemical reaction with polypyrrole films (chemical glue) around it.

A novel idea is to grow large crystals on transparent electrodes (ITO for instance) by following the strategy previously described. Then a new solution of protein, precipitant, and additives can be added to the crystal fixed on the electrode. This tiny crystal can reach good sizes for structural research by neutron diffraction techniques. At the same time, having this type of fixed crystals on ITO or HOPG, we can say that an electron-transfer biosensor has been obtained.

47.7 3-D Characterization via X-Ray Diffraction and Related Methods

Protein crystallography is a part of the solid-state sciences which aims to solve the three-dimensional structure of biological macromolecules by means of the x-ray diffraction of single crystals [47.162–166]. The state of the art in the three-dimensional (3-D) structure of several proteins, nucleic acids, and polysaccharides is collected in the protein databank (PDB, initially administrated in Brookhaven National Laboratory, whose updated website is http://www.rcsb.org). This database is collecting all x-ray diffraction information for most of the biological macromolecules. A review of a variety and use of several databases has recently been presented by *Einspahr* [47.167].

The appropriate way to solve the 3-D structure begins by performing precise x-ray data collection. In order to perform this we must work with high-quality single crystals. As pointed out in Sect. 47.6, the better the crystal, the higher the resolution limit. The vast majority of proteins need to have a certain degree of purity to be crystallized. However, crystallization also depends on the molecular weight of the protein, the type of the protein, and even the method of crystal growth used to obtain the crystals. The first well-known case is the protein horse spleen ferritin. In order to produce wonderful cubic-octahedral single crystals, horse spleen ferritin must be purified to the extent of having just monomers in the solution (Fig. 47.6a,b). Otherwise the presence of a small amount of impurities will poison some crystal growth sectors, producing dendritic growth such as that shown in Fig. 47.6c. It is important to emphasize that nowadays we can also use twin crystals in order to obtain the 3-D structure of almost any protein. It is clear that, using higher-quality single crystals, we can produce marvelous electron density maps or a wonderful structure (Fig. 47.7) of almost any protein. This figure shows the 3-D structure of cytochrome c from bovine heart obtained at 1.5 Å resolution [47.161]. It took that group nearly 2 years to crystallize this protein due to the existence of isoforms (which made the crystallization process difficult), only one of which was crystallized (native cytochrome). The only way in which we could produce suitable crystals of cytochrome c for x-ray diffraction was by means of the microseeding technique. The crystallization of membrane proteins, macromolecular complexes, and large assemblies is still a challenge.

Behind these beautiful structures is still a challenge: crystallization of membrane proteins and macromolecular complexes and assemblies. In this regard, in the near future, as mentioned in Sect. 47.5, the existence of robots and high-throughput techniques will help greatly. The real challenge will be how to predict accurate 3-D structure from de novo proteins, based on powerful databases of the three-dimensional structures of many biological macromolecules.

Finally, neutron-diffraction protein crystallography methods are also becoming promising powerful tools for 3-D structural characterization. Neutron diffraction provides an experimental approach for directly locating H atoms and hydration in proteins, a technique complementary to ultrahigh-resolution x-ray diffraction [47.168]. Recently, technical aspects as well as

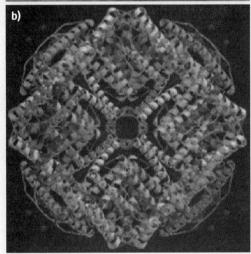

Fig. 47.6 (a) Crystal of ferritin grown by the batch method using only a purified fraction of monomers. (b) Three-dimensional structure of horse spleen ferritin showing the variety of cubic elements of the symmetry. (c) Dendrite of ferritin containing a mixture of oligomers ◀ ▲

 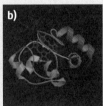

Fig. 47.7 (a) A perfect twin crystal of cytochrome c from bovine heart. (b) Image of the cytochrome structure obtained at high resolution (1.5 Å, PDB code: 2b4z) by *Mirkin* et al. 47.161

some potential applications and limitations of neutron protein crystallography have been reviewed [47.169].

New related methods coupling x-ray crystallography and scanning electron microscopy to obtain three-dimensional electron microscopy of macromolecular assemblies have also been published [47.170].

References

47.1 A. McPherson: Macromolecular crystallization in the structural genomics era, J. Struct. Biol. **142**, 1–2 (2003)

47.2 D. Roses: Genome-based pharmacogenetics and the pharmaceutical industry, Nat. Rev. Drug Discov. **1**, 541–549 (2002)

47.3 P. Kuhn, K. Wilson, M.G. Patch, R.C. Stevens: The genesis of high-throughput structure-based drug discovery using protein crystallography, Curr. Opin. Chem. Biol. **6**, 704–710 (2002)

47.4 J.L. DeLucas, T.L. Bray, L. Nagy, K. McCombs, N. Chernov, D. Hamrick, L. Cosenza, A. Belgovskiy, B. Stoops, A. Chait: Efficient protein crystallization, J. Struct. Biol. **142**, 188–206 (2003)

47.5 R.J. Davey: The role of the solvent in crystal growth from solution, J. Cryst. Growth **76**, 637–644 (1986)

47.6 A.G.W. Wilson: Predicting protein crystallization from a dilute solution property, Acta Crystallogr. D **50**, 361–365 (1994)

47.7 V. Mikol, E. Hirsch, R. Giegé: Diagnostic of precipitant for biomacromolecule crystallization by quasi-elastic light-scattering, J. Mol. Biol. **213**, 187–195 (1990)

47.8 W.W. Wilson: Light scattering as a diagnostic for protein crystal growth – A practical approach, J. Struct. Biol. **142**, 56–65 (2003)

47.9 C.N. Nanev: Protein crystal nucleation, Cryst. Res. Technol. **42**, 4–12 (2007)

47.10 A.A. Chernov: *Modern Crystallography III, Crystal Growth* (Springer, Berlin Heidelberg 1984)

47.11 S. Sarig: *Handbook of Crystal Growth*, Vol. 2B, ed. by D.T.J. Hurle (North-Holland, Amsterdam 1994)

47.12 P. Bennema: Crystal growth from solution – Theory and experiment, J. Cryst. Growth **24**, 76–83 (1974)

47.13 D.T.J. Hurle: *Handbook of Crystal Growth*, Vol. 1B (North-Holland, Amsterdam 1994)

47.14 D.T.J. Hurle: *Handbook of Crystal Growth*, Vol. 2A (North-Holland, Amsterdam 1994)

47.15 F.E. Neumann: Über die optischen Eigenschaften der hemiprismatischen oder zwei- und eingliedrigen Krystalle, Ann. Physik **111**, 81–95 (1835), in German

47.16 F. Rosenberger: Inorganic and protein crystal growth – Similarities and differences, J. Cryst. Growth **76**, 618–636 (1986)

47.17 P.S. Cheng, P.J. Shlichta, W.R. Wilcox, R.A. Lefever: Convection phenomena during the growth of sodium chlorate crystals from solution, J. Cryst. Growth, **47**, 43–60 (1979)

47.18 A. McPherson: Macromolecular crystal growth in microgravity, Crystallogr. Rev. **6**, 157–308 (1996)

47.19 F. Rosenberger: *Fundamentals of Crystal Growth I, Macroscopic Equilibrium Concepts* (Springer, Berlin Heidelberg 1979)

47.20 T.M. Bergfors: Protein crystallization (Int. Univ. Line, La Jolla 1999)

47.21 A. Ducruix, R. Giegé: *Crystallization of Nucleic Acids and Proteins, a Practical Approach*, 2nd edn. (IRL, Oxford 1999)

47.22 A. McPherson: *Crystallization of Biological Macromolecules* (Cold Spring Harbor Laboratory, New York 1999)

47.23 R. Giegé, A. McPherson: General methods. In: *International Tables for Crystallography*, Vol. F, ed. by M.G. Rossmann, E. Arnold (IUCr, Kluwer Academic, Boston 2001)

47.24 A. D'Arcy, C. Elmore, M. Stihle, J.E. Johnston: A novel approach to crystallising proteins under oil, J. Cryst. Growth **168**, 175–180 (1992)

47.25 E.P.K. Hade, C. Tanford: Isopiestic compositions as a measure of preferential interactions of macromolecules in two-component solvents. Application to proteins in concentrated aqueous cesium chloride and guanidine hydrochloride, J. Am. Chem. Soc. **89**, 5034–5040 (1967)

47.26 J.R. Luft, J. Wolfley, I. Jurisica, J. Lasgow, S. Fortier, G.T. DeTitta: Macromolecular crystallization in a high throughput laboratory – The search phase, J. Cryst. Growth **232**, 591–595 (2001)

47.27 A. McPherson, J. Geller, A. Rich: Crystallographic studies on concanavalin B, Biochem. Biophys. Res. Commun. **57**, 494–499 (1974)

47.28 B.H. Weber, P.E. Goodkin: A modified microdiffusion procedure for the growth of single protein crystals by concentration-gradient equilibrium dialysis, Arch. Biochem. Biophys. **141**, 489–498 (1970)

47.29 F.R. Salemme: A free interface diffusion technique for the crystallization of proteins for X-ray crystallography, Arch. Biochem. Biophys. **151**, 533–539 (1972)

47.30 B. Lorber, R. Giegé: Nucleation and growth of thaumatin crystals within a gel under microgravity on STS-95 mission vs. under Earth's gravity, J. Cryst. Growth **231**, 252–261 (2001)

47.31 M. Ataka, E. Katoh, N.I. Wakayama: Magnetic orientation as a tool to study the initial stage of crystallization of lysozyme, J. Cryst. Growth **173**, 592–596 (1997)

47.32 N.I. Wakayama, M. Ataka, H. Abe: Effect of a magnetic field gradient on the crystallization of hen lysozyme, J. Cryst. Growth **178**, 653–656 (1997)

47.33 J.M.. García-Ruíz, M.L. Novella, R. Moreno, J.A. Gavira: Agarose as crystallization media for proteins. I: Transport processes, J. Cryst. Growth **232**, 165–172 (2001)

47.34 M. Pusey, W.K. Witherow, R. Naumann: Preliminary investigations into solutal flow about growing tetragonal lysozyme crystals, J. Cryst. Growth **90**, 105–111 (1988)

47.35 M. Beth, H. Broom, W.K. Witherow, R.S. Snyder, D.C. Carter: Preliminary observations of the effect of solutal convection on crystal morphology, J. Cryst. Growth, **90**, 130–135 (1988)

47.36 J.K. Baird, E.J. Meehan, A.L. Xidis, S.B. Howard: Convective diffusion in protein crystal growth, J. Cryst. Growth **76**, 694–700 (1986)

47.37 H. Lin, F. Rosenberger, J.L.D. Alexander, A. Nadarajah: Convective-diffusive transport in protein crystal growth, J. Cryst. Growth **151**, 153–162 (1995)

47.38 Y.P. Wang, Y. Han, J.S. Pan, K.Y. Wang, R.C. Bi: Protein crystal growth in microgravity using a liquid/liquid diffusion method, Microgravity Sci. Technol. **9**, 281–283 (1996)

47.39 W. Littke, C. John: Protein single crystal growth under microgravity, Science **225**, 203–204 (1984)

47.40 A. McPherson: *Methods of Biochemical Analysis*, Vol. 23, ed. by D. Glick (Academic, New York 1976)

47.41 A. McPherson: *The Preparation and Analysis of Protein Crystals* (Wiley, New York 1982)

47.42 J.L. DeLucas, C.D. Smith, H.W. Smith, V.K. Senagdi, S.E. Senadhi, S.E. Ealick, C.E. Bugg, D.C. Carter, R.S. Snyder, P.C. Weber, F.R. Salemme, D.H. Ohlendorf, H.M. Einspahr, L. Clancy, M.A. Navia, B. McKeever, T.L. Nagabhushan, G. Nelson, Y.S. Babu,

47.42 A. McPherson, S. Koszelak, D. Stammers, K. Powell, G. Darby: Protein crystal growth in microgravity, Science **246**, 651–654 (1989)

47.43 R.S. Snyder, K. Fuhrmann, H.U. Walter: Protein crystallization facilities for microgravity experiments, J. Cryst. Growth **110**, 333–338 (1991)

47.44 M.M. Long, L.J. DeLucas, C. Smith, M. Carson, K. Moore, M.D. Harrington, D.J. Pilion, S.P. Bishop, W.M. Rosenblum, R.J. Naumann, A. Chait, J. Prahl, C.E. Bugg: Protein crystals growth in microgravity-temperature induced large scale crystallization of insulin, Microgravity Sci. Technol. **7**, 196–202 (1994)

47.45 J.L. DeLucas, F.L. Suddath, R. Snyder, R. Naumann, M.B. Broom, M. Pusey, V. Yost, B. Herren, D. Carter, B. Nelson, E.J. Meehan, A. McPherson, C.E. Bugg: Preliminary investigations of protein crystal growth using the space shuttle, J. Cryst. Growth **76**, 681–693 (1986)

47.46 J.L. DeLucas, M.M. Long, K.M. Moore, W.M. Rosenblum, T.L. Bray, C. Smith, M. Carson, S.V.L. Narayana, D. Carter, A.D. Clark Jr., R.G. Nanni, J. Ding, A. Jacobo-Molina, G. Kamer, S.H. Hughes, E. Arnold, H.M. Einspahr, L.L. Clancy, G.S.J. Rao, P.F. Cook, B.G. Harris, S.H. Munson, B.C. Finzel, A. McPherson, P.C. Weber, F. Lewandowski, T.L. Nagabhushan, P.P. Trotta, P. Reichert, M.A. Navia, K.P. Wilson, J.A. Thomson, R.R. Richards, K.D. Bowersox, C.J. Meade, E.S. Baker, S.P. Bishop, B.J. Dunbar, E. Trinh, J. Prahl, A. Sacco Jr., C.E. Bugg: Recent results and new hardware developments for protein crystal growth in microgravity, J. Cryst. Growth **135**, 183–195 (1994)

47.47 J.R. Helliwell, E. Snell, S. Weisgerber: Proc. 9th Europ. Symp. Gravity Depend. Phenom. Phys. Sci. (Berlin 1995)

47.48 E.H. Snell, S. Weisgerber, J.R. Helliwell: Improvements in lysozyme protein crystal perfection through microgravity growth, Acta Crystallogr. D **51**, 1099–1102 (1995)

47.49 R. Boistelle, J.P. Astier: Crystallization mechanisms in solution, J. Cryst. Growth **90**, 14–30 (1988)

47.50 H.K. Henisch: *Crystals in Gels and Liesegang Rings* (Cambridge Univ. Press, Cambridge 1988)

47.51 K.-T. Wilke: *Kristallzüchtung* (Verlag Harri Deutsch, Frankfurt/Main 1988), in German

47.52 P.S. Chen, P.J. Schlichta, W.R. Wilcox, R.A. Lefever: Convection phenomena during the growth of sodium chlorate crystals from solution, J. Cryst. Growth **47**, 43–60 (1979)

47.53 M.C. Robert, F. Lefaucheux: Crystal growth in gels: Principle and applications, J. Cryst. Growth **90**, 358–367 (1988)

47.54 B. Rubin: The growth of single crystals by controlled diffusion in silica gel, AIChE J. **15**, 206–208 (1969)

47.55 M.C. Robert, F. Lefaucheux, B. Jannot, G. Godefroy, E. Garnier: A comparative study of gel grown and space grown lead hydrogen phosphate crystals, J. Cryst. Growth **88**, 499–510 (1988)

47.56 J.M. García-Ruiz, O. Fermín, M.L. Novella, J.A. Gavira, C. Sauter, O. Vidal: A supersaturation wave of protein crystallization, J. Cryst. Growth **232**, 149–155 (2001)

47.57 B.W. Low, F.M. Richards: Measurements of the density, composition and related unit cell dimensions of some protein crystals, J. Am. Chem. Soc. **76**, 2511–2518 (1954)

47.58 B. Lorber, C. Sauter, M.C. Robert, B. Capelle, R. Giegé: Crystallization within agarose gel in microgravity improves the quality of thaumatin crystals, Acta Crystallogr. D **55**, 1491–1494 (1999)

47.59 D. Maes, L.A. Gonzalez-Ramirez, J. Lopez-Jaramillo, B. Yu, H. De Bondt, I. Zegers, E. Afonina, J.M. García-Ruiz, S. Gulnik: Structural study of the type II 3-dehydroquinate dehydratase from *Actinobacillus pleuropneumoniae*, Acta Crystallogr. D **60**, 463–471 (2004)

47.60 J.M. García-Ruiz: Counterdiffusion methods for macromolecular crystallization, Methods Enzymol. **368**, 130–154 (2003)

47.61 F. Otálora, J.M. García-Ruiz, A. Moreno: Protein crystal quality studies using rod-shaped crystals, J. Cryst. Growth **168**, 93–98 (1996)

47.62 A. McPherson, A.J. Makin, Y.G. Kuznetsov, S. Koszelak, M. Wells, G. Jenkins, G. Howard, J. Lawson: The effects of microgravity on protein crystallization: evidence for concentration gradients around growing crystals, J. Cryst. Growth **196**, 572–586 (1988)

47.63 F. Otálora, J.M. García-Ruiz, L. Carotenuto, D. Castagnolo, M.L. Novella, A.A. Chernov: Lysozyme crystal growth kinetics in microgravity, Acta Crystallogr. D **58**, 1681–1689 (2002)

47.64 M. Zeppezauer, H. Eklund, E.S. Zeppezauer: Micro diffusion cells for the growth of single protein crystals by means of equilibrium dialysis, Arch. Biochem. Biophys. **126**, 564–573 (1968)

47.65 A. Yonath, J. Müssig, H.G. Witlmann: Parameters for crystal growth of ribosomal subunits, J. Cell. Biochem. **19**, 145–155 (1982)

47.66 J.M. García-Ruíz: The uses of crystal growth in gels and other diffusing-reacting systems, Key Eng. Mater. **58**, 87–106 (1991)

47.67 J.M. García-Ruíz, A. Moreno, C. Viedma, M. Coll: Crystal quality of lysozyme single crystals grown by the gel acupuncture method, Mater. Res. Bull. **28**, 541–546 (1993)

47.68 J.M. García-Ruiz, A. Moreno: Investigations on protein crystal growth by the gel acupuncture method, Acta Crystallogr. D **50**, 484–490 (1994)

47.69 V.M. Bolaños-García: The use of oil in a counter-diffusive system allows to control nucleation and coarsening during protein crystallization, J. Cryst. Growth **253**, 517–523 (2003)

47.70 J.M. García-Ruíz, A. Moreno, D. Rondón, F. Otálora, F. Zauscher: Teaching protein crystallization by the

47.71 J.D. Ng, J.A. Gavira, J.M. García-Ruíz: Protein crystallization by capillary counterdiffusion for applied crystallographic structure determination, J. Struct. Biol. **142**, 218–231 (2003)

47.72 J.A. Gavira, D. Toh, J. Lopez-Jaramillo, J.M. García-Ruíz, J.D. Ng: *Ab initio* crystallographic structure determination of insulin from protein to electron density without crystal handling, Acta Crystallogr. D **58**, 1147–1154 (2002)

47.73 C. Biertümpfel, J. Basquin, D. Suck, C. Sauter: Crystallization of biological macromolecules using agarose gel, Acta Crystallogr. D **58**, 1657–1659 (2002)

47.74 N. Mirkin, B.A. Frontana-Uribe, A. Rodriguez-Romero, A. Hernandez-Santoyo, A. Moreno: The influence of an internal electric field upon protein crystallization using the gel-acupuncture method, Acta Crystallogr. D **59**, 1533–1538 (2003)

47.75 M. Taleb, C. Didierjean, C. Jelsch, J.P. Mangeot, A. Aubry: Equilibrium kinetics of lysozyme crystallization under an external electric field, J. Cryst. Growth **232**, 250–255 (2001)

47.76 C. Nanev, A. Penkova: Nucleation of lysozyme crystals under external electric and ultrasonic fields, J. Cryst. Growth **232**, 285–293 (2001)

47.77 M.I. Al-Haq, E. Lebrasseur, H. Tsuchiya, T. Torii: Protein crystallization under an electric field, Crystallogr. Rev. **13**, 29–64 (2007)

47.78 E. Nieto-Mendoza, B. Frontana-Uribe, G. Sazaki, A. Moreno: Investigations on electromigration phenomena for protein crystallization using crystal growth cells with multiple electrodes, effect of the potential control, J. Cryst. Growth **275**, 1443–1452 (2005)

47.79 A. Moreno, G. Sazaki: The use of a new ad hoc growth cell with parallel electrodes for the nucleation control of lysozyme, J. Cryst. Growth **264**, 438–444 (2004)

47.80 G. Sazaki, A. Moreno, K. Nakajima: Novel coupling effects of the magnetic and electric fields on protein crystallization, J. Cryst. Growth **262**, 499–502 (2004)

47.81 A. Penkova, O. Gliko, I.L.D. Feyzim, V. Hodjaoglu, C. Nanev, P.G. Vekilov: Enhancement and suppression of protein crystal nucleation due to electrically driven convection, J. Cryst. Growth **275**, e1527–e1532 (2005)

47.82 G. Sazaki, E. Yoshida, H. Komatsu, T. Nakada, S. Miyashita, K. Watanabe: Effects of a magnetic field on the nucleation and growth of protein crystals, J. Cryst. Growth **173**, 231–234 (1997)

47.83 Y. Suzuki, S. Miyashita, G. Sazaki, T. Nakada, T. Sawada, H. Komatsu: Effects of pressure on growth kinetics of tetragonal lysozyme crystals, J. Cryst. Growth **208**, 638–644 (2000)

47.84 A. Kadri, G. Jenner, M. Damak, B. Lorber, R. Giegé: Crystallogenesis studies of proteins in agarose gel-gel acupuncture technique, J. Chem. Educ. **75**, 442–446 (1998)

47.85 Y. Mori, K. Takano, H. Adachi, T. Inoue, S. Murakami, H. Matsumura, T. Sasaki: Protein crystallization using femto-second laser irradiation and solution-stirring, Proc. 11th Int. Conf. Cryst. Biol. Macromol. (Quebec City 2006)

47.86 A. McPherson, P. Shlichta: The use of heterogeneous and epitaxial nucleants to promote growth of protein crystals, J. Cryst. Growth **90**, 47–50 (1988)

47.87 T.E. Paxton, A. Sambanis, R.W. Rousseau: Mineral substrates as heterogeneous nucleants in the crystallization of proteins, J. Cryst. Growth **198/199**, 656–660 (1999)

47.88 N.E. Chayen, E. Saridakis, R. El-Bahar, Y. Nemirovsky: Porous silicon: An effective nucleation-inducing material for protein crystallization, J. Mol. Biol. **312**, 591–595 (2001)

47.89 N.E. Chayen, L. Hench: (Imperial College Innovations Limited, UK) Mesoporous glass as nucleant for macromolecule crystallisation, Patent No. W02004041847 (2003)

47.90 R.P. Sear: Protein crystals and charged surfaces: Interactions and heterogeneous nucleation, Phys. Rev. E **67**, 061907/1–061907/7 (2003)

47.91 N.E. Chayen, E. Saridakis, R.P. Sear: Experiment and theory for heterogeneous nucleation of protein crystals in a porous medium, Proc. Natl. Acad. Sci. USA **103**, 597–601 (2006)

47.92 W.W. Wilson: Monitoring crystallization experiments using dynamic light scattering: Assaying and monitoring protein crystallization in solution, Methods Companion Methods Enzymol. **1**, 110–117 (1990)

47.93 G. Juárez-Martínez, C. Garza, R. Castillo, A. Moreno: A dynamic light scattering investigation of the nucleation and growth of thaumatin crystals, J. Cryst. Growth **232**, 119–131 (2001)

47.94 N. Chayen, M. Dieckmann, K. Dierks, P. Fromme: Size and shape determination of proteins in solution by a noninvasive depolarized dynamic light scattering instrument, Ann N. Y. Acad. Sci. (Transport Phenomena in Microgravity), **1027**, 20–27 (2004)

47.95 C. Chin, J.B. Dence, J.C. Warren: Crystallization of human placental estradiol 17β-dehydrogenase. A new method for crystallizing labile enzymes, J. Biol. Chem. **251**, 3700–3705 (1976)

47.96 M.C. Vaney, I. Broutin, P. Retailleau, A. Douangmath, S. Lafont, C. Hamiaux, T. Prangé, A. Ducruix, M. Riès-Kautt: Structural effects of monovalent anions on polymorphic lysozyme crystals, Acta Crystallogr. D **57**, 929–940 (2001)

47.97 A. McPherson, J. Weickmann: X-ray analysis of new crystal forms of the sweet protein thaumatin, J. Biomol. Struct. Dyn. **7**, 1053–1060 (1990)

47.98 J. Qi, N.I. Wakayama, M. Ataka: Magnetic suppression of convection in protein crystal growth processes, J. Cryst. Growth **232**, 132–137 (2001)

47.99 S.X. Lin, M. Zhou, A. Azzi, G.J. Xu, N.I. Wakayama, M. Ataka: Magnet used for protein crystallization: novel attempts to improve the crystal quality, Biophys. Res. Commun. **275**, 274–278 (2000)

47.100 T. Sato, Y. Yamada, S. Saijo, T. Hori, R. Hirose, N. Tanaka, G. Sazaki, K. Nakajima, N. Igarashi, M. Tanaka, Y. Matsuura: Enhancement in the perfection of orthorhombic lysozyme crystals grown in a high magnetic field (10 T), Acta Crystallogr. D **56**, 1079–1083 (2000)

47.101 C.W. Zhong, N.I. Wakayama: Effect of a high magnetic field on the viscosity of an aqueous solution of protein, J. Cryst. Growth **226**, 327–332 (2001)

47.102 L. Wang, C.W. Zhong, N.I. Wakayama: Damping of natural convection in the aqueous protein solutions by the application of high magnetic fields, J. Cryst. Growth **237**, 312–316 (2002)

47.103 D.C. Yin, N.I. Wakayama, Y. Inatomi, W.D. Huang, K. Kuribayashi: Strong magnetic field effect on the dissolution process of tetragonal lysozyme crystals, Adv. Space Res. **32**, 217–223 (2003)

47.104 D. Lübbert, A. Meents, E. Weckert: Accurate rocking-curve measurements on protein crystals grown in a homogeneous magnetic field of 2.4 T, Acta Crystallogr. D **60**, 987–998 (2004)

47.105 A. Moreno, B. Quiroz-García, F. Yokaichiya, V. Stojanoff, P. Rudolph: Protein crystal growth in gels and stationary magnetic fields, Cryst. Res. Technol. **42**, 231–236 (2007)

47.106 T. Sato, Y. Yamada, S. Saijo, T. Hori, R. Hirose, N. Tanaka, G. Sazaki, K. Nakajima, N. Igarashi, M. Tanaka, Y. Matsuura: Improvement in diffraction maxima in orthorhombic HEWL crystal grown under high magnetic field, J. Cryst. Growth **232**, 229–236 (2001)

47.107 T. Terwilliger: Structural genomics in America, Nat. Struct. Biol. **7**, 935–939 (2000)

47.108 R. Hui, A. Edwards: High-throughput protein crystallization, J. Struct. Biol. **142**, 154–161 (2003)

47.109 J.-W. de Gier, J. Luirink: Biogenesis of inner membrane proteins in *Escherichia coli*, J. Mol. Microbiol. **40**, 314–322 (2001)

47.110 S.A. Lesley: High-throughput proteomics: Protein expression and purification in the postgenomic world, Protein Expr. Purif. **22**, 159–164 (2001)

47.111 Stephen White: http://blanco.biomol.uci.edu/Membrane_Proteins_xtal.html (2009)

47.112 R. Grisshammer, C.G. Tate: Overexpression of integral membrane proteins for structural studies, Q. Rev. Biophys. **28**, 315–422 (1995)

47.113 A. Arora, D. Rinehart, G. Szabo, L.K. Tamm: Refolded outer membrane protein A of *Escherichia coli* forms ion channels with two conductance states in planar lipid bilayers, J. Biol. Chem. **275**, 1594–1600 (2000)

47.114 M. Müller, H.G. Koch, K. Beck, U. Schäfer: Protein traffic in bacteria: Multiple routes from the ribosome to and across the membrane, Prog. Nucleic Acid Res. Mol. Biol. **66**, 107–157 (2001)

47.115 G. Chang, R.H. Spencer, A.T. Lee, M.T. Barclay, D.C. Rees: Structure of the MscL homolog from *Mycobacterium tuberculosis*: A gated mechanosensitive ion channel, Science **282**, 2220–2226 (1998)

47.116 C.G. Tate: Overexpression of mammalian integral membrane proteins for structural studies, FEBS Letters **504**, 94–98 (2001)

47.117 K.E.S. Matlack, W. Mothes, T.A. Rapoport: Protein translocation: Tunnel vision, Cell **92**, 381–390 (1998)

47.118 J.C. Otto, D.L. De Witt, W.L. Smith: N-glycosylation of prostaglandin endoperoxide synthases-1 and -2 and their orientations in the endoplasmic reticulum, J. Biol. Chem. **268**, 18234–18242 (1993)

47.119 M.P. Patricelli, H.A. Lashuel, D.K. Giang, J.W. Kelly, B.F. Cravatt: Comparative characterization of a wild type and transmembrane domain-deleted fatty acid amide hydrolase: Identification of the transmembrane domain as a site for oligomerization, Biochemistry **37**, 15177–15187 (1998)

47.120 M.H. Bracey, M.A. Hanson, K.R. Masuda, R.C. Stevens, B.F. Cravatt: Structural adaptations in a membrane enzyme that terminates endocannabinoid signaling, Science **298**, 1793–1796 (2002)

47.121 G.J. Turner, R. Reusch, A.M. Winter-Vann, L. Martínez, M.C. Betlach: Heterologous gene expression in a membrane-protein-specific system, Protein Expr. Purif. **17**, 312–323 (1999)

47.122 J.L. Cereghino, J.M. Cregg: Heterologous protein expression in the methylotrophic yeast *Pichia pastoris*, FEMS Microbiol. Rev. **24**, 45–66 (2000)

47.123 K. Sreekrishna, R.G. Brankamp, K.E. Kropp, D.T. Blankenship, J.T. Tsay, P.L. Smith, J.D. Wierschke, A. Subramaniam, L.A. Birkenberger: Strategies for optimal synthesis and secretion of heterologous proteins in the methylotrophic yeast *Pichia pastoris*, Gene **190**, 55–62 (1997)

47.124 P.J. Loll: Membrane protein structural biology: The high throughput challenge, J. Struct. Biol. **142**, 144–153 (2003)

47.125 J.P. Rosenbusch: Stability of membrane proteins: relevance for the selection of appropriate methods for high-resolution structure determinations, J. Struct. Biol. **136**, 144–157 (2001)

47.126 E.M. Landau, J.P. Rosenbusch: Lipidic cubic phases: A novel concept for the crystallization of membrane proteins, Proc. Natl. Acad. Sci. USA **93**, 14532–14535 (1996)

47.127 A. Cheng, B. Hummel, H. Qiu, M. Caffrey: A simple mechanical mixer for small viscous lipid-containing samples, Chem. Phys. Lipids **95**, 11–21 (1998)

47.128 S.D. Durbin, W.E. Carlson: Lysozyme crystal growth studied by atomic force microscopy, J. Cryst. Growth **122**, 71–79 (1992)

47.129 A.J. Malkin, Y.G. Kuznetsov, A. McPherson: Incorporation of microcrystals by growing protein and virus crystals, Proteins **24**, 247–252 (1996)

47.130 D.A. Walters, B.L. Smith, A.M. Belcher, G.T. Paloczi, G.D. Stucky, D.E. Morse, P.K. Hansma: Modification of calcite crystal growth by abalone shell proteins: an atomic force microscope study, Biophys. J. **72**, 1425–1433 (1997)

47.131 H. Li, A. Nadarajah, M.L. Pusey: Determining the molecular-growth mechanisms of protein crystal faces by atomic force microscopy, Acta Crystallogr. D **55**, 1036–1045 (1999)

47.132 Y.F. Dufrene: Application of atomic force microscopy to microbial surfaces: From reconstituted cell surface layers to living cells, Micron **32**, 153–165 (2001)

47.133 M. Plomp, A. McPherson, A.J. Malkin: Repair of impurity-poisoned protein crystal surfaces, Proteins **50**, 486–495 (2003)

47.134 A.J. Malkin, R.E. Thorne: Growth and disorder of macromolecular crystals: Insights from atomic force microscopy and X-ray diffraction studies, Methods **34**, 273–299 (2004)

47.135 S. Hiroyuki: Modification, characterization and handling of protein molecules as the first step to bioelectronic devices, Electron. Biotechnol. Adv. (EL.B.A.) Forum Ser. **2**, 157–174 (1996)

47.136 I. Reviakine, W. Bergsma-Schutter, A. Brisson: Growth of protein 2-D crystals on supported planar lipid bilayers imaged in situ by AFM, J. Struct. Biol. **121**, 356–361 (1998)

47.137 S.-T. Yau, B.R. Thomas, P.G. Vekilov: Molecular mechanisms of crystallization and defect formation, Phys. Rev. Lett. **85**, 353–356 (2000)

47.138 A.P. Wheeler, C.S. Sikes: Proteins from oyster shell: biomineralization regulators and commercial polymer analogs, Mater. Res. Soc. Symp. Proc. **599**, 209–224 (2000)

47.139 H. Kim, R.M. Garavito, R. Lal: Atomic force microscopy of the three-dimensional crystal of membrane protein, OmpC porin, Colloids Surf. B: Biointerfaces **19**, 347–355 (2000)

47.140 A. Nadarajah, H. Li, J.H. Konnert, M.L. Pusey: New AFM techniques for investigating molecular growth mechanisms of protein crystals, Proc. SPIE **4098**, 31–39 (2000)

47.141 A. McPherson: Macromolecular crystal structure and properties as revealed by atomic force microscopy, NATO Sci. Ser. I: Life Behav. Sci. **325**, 1–8 (2001)

47.142 A. McPherson, A.J. Malkin, Y.G. Kuznetsov: Atomic force microscopy in the study of macromolecular crystal growth, Annu. Rev. Biophys. Biomol. Struct., **29**, 361–410 (2000)

47.143 J. Yang, L.K. Tamm, T.W. Tillack, Z. Shao: New approach for atomic-force microscopy of membrane proteins. The imaging of cholera toxin, J. Mol. Biol. **229**, 286–290 (1993)

47.144 P. Hallett, G. Offer, M.J. Miles: Atomic force microscopy of the myosin molecule, Biophys. J. **68**, 1604–1606 (1995)

47.145 D. Pang, S. Yoo, W.S. Dynan, M. Jung, A. Dritschilo: Ku proteins join DNA fragments as shown by atomic force microscopy, Cancer Res. **57**, 1412–1415 (1997)

47.146 K.K. Chittur: Proteins on surfaces: methodologies for surface preparation and engineering protein function, Surfact. Sci. Ser. **75**, 143–179 (1998)

47.147 J. Cao, D.K. Pham, L. Tonge, D.V. Nicolau: Simulation of the force-distance curves of atomic force microscopy for proteins by the Connolly surface approach, Proc. SPIE **4590**, 187–194 (2001)

47.148 W. Huang, S. Taylor, K. Fu, Y. Lin, D. Zhang, T.W. Hanks, A.M. Rao, Y.-P. Sun: Attaching proteins to carbon nanotubes via diimide-activated amidation, Nano Lett. **2**, 311–314 (2002)

47.149 J. Torres, T.J. Stevens, M. Samso: Membrane proteins: the "Wild West" of structural biology, Trends Biochem. Sci. **28**, 137–144 (2003)

47.150 A. Tulpar, D.B. Henderson, M. Mao, B. Caba, R.M. Davis, K.E. Van Cott, W.A. Ducker: Unnatural proteins for the control of surface forces, Langmuir **21**, 1497–1506 (2005)

47.151 P.L. Silva: Imaging proteins with atomic force microscopy: An overview, Curr. Protein Peptide Sci. **6**, 387–395 (2005)

47.152 E.A. Stura, I.A. Wilson: Applications of the streak seeding technique in protein crystallization, J. Cryst. Growth **110**, 270–282 (1991)

47.153 Y. Zhang, T. Fu, H. Li, K. Xu, K. State: Co-deposition of electocrystallized calcium phosphate and protein as biocoatings on the prosthetic materials, Guisuanyan Xuebao **28**, 379–380, 384 (2000), in Chinese

47.154 A. Moreno, M. Rivera: Conceptions and first results on the electrocrystallization behaviour of ferritin, Acta Crystallogr. D **61**, 1678–1681 (2005)

47.155 Y.H. Chen, J.Y. Wu, Y.C. Chung: Preparation of polyaniline-modified electrodes containing sulfonated polyelectrolytes using layer-by-layer techniques, Biosens. Bioelectron. **22**, 489–494 (2006)

47.156 J.P.H. Perez, E. Lopez-Cabarcos, B. Lopez-Ruiz: The application of methacrylate-based polymers to enzyme biosensors, Biomol. Eng. **23**, 233–245 (2006)

47.157 A. Baba, W. Knoll, R. Advincula: Simultaneous in situ electrochemical, surface plasmon optical, and atomic force microscopy measurements: Investigations of conjugated polymer electropolymerization, Rev. Sci. Instrum. **77**, 064101 (2006)

47.158 G.A. Álvarez-Romero, E. Garfias-García, M.T. Ramírez-Silva, C. Galán-Vidal, M. Romero-Romo,

M. Palomar-Pardavé: Electrochemical and AFM characterization of the electropolymerization of pyrrole over a graphite–epoxy resin solid composite electrode, in the presence of different anions, Appl. Surf. Sci. **252**, 5783–5792 (2006)

47.159 T. Hernández-Pérez, N. Mirkin, A. Moreno, M. Rivera: In situ immobilization of catalase monocrystals on HOPG by the voltammetric growth of polypyrrole films for AFM investigations, Electrochem. Solid-State Lett. **5**, 37–39 (2002)

47.160 F. Acosta, D. Eid, L. Marín-García, B.A. Frontana-Uribe, A. Moreno: From cytochrome c crystals to a solid-state electron-transfer device, Cryst. Growth. Des. **7**, 2187–2191 (2008)

47.161 N. Mirkin, J. Jaconcic, V. Stojanoff, A. Moreno: High resolution X-ray crystallographic structure of cytochrome c from bovine heart and its application to the design of an electron transfer biosensor, Proteins: Structure, Function, Bioinformatics **70**, 83–92 (2008)

47.162 T.L. Blundell, L.N. Johnson: *Protein Crystallography* (Oxford Univ. Press, Oxford 1976)

47.163 C. Giacovazzo, C. Giazovazzo, H.L. Monaco, G. Artioli, D. Viterbo, G. Ferraris: *Fundamentals of Crystallography* (Oxford Univ. Press, Oxford 2002)

47.164 D.M. Blow: *Outline of Crystallography for Biologists* (Oxford Univ. Press, Oxford 2002)

47.165 A. McPherson: *Macromolecular Crystallography* (Wiley, New York 2002)

47.166 D.E. McRee: *Practical Protein Crystallography* (Academic, New York 1999)

47.167 H.M. Einspahr: A functioning crystallization database: What do we want and how do we get it? Int. Sch. Biological Cryst. (Granada 2006)

47.168 N. Niimura: *Neutron Protein Crystallography: Hydrogen and Hydration in Proteins. Neutron Scattering in Biology* (Springer, Berlin Heidelberg 2006) pp. 43–62

47.169 D.A.A. Myles: Neutron protein crystallography: current status and brighter future, Curr. Opin. Struct. Biol. **16**, 630–637 (2006)

47.170 R.M. Glaeser, K. Downing, D. DeRosier, W. Chiu, J. Frank: *Electron Crystallography of Biological Macromolecules* (Oxford Univ. Press, Oxford 2007)

1606

48. Crystallization from Gels

S. Narayana Kalkura, Subramanian Natarajan

- 48.1 **Gel Growth in Crystal Deposition Diseases** 1608
  - 48.1.1 Gel Growth of Crystals 1608
  - 48.1.2 Types of Gels 1608
  - 48.1.3 Mechanism of Gelling 1609
- 48.2 **Experimental Methods** 1609
  - 48.2.1 Chemical Reaction 1609
  - 48.2.2 Complex Dilution 1609
  - 48.2.3 Solubility Reduction 1610
  - 48.2.4 Chemical Reduction 1610
  - 48.2.5 Electrochemical/Electrolysis 1610
  - 48.2.6 Crystal Growth in the Presence of a Magnetic Field 1610
  - 48.2.7 Nucleation Control 1610
- 48.3 **Pattern Formation in Gel Systems** 1610
- 48.4 **Crystals Grown Using Gel Technique** 1611
  - 48.4.1 Advantages of Crystallization in Gels 1613
- 48.5 **Application in Crystal Deposition Diseases** 1614
  - 48.5.1 Crystal Deposition Diseases 1614
  - 48.5.2 Significance of In Vitro Crystallization 1614
  - 48.5.3 Crystallization of the Constituents of Crystal Deposits 1616
- 48.6 **Crystal-Deposition-Related Diseases** 1616
  - 48.6.1 Urinary Stone Disease 1616
  - 48.6.2 Theories of Urinary Stone Formation 1616
  - 48.6.3 Role of Trace Elements in Urinary Stone Formation 1617
- 48.7 **Calcium Oxalate** 1617
  - 48.7.1 Crystallization of Calcium Oxalate 1617
  - 48.7.2 Effect of Trace Elements 1618
  - 48.7.3 Effect of Tartaric and Citric Acids. 1618
  - 48.7.4 Effect of the Extracts of Cereals, Plants, and Fruits 1618
- 48.8 **Calcium Phosphates** 1619
- 48.9 **Hydroxyapatite (HAP)** 1620
- 48.10 **Dicalcium Phosphate Dihydrate (DCPD)** ... 1620
  - 48.10.1 Effect of Additives on Crystallization of Calcium Phosphates 1620
  - 48.10.2 Effect of Some Extracts of Cereals, Plants, and Fruits and Tartaric Acid 1622
  - 48.10.3 Calcium Hydrogen Phosphate Pentahydrate (Octacalcium Phosphate, OCP) 1622
  - 48.10.4 Magnesium Ammonium Phosphate Hexahydrate (MAP) and Magnesium Hydrogen Phosphate Trihydrate (MHP) 1622
- 48.11 **Calcium Sulfate** 1623
- 48.12 **Uric Acid and Monosodium Urate Monohydrate** 1623
- 48.13 **L-Cystine** 1624
- 48.14 **L-Tyrosine, Hippuric Acid, and Ciprofloxacin** 1625
- 48.15 **Atherosclerosis and Gallstones** 1625
  - 48.15.1 Crystal Growth in Bile 1625
  - 48.15.2 Cholesterol and Related Steroids 1626
  - 48.15.3 Cholic Acid 1627
- 48.16 **Crystallization of Hormones: Progesterone and Testosterone** 1628
- 48.17 **Pancreatitis** 1628
  - 48.17.1 Calcium Carbonate 1629
- 48.18 **Conclusions** 1629
- **References** 1630

Among the various crystallization techniques, crystallization in gels has found wide applications in the fields of biomineralization and macromolecular crystallization in addition to crystallizing materials having nonlinear optical, ferroelectric, ferromagnetic, and other properties. Furthermore, by using this method it is possible to grow single crystals with very high perfection that are difficult to grow by other techniques. The gel method of crystallization provides an ideal technique to study crystal deposition diseases, which could lead to better understanding of their etiology. This chapter focuses on crystallization in gels of compounds that are responsible for crystal deposition diseases. The introduction is followed by a description of the various gels used, the mechanism of gelling, and the fascinating phenomenon of Liesegang ring formation, along with various gel growth techniques. The importance and scope of study on crystal deposition diseases and the need for crystal growth experiments using gel media are stressed. The various crystal deposition diseases, viz. (1) urolithiasis, (2) gout or arthritis, (3) cholelithiasis and atherosclerosis, and (4) pancreatitis and details regarding the constituents of the crystal deposits responsible for the pathological mineralization are discussed. Brief accounts of the theories of the formation of urinary stones and gallstones and the role of trace elements in urinary stone formation are also given. The crystallization in gels of (1) the urinary stone constituents, viz. calcium oxalate, calcium phosphates, uric acid, cystine, etc., (2) the constituents of the gallstones, viz. cholesterol, calcium carbonate, etc., (3) the major constituent of the pancreatic calculi, viz., calcium carbonate, and (4) cholic acid, a steroidal hormone are presented. The effect of various organic and inorganic ions, trace elements, and extracts from cereals, herbs, and fruits on the crystallization of major urinary stone and gallstone constituents are described. In addition, tables of gel-grown organic and inorganic crystals are provided.

48.1 Gel Growth in Crystal Deposition Diseases

48.1.1 Gel Growth of Crystals

Recently, there has been increasing interest in crystal growth in gels on account of its suitability to grow crystals of biological macromolecules and in studies involving biomineralization. The gel method of crystal growth is probably the most simple and versatile technique, compared with other methods of crystal growth from solutions under ambient conditions [48.1]. In 1896, *Liesegang* first observed the periodic precipitation of slightly soluble salts in gelatin [48.2]. Later, this technique was used extensively to crystallize many organic, inorganic, and even biological macromolecules in various colloidal media. The gel method of crystallization is well described by *Henisch* [48.1, 3], *Arora* [48.4], and *Patel* and *Venkateswara Rao* [48.5] as well as by *Lefaucheux* and *Robert* [48.6]. The present chapter aims to review crystallization in gels with a special focus on crystallization of compounds causing crystal deposition diseases.

48.1.2 Types of Gels

Gel growth is a particular case of solution growth where the solution is trapped in a polymeric structure. The gel is a loosely linked polymer of a two-component system formed by the establishment of a three-dimensional system of cross-linkages between molecules of one of the components. The system as a whole is permeated by the other component as a continuous phase, giving a semisolid, generally rich in liquid. Gels can be prepared by a variety of techniques and materials. The most commonly used ones for crystallization are gels of silica, agar, gelatin, clay (bentonite), and polyacrylamide.

Silica gel is prepared [48.1] by mixing aqueous solution of sodium metasilicate and mineral or organic acid. Gelation takes place in times ranging from a few seconds to a few months, depending on the pH, temperature, and concentration of the gel solution. To form agar gel [48.7], agar-agar is dissolved in water (1–2% by weight) and boiled. When this solution cools down, gelation takes place. Gelatin gel is prepared [48.8] by dissolving gelatin in water, stirring at a constant temperature of 50 °C for 1 h and cooling to room temperature. A small quantity of formaldehyde is added to strengthen the gel. To prepare clay gel, powdered clay is slowly sifted on rapidly stirred water making a blend until about 9% clay has been added; the gel sets immediately. Polyacrylamide gel [48.9] is prepared by dissolving 3.99 wt % acrylamide and 0.02 wt % of a cross-linking agent in water. The solution is bubbled with nitrogen and degassed by lowering the pressure. This results in a rigid transparent gel. Crystallization has also been attempted using other gels such as pectin, polyethylene oxide (PEO), polyvinyl alcohol (PVA), and tetramethoxysilane (TMOS) [48.1, 3–7].

48.1.3 Mechanism of Gelling

Gels can be formed by cooling of a sol, by chemical reaction or by addition of a precipitating agent or incompatible solvents. The time taken for the gelling process varies from minutes to days, depending on the nature of the reagents and its temperature, pH, and history. The mechanical properties of fully developed gels can vary, depending on the gel density.

In the case of silica gels, when sodium metasilicate (Na_2SiO_3) is dissolved in water, monosilicic acid and sodium hydroxide are produced initially as per the following reaction

$$Na_2SiO_3 + 3H_2O \rightarrow H_4SiO_4 + 2NaOH . \qquad (48.1)$$

Later on, monosilicic acid polymerizes with the liberation of water. The process of polymerization continues, until a three-dimensional network of Si-O links is established. The hydrogen ion concentration (pH) plays a vital role in the gelling process. During polymerization, it is known that two types of ions, viz. $H_3SiO_4^-$ and $H_2SiO_4^{2-}$, are produced, whose relative amounts depend on the pH. The formation of the more reactive $H_2SiO_4^{2-}$ is favored at high pH values. However, higher charge implies a greater degree of mutual repulsion. $H_3SiO_4^-$ is favored at low pH values and is responsible for the sharp increase in viscosity. Very high or very low pH values inhibit gelation. Gelling mechanism regarding agar and TMOS gels are reviewed by *Lefaucheux* and *Robert* [48.6].

48.2 Experimental Methods

Crystal growth methods in gels fall into the following classes: chemical reaction, complex dilution, reduction of solubility, chemical reduction, and electrochemical techniques [48.5].

48.2.1 Chemical Reaction

This method is suitable for crystals which are mostly insoluble or sparingly soluble in water and which decompose before melting. Here, two soluble reactants are allowed to react inside the gel medium by incorporation of one of the reactants (I) in the gel, whereas the other reactant (II), which is used as the supernatant, diffuses into the gel medium (Fig. 48.1). The reaction inside the gel leads to the formation of an insoluble or sparingly soluble crystalline product. The basic requirements of this method are:

1. The gel must remain stable in the presence of reacting solutions
2. It must not react either with the solutions or with the product.

U-tubes are used, where the two reactants are allowed to react by diffusion into an inactive gel. This technique can also be adopted to crystallize compounds which have poor aqueous and organic solubility (uric acid and cystine) using the displacement reaction method [48.10, 11].

48.2.2 Complex Dilution

In this method, the material to be crystallized is first complexed in some reagents which enhance its solubility. It is then allowed to diffuse into a gel which is free from active reagents. As the complex solution diffuses through the gel it gets diluted. This results in a high supersaturation of the material to be crystallized and

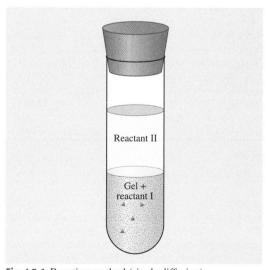

Fig. 48.1 Reaction method (single diffusion)

hence nucleation and subsequent crystallization occurs. Crystals of mercuric sulfide, cuprous, and silver halides and selenium have been grown using this method [48.5].

48.2.3 Solubility Reduction

This method is generally used for crystallization of highly water-soluble substances. The substance to be grown is dissolved in water and incorporated in the gel before gelation. After the gel has been set, a solution that reduces the solubility of the substance (solute) is added as a supernatant solution to induce crystallization. Compounds which have low aqueous solubility can also be crystallized using this technique. Steroids which have low aqueous solubility have been crystallized using this technique by reducing the water content in the gel by incorporating an organic solvent [48.12]. During this process of crystallization, crystals can be observed in the supernatant solution above the gel due to the reverse diffusion of the precipitating solvent [48.13, 14].

48.2.4 Chemical Reduction

This method is particularly suitable for growing metallic crystals (Cu, Au, and Ni). The metallic salt is incorporated with the gel and an aqueous solution of a reducing agent is slowly allowed to diffuse through the gel, where chemical reduction takes place to form metallic crystals [48.5].

48.2.5 Electrochemical/Electrolysis

Gels can be used to crystallize metals by electrolysis. *George* and *Vaidyan* [48.15] reported growth of single crystals and dendrites of silver. Recently, *Muzikar* et al. [48.16] reported crystallization of microcrystals of gold. Here, a small current was passed between two gold electrodes through a silica gel doped with $HAuCl_4$. Single crystals of gold of sizes ranging from hundreds of nanometer to hundreds of micrometer were grown by this technique. *Muzikar* et al. [48.17] have also crystallized platinum particles and a platinum complex by an identical procedure.

48.2.6 Crystal Growth in the Presence of a Magnetic Field

The effect of a magnetic field (up to 3 T) on the crystallization of calcium tartrate [48.18] and strontium tartrate [48.19] has been reported. The magnetic field has the effect of reducing the number of crystals formed but helps in the growth of a few, larger crystals. Crystallization of cholesterol in gels in a magnetic field was studied by *Sundaram* et al. [48.20, 21]; it was found that the presence of the magnetic field reduced the nucleation time and number of crystals but that there was an increase in the size of the crystals grown.

48.2.7 Nucleation Control

During crystallization in gels, the growing crystals compete with one another for the solute atoms, leading to a reduction in crystal size and perfection. Hence, nucleation has to be suppressed until only a few crystals are formed. The use of a particular combination of reactants, to grow the required crystals, has been found to reduce the nucleation density [48.22]. Acid-set gels, choosing a particular acid (based on experiments with various acids), is known to yield larger crystals [48.22]. It has been reported that changing the gel structure by varying one or more of the parameters, viz. pH and density, and by gel aging, can decrease the number of nucleation centers [48.1, 3–7]. The use of an intermediate neutral gel is also found to slow down the reaction and thereby reduce the number of nucleated crystals [48.23]. In another method, the concentration of the diffusing reactant is initially kept below the level at which nucleation is known to occur and then increased gradually in a series of small steps. Crystals grown by this method are more perfect and larger than those grown otherwise.

48.3 Pattern Formation in Gel Systems

Pattern formation is widespread in nature and can be found in structures ranging from agate rocks and gold veins to the growth of bacterial rings in agar and gallstones [48.24–26]. A specific example discussed below is the Liesegang ring structure, discovered in 1896 [48.2]. When coprecipitated ions interdiffuse in a gelatinous medium, the sparingly soluble salt may precipitate discontinuously in a spectacular pattern of parallel bands. The pleasing appearance of Liesegang bands, as well as their spatiotemporal distribution, has elicited a steady proliferation of publications on the subject [48.3, 27–29]. It is quite easy to produce the

Fig. 48.2 Disc and helical Liesegang rings of calcium phosphates in silica gel (after [48.32])

patterns in the laboratory. Liesegang rings can be produced in gels by diffusing one of the reactants into an inert gel medium containing another reactant. The simplest and usual way to perform a Liesegang ring experiment is to fill a tube with an inert semisolid medium which contains one of the reactants (B), called the inner electrolyte. The other reactant (A), referred to as the outer electrolyte, is poured over the gel column as the supernatant solution. Usually the concentration level of the outer electrolyte is maintained high to minimize the possible loss of ions due to evaporation and other such transport mechanisms during the experimental processes. It involves the formation of concentric laminated rings or bands clearly separated in the direction perpendicular to the motion of the front [48.27]. The structures, which are often rings in circular geometries and bands in linear geometries, are formed by the nonuniform spatial distribution of crystals in a precipitation reaction in a gel [48.28, 29]. Recently, *George* and *Varghese* reported the formation of triplet Liesegang ring patterns [48.30]. *Terada* et al. [48.31] described the formation of sheets and helices of strontium carbonate in silica gel. Helical rings of calcium phosphates (Fig. 48.2) were formed in silica gels by the diffusion of calcium ions into a phosphate-containing gel [48.32].

Though in vivo occurrence is rare, chemical concentration, geometry of the reactant containers or vessels, temperature, pH, and the presence of impurities are the main factors that influence the formation of Liesegang rings [48.33]. The role of the gel during the formation of Liesegang rings is essentially passive, i.e., to prevent convection of solutions and sedimentation of the precipitate. The dynamics of the banding are very complex, involving the coupling of diffusion and precipitation processes in a nonequilibrium regime.

Theoretical models explaining Liesegang patterns fall into two broad categories. The first one, called the prenucleation model, is based on the classical feedback cycle of supersaturation, precipitation, and depletion as originally proposed by *Ostwald* [48.34]. In the second theory, the so-called postnucleation or competitive particle growth model, it is assumed that competition between growing particles can, by itself, produce periodic precipitation structures, even in the absence of strong external gradients. It is worth mentioning that all the theories share the assumption that the precipitate appears as the system passes through some nucleation or coagulation thresholds. However, the theories differ in their pre- or postnucleation assumptions. The main unresolved problem in all these theories is the understanding of the mechanism behind the transformation of the diffusive reagents A and B into an immobile coagulant. *George* and *Varghese* proposed a new model [48.30, 35–38] based on the prime assumption that the boundary that separates the outer ions and the inner electrolyte virtually migrates in the positive direction of the advancement of the type A ions. *Izsák* and *Lagzi* [48.39] proposed a universal law, which is also valid in the case of various transport dynamics (purely diffusive, purely advective, and diffusion–advection cases). The mechanism responsible for these structures is not yet fully understood. The interest in precipitate patterning phenomena is growing because of the suitability of their underlying dynamics for modeling many self-organization processes [48.40, 41].

48.4 Crystals Grown Using Gel Technique

Recently, there have been several reports on the crystallization of inorganic, organic, organometallic compounds, and metals apart from biological macromolecules using gels. Some of the compounds (except biological macromolecules) which have been crystallized using gels are listed in Table 48.1a,b.

Table 48.1 (a) Organic/organometallic crystals grown in gels

Crystals grown	Gel used	Method adopted	Studies carried out	Reference
2-Amino-5-nitropyridinium dihydrogen phosphate	Silica	Single diffusion	Crystal growth and growth rate studies by holographic interferometry	[48.42]
Ammonium hydrogen D-tartarate	Silica	Single diffusion	Bulk growth, habit	[48.43]
Antimony thiourea-bromide	Silica	Single diffusion	Dielectric constant, refractive indices	[48.44]
Aspargine monohydrate	Agarose, carrageenan, gelatin	Single diffusion	Crystallization and habit modification	[48.45]
Barium oxalate	Agar-agar	Double diffusion	Growth and characterization	[48.46]
Benzil, phenyl phenol	Sephadex	Single diffusion	Growth and characterization	[48.47]
Bis-(1,3,5-benzenetricarboxylato) dipyridine, Zinc(II) nitrate	PEO	Double diffusion	Crystal growth	[48.48]
Cadmium tartarate	Silica	Double diffusion	Optical absorption	[48.49]
Calcium malate-decahydrate	Silica	Single diffusion	Growth and characterization	[48.50]
Calcium L-tartarate-tetrahydrate (Mn^{2+} doped)	Silica	Single diffusion	Doping and dielectric studies	[48.51]
(Sr^{2+} doped)			Growth and characterization	[48.52]
Cerium lanthanum-oxalate	Silica	Single diffusion	Growth and characterization	[48.53]
Cholic acid	Silica, TMOS	Single diffusion	Growth and characterization	[48.54]
Dysprosium-gadolinium oxalate	Silica	Single diffusion	Optical absorption and fluorescence studies	[48.55]
γ-Glycine	Silica	Single diffusion	NLO studies, dielectric constant, photoconductivity	[48.56]
Iron-manganese levo-tartarate (pure and mixed)	Silica	Single diffusion	Growth and characterization	[48.57]
Lead (II) n-octa-, n-nona- and n-decanoate	Silica	Double diffusion	Morphology	[48.58]
β-DL-Methionine	Silica	Single diffusion	Growth and characterization	[48.59]
3-Methyl-4-nitropyridine-1-oxide	TMOS	Single diffusion	X-ray topography, NLO experiments	[48.60]
L-Phenylalanine	Silica	Single diffusion	Growth and characterization	[48.61]
Rubidium hydrogen-tartrate, strontium tartrate-trihydrate and tetrahydrate	Silica	Single diffusion	Growth and characterization, effect of magnetic field, $(Pb)^{2+}$ doping laser scattering tomography, magnetic susceptibility measurements	[48.62] [48.19] [48.63]
Strontium malate trihydrate	Silica	Single diffusion	Growth and characterization	[48.64]

Table 48.1 (b) Inorganic crystals grown in gels

Crystals grown	Gel used	Method adopted	Studies carried out	Reference
Ammonium chloride	Agarose,	Single diffusion	Periodic roughening transitions, AFM, interference contrast microscopy	[48.65] [48.66]
	Agar, gelatin, pectin, PVA	Reduction of solubility by reducing the temperature	Morphology	
Barium nitrate, boric acid, potassium dichromate	Agar, gelatin, pectin, PVA	Reduction of solubility by reducing the temperature	Morphology	[48.66]
Barium iodate-monohydrate, calcium iodate (mono- and hexahydrates)	Silica	Single diffusion Double diffusion	Growth and characterization Growth, doping microtopography	[48.67] [48.68]
Calcium tungstate	Silica	Double diffusion	Nucleation behavior, crystal morphology	[48.69]
Copper iodide	Silica	Single diffusion (decomplexation)	Growth and characterization Influence of CuI-HI complex	[48.70] [48.71]
Gold	Silica	Electrochemical growth	Backscatter electron diffraction	[48.16]
Lead bromide	Silica	Double diffusion	Growth	[48.72]
Lead carbonate	Silica	Double diffusion	Topotactic relationships	[48.73]
Lead hydrogen-phosphate	TMOS	Single diffusion	Influence of a polycrystalline precipitation zone	[48.74]
Lead iodide	Silica	Single diffusion	Morphology, microstructure	[48.75]
β-Lithium ammonium-sulfate,	TMOS	Single diffusion	Morphology, epitaxic phenomena	[48.76]
β-Lithium sodium-sulfate, lithium sulfate-monohydrate	TMOS	Single diffusion	Epitaxic, intergrowth phenomena	[48.77]
Platinum, ammonium-hexachloroplatinate(IV)	Silica	Electrochemical	Growth	[48.17]
Potassium ferrocyanide-trihydrate	Silica	Solubility reduction	Growth and characterization	[48.78]
Sodium bromate	Agrose, gelatin, silica	Single diffusion	Habit changes in various gels	[48.79]
Sodium chlorate	Agarose	Single diffusion	Selective growth and distribution of crystalline enantiomers	[48.80]

48.4.1 Advantages of Crystallization in Gels

The gel is an ideal medium for diffusion reaction and it acts as a three-dimensional crucible holding the crystals in fixed positions without overlapping. It is a chemically inert, transparent system for growing good-quality crystals. It is possible to observe the entire growth process and the grown crystals can be harvested easily. Gel method allows effective control over factors such as density, concentration, and pH of the medium. Since crystals grow in the absence of convection, gel growth provides a good simulation of space growth experiments, as in the case of growth of protein crystals. To get valuable clues towards understanding the biomineralization process, extensive studies have been carried out on carbonates of calcium, barium, and strontium in silica gel medium [48.81–84]. Crystal growth in gels has been used to simulate the crystal growth process in sedimentary environments [48.85]. The simulation of pathological biomineralization in human organs using gel technique and the crystallization of biological macromolecules using gels are some of the recent advancements in this field.

48.5 Application in Crystal Deposition Diseases

Quite a variety of materials occur as crystals in living tissues and plants. Biomineralization is found extensively among living systems (eggs, mollusks, shells, pearls, corals, bone, and teeth). Biomineralization leads to crystals of uniform size and morphology with specific crystallographic orientation and properties [48.86]. The use of sophisticated methods of crystal identification including analytical electron microscopy has led to the discovery of crystals in the human body system. In organisms, some tissues calcify while others do not. The control over their nucleation and growth patterns and their location developed during evolution. Many of the mineral deposits assist in the regulation of the levels of free cations and anions in cellular systems. The presence of crystals in the human body produces both beneficial and harmful effects. Bone and teeth are composed of oriented nanocrystals of hydroxyapatite (HAP, $Ca_{10}(PO_4)_6(OH)_2$). Our sense of balance and acceleration is dependent upon the small calcite crystals present in the inner ear. Many crystal depositions in tissues are pathological, with unusual crystal forms and patterns of mineralization. The harmful effects of crystals result in the pathological deposition of crystals known as crystal deposition diseases.

48.5.1 Crystal Deposition Diseases

Crystal deposition diseases may be defined as pathological processes associated with the presence of microcrystals which contribute to tissue damage and cause pain and suffering [48.87]. The increasing incidence of crystal deposition diseases (Table 48.2) such as heart diseases, gout, gallstones, urinary stones, deposition in eyes, thyroid glands and bone marrow, etc., among our population has resulted in extensive research to understand their etiology and cure. Crystal nucleation frequently occurs in urinary tract, coronary artery tract, and gallbladder, resulting in formation of pathological urinary stones, atherosclerotic plaques, and gallstones, respectively. Although much technical advancements have been made in the area of treatment of crystal deposition diseases, relatively little is known about the mechanisms involved in the process of pathological crystallization. Crystal deposition diseases are the result of a complex sequence of events that give rise to diseases through simple mechanical effects such as blocking ducts or hardening or weakening of flexible tissues. The formation of crystalline deposits in vivo is a multifactorial disorder, some of the significant factors being age, sex, occupation, diet, fluid intake, geographical location, and climate.

48.5.2 Significance of In Vitro Crystallization

Extensive research work is being carried out on the physicochemical aspects of compounds that are responsible for diseases such as atherosclerosis, gallstones, arthritis, urinary stones, etc., to study their growth mechanism and to find factors that can inhibit their growth in our body system [48.87–99]. One of the principal aims in this area is first to identify the mechanism of crystal growth conditions prevalent in biological systems and then to devise means of inhibiting the unwanted crystal growth. Despite the enormous research aimed at obtaining thorough knowledge of the genesis of pathological mineralization, the mechanism of crystal deposition remains largely unexplained. The main reason for this is the fact that mineralization in vivo cannot be observed directly. Also, in vivo experiments are possible only to a limited extent. Thus, the need to understand the situations that give rise to crystal deposition diseases has necessitated in vitro investigations of the crystalline components that lead to pathological deposits. The crystal growth process in gels is essentially like crystallization from solution and consists of the following three basic steps [48.81]:

1. Attainment of supersaturation of the salt to be deposited

Table 48.2 Location of stones/crystal deposits and the major crystals involved in crystal deposition diseases (*stones or crystal deposits are heterogeneous mixtures, some of them with more than one crystalline phase held together by a matrix)

Diseases	Location of the deposits	Major crystals involved*
Acute and chronic gouty arthritis	Joints of hands and feet	Monosodium urate monohydrate, uric acid
Acute pseudo gout	Knees, wrists, and pelvis	Calcium pyrophosphate dihydrate
Acute calcific periarthritis	Shoulders, hip, and spine	Hydroxyapatite (HAP)
Atherosclerotic arteries and veins	Arteries and veins	Cholesterol
Calcific pancreatitis	Pancreas	Calcium carbonate
Cholelithiasis	Gallbladder	Cholesterol
Urolithiasis	Stones in kidney, ureter, bladder, and urethra	Calcium oxalate-monohydrate (COM), calcium oxalate dihydrate (COD), dicalcium-phosphate dihydrate (DCPD), magnesium hydrogen phosphate-trihydrate (MHP), magnesium ammonium-phosphate hexahydrate (MAP), uric acid, urates, cystine, xanthine

Table 48.3 Constituents of crystal deposits grown in gel

Crystals grown	Gel used	Method adopted	References
Calcium carbonate	Silica, polyacrylamide	Double diffusion	[48.82, 90–94]
Calcium oxalate-monohydrate (COM) Calcium oxalate dihydrate (COD)	Agar-agar, agarose, gelatin, silica	Single, double diffusion	[48.89, 95–99]
Calcium sulfate dihydrate	Silica	Double diffusion	[48.100, 101]
Cholesterol	Silica	Single diffusion	[48.12, 102, 103]
Cholesteryl acetate	Silica	Solubility reduction	[48.12]
L-Cystine	Silica	Single, double diffusion	[48.11, 104]
		Single, double diffusion	[48.98, 99, 105–110]
Hippuric acid	Silica	Double diffusion	[48.111]
Hydroxyapatite (HAP)	Silica	Single diffusion	[48.112, 113]
Magnesium hydrogen-phosphate trihydrate (MHP), magnesium ammonium-phosphate hexahydrate (MAP)	Silica	Single diffusion	[48.114]
Monosodium urate-monohydrate (MSUM)	Silica, TMOS	Single diffusion	[48.115]
Octacalcium phosphate (OCP)	Polyacrylamide	Double diffusion	[48.116]
β-Sitosterol	Silica	Single diffusion	[48.12]
L-Tyrosine	Silica	Single diffusion	[48.117]
Uric acid	Silica, TMOS	Single diffusion	[48.10]

2. Nucleation
3. Subsequent crystal growth.

Investigations of either solution or solid-phase phenomenon will yield useful information regarding the etiology of calculi formation. The growth of crystals in vivo will be dependent on the same factors of solubility, nucleation, and growth rate as in vitro. It is also possible to study the effects of various inhibitors (covering a wide range of molecular weights) which are present in biological fluids, by knowing the growth and dissolution of the stones and stone mineral phases. The in vitro experiments simulate the conditions of stone formation artificially to infer the general principles of calculogenesis, which can shed some light on crystallogenesis of pathological stones.

There is increasing interest in using gel as a crystallizing medium, as it acts as a medium for crystallization of biomolecules and to evaluate the processes of biomineralization in vitro [48.6, 12, 32, 88, 89]. This is because of their viscous nature, providing simulation of synovial fluid, cartilage, and other biological fluids. This method also provides a convenient technique for assessing the effect of various compounds (present in the human body) in altering crystallization parameters such as aggregation of crystals and also the morphology of the insoluble particles. Hence, the gel method of crystal growth is considered ideal for studying crystal deposition diseases [48.88, 89].

48.5.3 Crystallization of the Constituents of Crystal Deposits

Table 48.3 lists references (1994–2006) related to the gel growth of materials that are constituents of calculi occurring in crystal deposition diseases. The references to earlier research contributions of this category are available in previous reviews [48.1, 3–6].

48.6 Crystal-Deposition-Related Diseases

48.6.1 Urinary Stone Disease

Generally, urinary stones are called renal stones, ureteral stones, bladder stones, or urethral stones, depending on their specific site of growth (Table 48.2). Variations in urine flow and tube blockage can cause local fluctuations in pH and concentration, leading to supersaturation of urine and subsequently crystal nucleation, growth, and aggregation resulting in the formation of calculi of different shapes and sizes (Fig. 48.3 [48.118, 119]).

The ultimate control of urolithiasis requires proper application of both approaches of stone removal and drug therapy. Drug therapy is mainly aimed at inhibition of growth and dissolution of the existing stone and prevention of stone recurrence. Hence, it is important to understand the mechanism of stone formation and identification of the inhibitors of different crystalline materials present in urinary calculi.

48.6.2 Theories of Urinary Stone Formation

There are various theories to explain the actual mechanism by which urinary calculi develop [48.89, 120–124]. *Randal*, in 1937 [48.125], reviewed all the existing theories and described that there must be some lesion in the renal pelvis to which a calculus could remain anchored during its period of growth. *Anderson* and *McDonald*, in 1946 [48.126], improved this theory by suggesting that the small crystalline deposits in the kidney reach renal papilla, come into contact with urine, and subsequently grow into stones by deposition of urinary salts. *Butt* and *Hauser*, in 1952 [48.127], concluded that crystalline constituents of the urine separate out of the urine because of the shortage of protective colloids. Crystalline deposits can also occur as a result of excessive excretion or abnormal pH of the urine. *Vermeulon* and *Lyon*, in 1968 [48.128], proposed an advanced theory by explaining the different stages such as supersaturation, nucleation, growth, aggregation, and retention involved in the formation of the

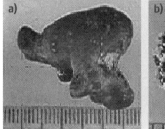

Fig. 48.3a,b COM urinary stones with (a) smooth surface (b) rough surface

stone. During the formation of urine, when it traverses the nephron, much of its water content is reabsorbed and urine gets supersaturated with respect to substances of limited solubility such as those found in stones. Molecular agglomerates of these substances having proper lattice orientation result in the formation of nuclei, or such nucleation can also be extraneous. When these nuclei attain critical size, they act as seeds and then grow into visible crystals. There are two different types of agglomeration, viz. primary and secondary agglomeration. Primary agglomeration is the growth of crystals on the surface or tips of crystals already formed. Secondary agglomeration is that which results from crystal-to-crystal collision. Primary agglomeration was recognized as a possible mechanism for the development of calculi [48.129, 130]. The conditions prevailing in the kidney are not conducive to the secondary agglomeration process, which is expected to play only a minor role in the formation of stones. The crystals, once formed, are retained inside the urinary tract and grow [48.89]. New small crystals originate upon the larger parent crystals and the matrix material binds crystalline material together, giving the stone its cohesive wholeness.

48.6.3 Role of Trace Elements in Urinary Stone Formation

The presence of certain trace elements may play a vital role in the formation of crystal deposition diseases. In recent years, the function of trace elements in the human body and the environment and their importance for medical practice have been increasingly recognized in the biomedical field. It has been suggested that some trace elements enhance the growth rate of deposits of crystalline compounds and that their high concentration in body fluids and tissues may be significant in this respect. One of the main problems is in determining whether nucleation is essentially homogeneous and takes place spontaneously from highly supersaturated fluids or whether it is heterogeneous and is initiated by some other agent such as trace elements (Mg, Pb, Ba, Fe, Sr, Zn, etc. [48.125, 127, 128, 131–133]). Impurities and diffusion limitations are suspected to play a major role in agglomeration. The process of aggregation through electrostatic forces would be expected to be controlled by the zeta potential of the particles, which is readily modified if any ion gets absorbed on the particles.

48.7 Calcium Oxalate

Calcium oxalate urolithiasis is the most common pathological condition associated with oxalate crystal deposition. Calcium oxalate crystalline deposits are also found in myocardium, bone marrow, and blood vessels. In urinary stones, calcium oxalate may exist as monohydrate (COM, $Ca(COO)_2 \cdot H_2O$, 43%) or as dihydrate (COD, $Ca(COO)_2 \cdot 2H_2O$, 20.5%). The formation of oxalate stone may be the result of supersaturation of the urine with its components, viz. calcium and oxalate, deficiency of inhibitors or the increase in concentration of promoters. Calcium oxalate forms mixed stones with calcium phosphates, uric acid, and magnesium ammonium phosphate hexahydrate (struvite, MAP, $MgNH_4PO_4 \cdot 6H_2O$).

48.7.1 Crystallization of Calcium Oxalate

COM was crystallized by *Girija* et al. [48.95] using calcium chloride and oxalic acid as the reactants in silica gel. Single-diffusion experiments yielded microcrystals of COM as seen in Fig. 48.4a. Crystallization using double diffusion yielded COM crystals of millimeter size. COM is a polymorphic crystal, and the crystals obtained in this experiment were of prismatic (single, twinned, and bunched) morphology. These crystals enabled the measurement of Vicker's and Knoop microhardness and the values were found to be 192 and 73 kg/mm², re-

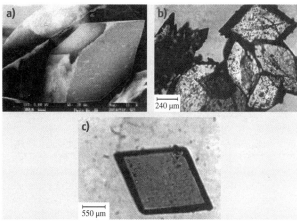

Fig. 48.4a–c COM single crystal in silica gel: (**a**) without any additives (**b**) in presence of cadmium, (**c**) in presence of zinc

spectively. *Achilles* et al. [48.89] designed a flow model of crystallization in gels, where crystal growth is performed from a flow of supersaturated urine at nearly constant supersaturation on COM particles in a gel matrix. Here, they used the discontinuous measurement of scattered light intensity by microphotometry to quantify various crystal growth parameters. These experiments to a large extent simulate the physiological conditions in which the pathological urinary crystals develop in vivo [48.134].

48.7.2 Effect of Trace Elements

Analyses of urinary stones showed the presence of more than 20 trace elements, and it is believed that these elements might play an important role in the formation of these stones [48.133]. *Girija* et al. [48.95] studied the effect of Mg, Cd, Zn, and Pb on the crystallization of COM at room temperature. There was significant influence of these ions on the process of primary agglomeration of COM.

The presence of the elements Mg and Zn, which are known to act as inhibitors of calcium oxalate urinary stones, was found to have less effect on crystal agglomeration and dendritic formation than the presence of the nephrotoxic elements Cd and Pb. Furthermore, Cd and Pb reduced the number of individual crystals and increased the tendency to form agglomerates of COM and dendrites of COD as seen in Fig. 48.4b. The incorporation of Cd and Pb was found to be high, the amount of incorporation following the order: Pb > Cd > Zn > Mg. The addition of Zn modified the morphology of the crystals from prismatic to rhombohedral (Fig. 48.4c [48.95]). Furthermore, *Petrova* et al. [48.97] crystallized COM in gelatin, agar-agar, agarose, and sodium silicate gels and studied the dissolution rates in the presence of Al and Fe. The rate of dissolution was found to increase in the presence of these ions, due to the adsorption of ions on the surface of the crystal, hindering further growth.

48.7.3 Effect of Tartaric and Citric Acids

Marickar and *Koshy* [48.135] crystallized COM, COD, dicalcium phosphate dihydrate (DCPD, brushite, $CaHPO_4 \cdot 2H_2O$), octacalcium phosphate (OCP, $Ca_8(HPO_4)_2(PO_4)_4 \cdot 5H_2O$), HAP, MAP, and magnesium hydrogen phosphate trihydrate (MHP, newberyite, $Mg(PO_3OH) \cdot 3(H_2O)$) by reaction method in silica gel. The effect of known inhibitors of crystallization, viz. tartaric acid, citric acid, and human urine (affected and unaffected), were studied. Tartaric and citric acid altered the crystal habit and inhibited growth. Urine from patients with stones modified the crystal habit. Normal urine samples produced reduction in the size of the crystals along with change in morphology. It is surmised from these observations that normal urine contains inhibitors of crystallization and that these were absent in the urine of certain patients affected by urinary stones.

48.7.4 Effect of the Extracts of Cereals, Plants, and Fruits

Natarajan et al. [48.99] grew crystals of COM, DCPD, and MAP in silica gel medium using reaction method, incorporating extracts of some cereals, plants, and fruits to screen these materials for their inhibitory role on crystallization of these compounds. Control experiments were carried out simultaneously to compare the growth of crystals without [48.98] and with the extracts. The concentration of the various substances incorporated in the gel was about 5 g/l. The generally expected inhibitory effects are the following:

1. No nucleation to lead to crystal growth (a powdery mass being produced)
2. Reduction in the number and size of the crystals
3. Change in the morphology of the crystals.

By carefully observing the shape, size, and approximate number of crystals obtained and also from the knowledge of their total mass, conclusions were derived regarding the inhibitory or promotive effect of the extracts incorporated. It was observed that the extracts of *Cocus nucifera*, *Citrus limon*, *Lycopersicon esculentum*, *Mimosa pudica*, *Musa sapientum*, *Hordeum vulgare*, *Tribulus terrestris*, and *Policos biflorus* produced very good inhibitory effects on COM crystallization. The extracts of *Mentha spicata*, *Raphanus sativus*, and *Vitis vinifera* had no effect on crystal growth. It was also interesting to observe that the extracts of *Borassus flabellifer* and *Ananas comosus* promoted the crystallization, producing larger crystals of COM crystals.

Irusan et al. [48.108] studied the effect of the herbal extracts *Phyllanthus niruri* and *Ocimum sanctum* on the crystallization of DCPD and found that, in the presence of these herbal additives, there was considerable reduction in the thickness of the crystals and also that the crystals became dendritic in nature. *Freitas* et al. [48.136] have also studied the effect of the aqueous extract of the folk medicinal plant *Phyllanthus niruri* on COM crystallization and concluded that the plant extract has an inhibitory effect on crystal

Table 48.4 Effect of cereal, plant, and fruit extracts on the crystallization of some urinary stone components

Cereal/plant/fruit Botanical name	Popular name	Effect of extracts on crystallization			Reference
		COM	DCPD	MAP	
Ananas comosus	Pineapple	Promoter	Inhibitor	No effect	[48.99]
Bergenia ligulata	Pashanbhed	Inhibitor	Not done	Not done	[48.137]
Borassus flabellifer	Palm juice	Promoter	Inhibitor	Inhibitor	[48.99]
Citrus limon	Lemon fruit	Inhibitor	Inhibitor	Inhibitor	[48.99]
Cocus nucifera	Tender coconut water	Inhibitor	Inhibitor	Inhibitor	[48.99]
Hordeum vulgare	Barley	Inhibitor	Inhibitor	Promoter	[48.99]
Lycopersicon esculentum	Tomato fruit	Inhibitor	Inhibitor	Inhibitor	[48.99]
Mentha spicata	Mint	No effect	Inhibitor	Inhibitor	[48.99]
Mimosa pudica	Touch-me-not plant	Inhibitor	Inhibitor	Inhibitor	[48.99]
Musa sapientum	Plantain stem	Inhibitor	No effect	Inhibitor	[48.99]
Oscimum sanctum	Thulasi	Not done	Inhibitor	Not done	[48.108]
Phylanthus niruri	Kizhanelli	Not done	Inhibitor	Not done	[48.108]
		Inhibitor	Not done	Not done	[48.136]
Policos biflorus	Horsegram	Inhibitor	Promoter	No effect	[48.99]
Rahanus sativus	Radish	No effect	Inhibitor	Inhibitor	[48.99]
Tamarindus indica	Tamarind fruit pulp	Promoter	Inhibitor	Inhibitor	[48.99]
		Not done	Inhibitor	Not done	
Tribulus terrestris	Nerringi	Inhibitor	Inhibitor	Promoter	[48.98]
		Inhibitor	Not done	Not done	[48.138]
Vitis vinifera	Grapefruit	No effect	Inhibitor	No effect	[48.99]

growth. Recently, *Joshi* et al. [48.137] have shown the inhibitory effect of two herbs, viz. *Tribulus terrestris* and *Bergemia liuqulata*, on the growth of COM crystals, in vitro. *Bergemia liuqulata* was found to be a better inhibitor of COM than was *Tribulus terrestris*. They infer that the presence of some macromolecules in these herbs seems to play an important role in inhibition of COM crystals. It may be pertinent to point out that the particular ingredient of the cereals/plants/fruits responsible for the inhibitory or promotive effect is yet to be identified. The observations of the various experiments described are summarized in Table 48.4.

48.8 Calcium Phosphates

Calcium phosphates are the major constituents of bone and teeth. Dental calculus, bursitis, arthritis etc. are the pathological conditions involving the deposition of calcium phosphate-like minerals. Calcium phosphate may mineralize in various phases, such as HAP, tricalcium phosphate (TCP, whitlockite, $Ca_3(PO_4)_2$), OCP, dicalcium phosphate (DCP, monetite), and DCPD, in order of increasing solubility. Investigations on the

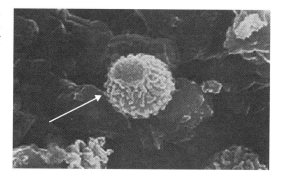

Fig. 48.5 Microstructure of a COM stone showing HAP as nidus of about 600 μm in diameter (indicated by an *arrow*) ▶

nucleation and growth of calcium phosphates such as DCPD, DCP, OCP, TCP, and HAP have received considerable attention in connection with the formation of metabolic and nonmetabolic stones [48.138–140]. Among these, DCPD plays an important role in the formation of caries lesions under acidified conditions. Needles of DCPD crystals were reported in the outer layer of calculi [48.141]. It also forms the basis for the nucleation of calcium oxalate crystals.

Tiny crystals of HAP sometimes form in and around joints and can cause inflammation around the articular cartilage and periarticular connective tissues. They have been described as a cause of inflammations of the shoulder [48.87]. HAP is also a constituent of the urinary calculi and is sometimes found in the nidus of COM stones (Fig. 48.5 [48.142]). HAP has also been detected in atherosclerotic lesions [48.143].

48.9 Hydroxyapatite (HAP)

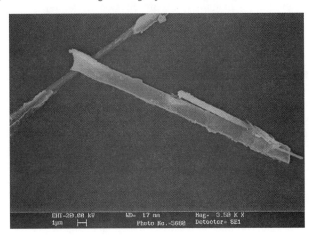

Fig. 48.6 Platy crystals of HAP grown in silica gel (after [48.144], © IOS) ◄

At present, there is no simple technique to crystallize large single crystals of HAP at ambient temperature. The crystal growth of HAP at ambient temperature and pressure conditions always yielded crystals of nanometer size. *Ashok* et al. [48.112] reported the crystallization of platy crystals of HAP, using single-test-tube diffusion technique in silica gel media at physiological temperature (37 °C). Discs of HAP containing platy crystals of size $27 \times 2\,\mu\text{m}$ were obtained (Fig. 48.6, [48.112, 144]) using low-temperature crystallization techniques using gels. *Girija* et al. [48.113] also crystallized HAP biomimetically on collagen gel to evaluate the biomineralization process.

48.10 Dicalcium Phosphate Dihydrate (DCPD)

There are several reports on the crystallization of DCPD in gels [48.109]. Spherulitic DCP and DCPD of different morphologies which have close resemblance to those found in pathological joints, stones, and dental calculi were crystallized in silica gel at physiological pH and temperature by *Sivakumar* et al. [48.107]. The crystals grown without any additives had characteristic platy and spherulitic morphology of DCPD and DCP, respectively (Figs. 48.7, 48.8a). Scanning electron photomicrographs of synthesized spherulitic DCP crystals revealed that the crystal consisted of radially arranged aggregation of rectangular rod-like spokes and rectangular platelets (Fig. 48.8b).

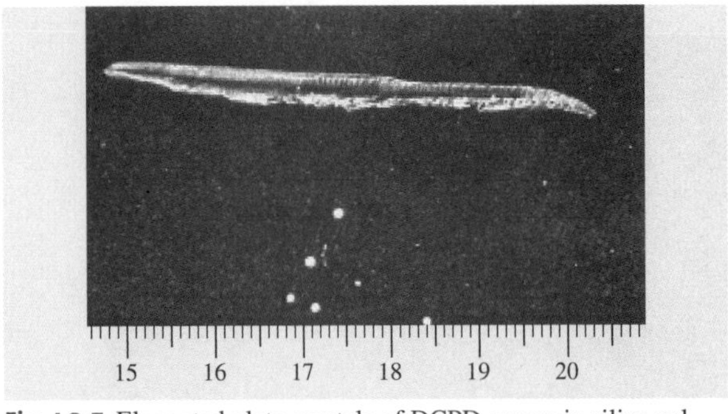

48.10.1 Effect of Additives on Crystallization of Calcium Phosphates

The influence of organic and inorganic ionic species (Mg/Ca, Cd, Zn, Pb, $P_2O_7^{4-}$, and citrate) on crystal-

Fig. 48.7 Elongated platy crystals of DCPD grown in silica gel

lization of DCP, DCPD, and magnesium phosphates have also been studied by *Sivakumar* et al. in silica gel [48.107, 114]. Incorporation of magnesium ions in the crystallizing medium led to a change in the size and morphology of crystals and the mechanical properties of the crystals (Fig. 48.9).

Even when the concentration of the calcium ions was considerably less than that of the magnesium ions, platy DCPD crystals were found to crystallize along with the MHP and MAP crystals (Fig. 48.10). The hardness of the MHP and MAP crystals increased with increasing calcium ions. The presence of magnesium ions in the crystallizing medium drastically reduced the hardness and also reduced the number of DCPD crystals. The magnesium ions decreased the hardness, whereas lead ions increased the hardness of the DCPD crystals. The incorporation of calcium in MHP and MAP crystals increased the hardness, whereas cadmium reduced the hardness. The presence of pyrophosphate and citrate ionic species in the crystallizing medium did not have any influence on the morphology, microstructure, and lattice parameters of DCP, DCPD, MHP, and MAP crystals. From in vitro crystallization using silica gels, it was concluded that Mg^{2+}, Zn^{2+}, $P_2O_7^{4-}$, and citrate ions inhibit growth of DCP, DCPD, and MAP crystals, whereas Cd^{2+} and Pb^{2+} accelerate the growth of crystals [48.107].

Anee et al. [48.110] studied the effect of Fe^{3+} ions on the crystallization of DCPD at various temperatures (27, 37, and 47 °C) using agarose gel medium. In these experiments, crystallization was inhibited in the absence of Fe^{3+} at 37 and 47 °C. Figure 48.11 shows iron-doped DCPD crystallized by single-diffusion technique at 27 °C. At 47 °C, spherulites of HAP were found to grow in the gel medium along with DCPD crystals. These crystals were of average length 1.33 mm

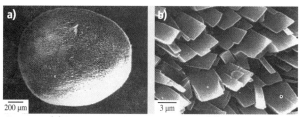

Fig. 48.8 (a) Spherulitic crystals of DCP and (b) its microstructure

Fig. 48.9 Change in habit of DCPD when crystallized in silica gel with magnesium ions

and breadth of 4.28 μm, forming spherulites of more than 2.7 mm in diameter. Previous studies reported crystallization of needle-like morphology of HAP, of only micro- or nanometer size [48.111, 145, 146]. These studies [48.110] showed that Fe^{3+} played a significant role in the crystallization of phase-pure HAP at physiological pH. Furthermore, this technique provided an easier method to synthesize large crystals of HAP, at a low temperature.

Combined influence of an organic and an inorganic ion (cobalt and malic acid) on crystallization of DCPD

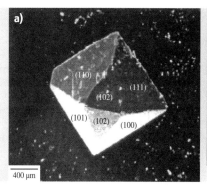
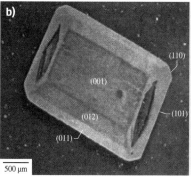

Fig. 48.10a,b Crystals of (a) newberyite (MHP) and (b) struvite (MAP) grown in silica gel

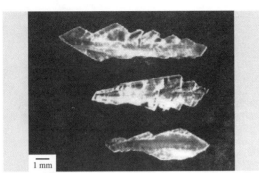

Fig. 48.11 DCPD crystals grown in agarose medium in presence of Fe^{3+}

crystallization of DCPD in silica gels, due to the formation of calcium tartrate in preference to DCPD, as reported by *Joseph* et al. [48.138].

48.10.3 Calcium Hydrogen Phosphate Pentahydrate (Octacalcium Phosphate, OCP)

Iijima and *Moriwaki* [48.116] crystallized OCP in polyacrylamide gels; their experiments revealed that, in addition to other enamel extracellular matrix proteins, amelogenin nanospores are directly involved in controlling the morphology of crystals during enamel mineralization. *Marickar* and *Koshy* [48.135] also reported growth of OCP crystals in silica gel.

48.10.4 Magnesium Ammonium Phosphate Hexahydrate (MAP) and Magnesium Hydrogen Phosphate Trihydrate (MHP)

MAP and MHP are two major crystalline constituents of nonmetabolic or infection-induced urinary stones [48.139, 140]. These stones arise from infection produced by urea-splitting organisms such as *Proteus mirabilis*. The urease-catalyzed hydrolysis of urea generates ammonia, which elevates urine pH and causes precipitation of Mg^{2+}, as struvite. Though surgical treatment gives an immediate relief, it is usually followed by relapse of infection and recurrence of stone formation. The recurrence rate of struvite and newberyite stone formation is high compared with other urinary stone constituents [48.148].

MAP and MHP were crystallized through the direct crystallization method in silica gel medium [48.114] as mentioned in Sect. 48.10.1; the presence of Mg/Ca, Zn, $P_2O_7^{4-}$, and citrate was found to inhibit the growth of MAP and MHP, whereas Pb and Cd promoted growth.

The effect of the extracts of several cereals, plants, and fruits on the crystallization of MAP was studied by *Natarajan* et al. [48.99]. As evident from the Table 48.4, all the extracts screened (except *Vitis vinifera*, *Policos biflorus*, *Ananas comosus*, *Hordeum vulgare*, and *Tribulus terrestris*) played inhibitory roles in the growth of MAP crystals. *Hordeum vulgare* and *Tribulus terrestris* promoted, whereas *Vitis vinifera*, *Policos biflorus*, and *Ananas comosus* had no effect on crystallization of MAP.

in agarose gel was reported by *Anee* et al. [48.105]. Morphological changes were noticed when the growth assay was doped with cobalt and malic acid. Even at very low concentrations, malic acid exerted an influence on crystal morphology, producing elongated crystals. Increase in the malic acid concentration in the crystallization media promoted entry of Co^{2+} into the DCPD crystals. Hence, malic acid is expected to be a mediator for transporting Co^{2+} from the supernatant liquid into the gel where crystallization occurred. However, no crystallization occurred in the presence of malic acid alone. Presence of cobalt promoted crystallization, whereas malic acid modified the morphology of DCPD crystals. The molecules of organic additives can bind specifically to certain crystal planes during growth and modify the morphology of the crystals [48.147]. These types of studies help in understanding the process by which organic molecules control the biomineralization process leading to crystal deposition.

48.10.2 Effect of Some Extracts of Cereals, Plants, and Fruits and Tartaric Acid

Natarajan et al. [48.99] have also grown crystals of DCPD using reactants in silica gel and experimented with extracts of various cereals, plants, and fruits as possible inhibitors, as described in Table 48.4. It is of interest to observe that all the extracts tested, except *Policos biflorus* and *Musa sapientum*, inhibited crystallization of DCPD. The extract of *Policos biflorus* showed a promotive effect, whereas *Musa sapientum* had no effect on the crystal growth. Tartaric acid and tamarind (*Tamarindus indica*) were found to inhibit the

48.11 Calcium Sulfate

Naturally occurring mineral, gypsum (calcium sulfate dihydrate ($CaSO_4 \cdot 2H_2O$, CSD)) is one of the rare constituents of urinary stones. Gypsum also finds applications in dentistry and pottery, apart from its major use in the medical field to immobilize broken limbs after the bones have been set. Growth of gypsum crystals in silica gel has been reported [48.100, 101]. *Kumareson* and *Devanarayanan* [48.101] crystallized CSD using double-diffusion technique in silica gel and further studied the influence of additives such as citric acid, borax, and barium chloride. It was found that the presence of these additives increased the perfection of the grown crystals.

48.12 Uric Acid and Monosodium Urate Monohydrate

Uric acid ($C_5H_4N_4O_3$) is the major end-product of purine metabolism (Fig. 48.12). The classic example of a crystal deposition disease is gout. Gout is a very painful inflammatory condition associated with deposition of monosodium urate crystals [48.149] in various tissues of the body. Crystals of uric acid anhydrate and dihydrate as well as its sodium salts have been found in the renal tract, and interstitial tissues of the kidney and its collection ducts. Uric acid anhydrate and dihydrate are often found as a constituent of urinary calculi and have a high incidence in nephrolithiasis resulting in severe kidney damage [48.150, 151]. The solubility of uric acid changes dramatically with pH and, in acidic urine when supersaturated with uric acid, it is deposited as uric acid crystals [48.152]. It is found to occur in pure form or in association with calcium oxalates, HAP, struvite, and sodium and ammonium urates. In addition, it has been suggested that uric acid crystals could act as a support for heterogeneous nucleation of other crystalline species [48.153, 154].

Uric acid is insoluble in alcohol, ether, and most common organic solvents and, owing to its extremely low solubility in water, it would be difficult to prepare a pure specimen because of the large volume of solvent needed. As uric acid has very poor aqueous and organic solubility, it is very difficult to crystallize them in an aqueous medium such as silica gel. *Kalkura* et al. [48.10] crystallized uric acid dihydrate in TMOS and silica gels using single- and double-diffusion techniques (Fig. 48.13). In this case, a solution of uric acid in sodium hydroxide was allowed to diffuse into the gel to produce a displacement reaction with dilute HCl inside the gel medium, producing uric acid crystals. These experiments showed that uric acid, which has a low aqueous and organic solubility, can be crystallized in an aqueous medium such as silica gel and in TMOS without the use of bulk solvents.

Monosodium urate monohydrate (MSUM) is the salt of singly ionized state of uric acid, and a component of urinary stones. It has also been identified in articular tissues of patients managed with long-term dialysis [48.155]. The presence of characteristic needle or spherulitic forms of MSUM in synovial fluid and within synovial leukocytes is recognized as a strong indication of the presence of gouty arthritis [48.156]. The synthetic crystals of MSUM were needle shaped, whereas the crystals grown in the presence of serum, synovial fluid, and components thereof closely resembled the morphology of the MSUM crystals present in the articular cartilage in vivo [48.157]. A saturated solution of

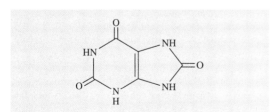

Fig. 48.12 Molecular structure of uric acid

Fig. 48.13 Crystals of uric acid grown in TMOS gels

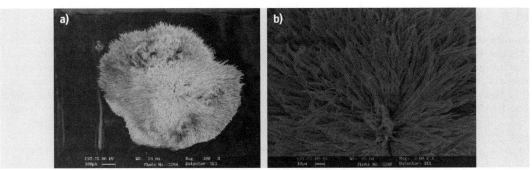

Fig. 48.14 (a) Spherulites of sodium urate monohydrate crystallized in TMOS gels, (b) magnified image

sodium urate in the presence of urate seeds at pH greater than 6.0 will form needle-shaped sodium urate crystals.

MSUM, which resembled the crystals found in synovial fluid of gouty patients, has been crystallized fairly easily by *Kalkura* et al. [48.115] without resorting to any heating of the solutions in a wide range of pH (3–10) using TMOS and silica gels compared with the other existing standard method, viz. slow cooling of a NaOH–uric acid solution kept at pH of 8.9 from 60 to 4 °C, as seen in Fig. 48.14 [48.115, 154].

48.13 L-Cystine

Crystallization of cystine ($C_6H_{12}N_2O_4S_2$, Fig. 48.15), a dibasic sulfur-containing amino acid present in the body, results in disorders such as cystinuria and cystinosis. Cystinuria is characterized by excretion of large quantities of the above amino acid in urine. Cystinosis results in the formation of cystine calculi and could even lead to kidney damage. Cystine crystals could also occur in liver, spleen, thyroid glands, bone marrow, and ocular tissues [48.158]. Cystine can act as a seed and favors the crystallization of MAP [48.159], a urinary stone constituent with a high degree of recurrence [48.160].

Since cystine has low aqueous and organic solubility, it is difficult to crystallize using conventional methods of gel technique. Earlier, cystine crystallization was carried out in solution by dissolving it in an alkaline solution (ammonia) and then reducing it

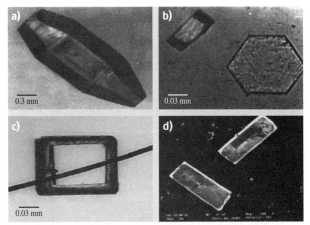

Fig. 48.16a–d Different morphologies of crystals of cystine crystallized from silica gels: (a) Bipyramidal (after [48.11], © Springer 1995), (b) hexagonal (after [48.11], © Springer 1995), (c) hollow crystal with a thread passing through it, (d) rectangular crystals

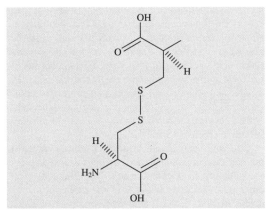

Fig. 48.15 Molecular structure of cystine

using a weak acid, such as acetic acid. In a novel technique, reported by *Girija* et al. [48.11], cystine was dissolved directly in silica gel and the crystals of cystine appeared without the diffusion of any supernatant solution into the gel medium. Here, acetic acid used for acidifying the silica gel also played the role of the reduction agent to produce crystals. Single, twinned, and bunched hexagonal, cubic, rectangular, bipyramidal, and hollow morphologies of cystine crystals (Fig. 48.16a–d) were obtained by this technique [48.11, 104]. Cystine crystallized in the presence of ascorbic acid exhibited tetragonal structure and in addition had hourglass-type inclusion in crystals of rectangular morphology [48.161].

48.14 L-Tyrosine, Hippuric Acid, and Ciprofloxacin

Some components that are rarely found in urinary stones include L-tyrosine ($C_9H_{11}NO_3$), hippuric acid, and ciprofloxacin ($C_{17}H_{18}FN_3O_3$). Tyrosine is an amino acid present in living organisms. Hippuric acid ($C_6H_5CONHCH_2COOH$) is a colorless crystal obtained from urine of domestic animals and humans [48.162]. It is believed to be a natural regulator of urinary saturation with regard to calcium oxalate crystallization, which is the most frequent chemical constituent of all the urinary stones. Also, hippuric acid is a potential nonlinear optical (NLO) material. Ciprofloxacin is an antibiotic drug used for a wide range of infections. Drugs such as ciprofloxacin and triamterene induce calculi and they, along with many other drug metabolites, represent 1–2% of all renal calculi [48.160]. Ciprofloxacin crystals form rarely in humans. *Ramachandran* and *Natarajan* [48.111, 117, 163] reported

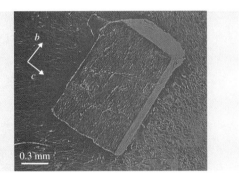

Fig. 48.17 Scanning electron micrograph of a crystal of ciprofloxacin

the crystal growth of L-tyrosine, hippuric acid, and ciprofloxacin (Fig. 48.17) in silica gel.

48.15 Atherosclerosis and Gallstones

Cardiovascular disease is one of the leading causes of death in humans. Cardiovascular diseases include such conditions as heart attacks, strokes, high blood pressure, and heart failure [48.165]. The main cause of these diseases is atherosclerosis (thickening of the artery due to the fatty deposits which are largely composed of cholesterol). Atherosclerotic plaques normally consist of lipids such as cholesterol, cholesteryl esters, and phospholipids [48.166]. These deposits interfere with normal blood flow and may result either in depriving some areas of adequate blood supply, in the formation of a blood clot or in the rupture of blood vessels.

48.15.1 Crystal Growth in Bile

Crystal growth frequently occurs in bile, usually resulting in the formation of gallstones. Generally, gallstones

Table 48.5 Constituents of gallstones and their relative abundance (%) (after [48.164], © Elsevier 1981)

Anhydrous cholesterol	$C_{27}H_{46}O$	52.3
Cholesterol monohydrate	$C_{27}H_{46}O \cdot H_2O$	16.0
Vaterite	$CaCO_3$	6.4
Calcium palmitate	$CH_3(CH_2)_{11}(COO)_2Ca$	5.9
Aragonite	$CaCO_3$	4.6
Calcite	$CaCO_3$	4.1
Cholesterol II	$C_{27}H_{46}O$	2.7
Apatite	$Ca_{10}(PO_4)_6(OH)_2$	2.4
Whitlockite	$\beta\text{-}Ca_3(PO_4)_2$	0.4
Palmitic acid	$CH_3(CH_2)_{14}COOH$	0.1
Artefacts, unidentified material	–	3.3

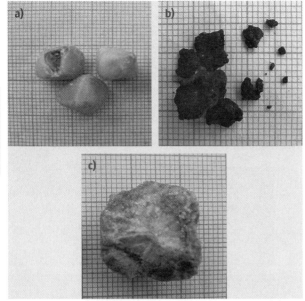

Fig. 48.18a–c Types of gallbladder stones: (**a**) cholesterol, (**b**) pigment, (**c**) mixed type

vary from patient to patient in number, size, color, and shape, as well as in composition (Fig. 48.18a–c, [48.164, 167]). Cholesterol is the most common constituent but certain calcium salts, notably calcium carbonate, are frequently present (Table 48.5). As cholesterol has low aqueous solubility, it is either solubilized in bile salt micelles or associated with different phospholipids, vesicles or with lipoprotein particles as free cholesterol or as cholesterol fatty-acid esters [48.168].

Bile secreted by the liver becomes supersaturated with cholesterol. Such abnormal bile contains an excess of cholesterol relative to the solubilizing agents, bile salts, and the phospholipid lecithin [48.169]. The liver, perhaps as a result of genetic programming, produces supersaturated bile by decreased secretion of bile salts or increased secretion of cholesterol, or both. The excess cholesterol precipitates out of the solution as solid monohydrated cholesterol microcrystals [48.170]. The cholesterol microcrystals precipitated from the bile are retained and subsequently aggregate and grow into macroscopic stones (Fig. 48.18a, [48.171]).

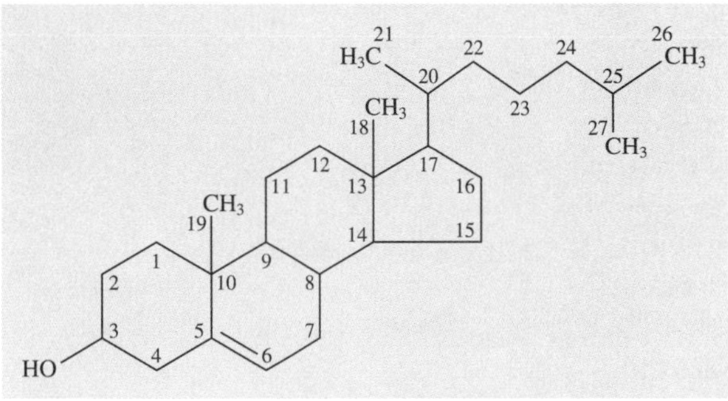

A possible explanation for the lack of cholesterol crystallization in some supersaturated bile is that a nucleating factor is required to seed crystal formation. Evidence has been obtained from bile-mixing experiments that gallbladder bile contains a nucleating factor that could be specific to the gallbladder or bile of cholesterol gallstone patients. Nucleation of cholesterol has been recognized as an important step in gallstone formation but few studies have been performed to examine the ability of potential nucleating agents to influence cholesterol crystal formation. Some of the factors which contribute to crystallization in bile may be crystal aggregation, overgrowth, and epitaxy [48.172].

48.15.2 Cholesterol and Related Steroids

Steroids are an important class of chemical compounds found in virtually all forms of plant and animal life. They are also extensively used in the treatment of ailments ranging from coronary insufficiency to endocrine hormone alterations. Cholesterol ($C_{27}H_{46}O$) is one of the most abundant steroids found in the animal kingdom. It is found in brain, nerve tissues, cell membranes, and gallstones. Plant sterols are structurally similar to cholesterol, and β-sitosterol ($C_{29}H_{50}O$) is the most common plant steroid. β-Sitosterol is a constituent of the gallbladder bile and gallstones. Cholesterol is supposed to be the causative agent for coronary heart diseases and gallstones. Cholesterol and similar compounds are the precursors of the steroid hormones, which play vital roles in human metabolism.

Cholesterol crystals are also seen in rheumatoid nodules, tophi, unicameral bone cysts, and even in isolated cholesterol granulomas of the skin [48.173]. When the cholesterol concentration exceeds the solubility limit in the lipid bilayers of micelles, deposition of cholesterol takes place. This leads to nucleation of crystalline cholesterol and subsequently to atherosclerosis and gallstones. The cholesterol has a ring system as illustrated in Fig. 48.19. It consists of three six-membered rings and a five-membered ring. The interesting fea-

Fig. 48.19 Molecular structure of cholesterol ring

ture of the cholesterol crystals is the bilayer nature of its structure, with a molecular arrangement generally similar to that of cholesterol in biological membrane [48.174].

The physicochemical properties of cholesterol are important to its necessary functions on membranes. Many of the features of cholesterol are also found in its physiologically important derivatives. Cholesterol plays a significant role in the synthesis of essential steroids such as bile acids, sex hormones produced in special glands (ovaries, corpus leutum, placenta, and testis), adrenocortical hormones and vitamin D [48.158]. An esterified form of cholesterol plays an important role in atherosclerosis and cholelithiasis. Cholesteryl esters are precursors for the cyclic formation of free cholesterol synthesis. In addition to cholesterol monohydrate, anhydrous form of cholesterol is also present in freshly removed human gallstones, whereas atherosclerotic plaques contain only monohydrated cholesterol [48.103].

Kalkura and *Devanarayanan* have shown that cholesterol and related steroids such as cholesteryl acetate and β-sitosterol, which has a low aqueous solubility, can be crystallized in an aqueous silica gel medium without cholesterol being precipitated [48.12]. The single-test-tube diffusion method was employed to grow crystals of cholesterol in silica gel medium by solubility reduction technique. Cholesterol and cholesteryl acetate crystals grown in vitro are shown in Fig. 48.20 [48.12]. Cholesterol and β-sitosterol formed anhydrous crystals in the absence of water and monohydrated crystals in the presence of water. Platy and fibrous crystals of cholesterol monohydrate were obtained when crystallized in silica gels. This technique has also been used to study the effect of various solvents and the extracts from medicinal plants on the crystallization [48.102, 175]. The extracts of some Indian medicinal plants, viz. *Commiphora mughul, Aegle marmeleos, Cynoden dactylon, Musa paradisiaca, Polygala javana, Alphinia officinarum*, and *Solafolium* were used as additives to study their effect on the crystallization behavior of cholesterol. It was found that many of these herbs have an inhibitory effect on crystallization in terms of nucleation, crystal size or habit modification. Trace elements also seem to play an important role in the pathological mineralization of gallstones [48.176–178].

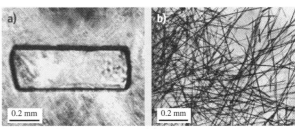

Fig. 48.20 (a) Cholesterol and (b) cholesteryl acetate crystallized in silica gel

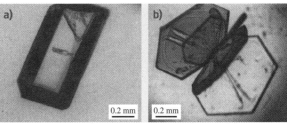

Fig. 48.21a,b Cholic acid crystallized in (a) silica, (b) TMOS gels (after [48.54], © Akademie Verlag 1997)

48.15.3 Cholic Acid

Cholic acid ($C_{24}H_{40}O_5$) is one of the bile acids and is responsible for many physiological functions, including promotion of lipid diffusion through intestinal mucosa, resorption of drugs, vitamins, and hormones, activation/inhibition of various enzyme reactions, and excretion of cholesterol [48.175]. Steroidal cholic acid and its derivatives can serve as host molecules to form inclusion compounds with a wide variety of organic substances. Cholic acid is also a promising candidate to be used as a building block in biomimetic/molecular recognition chemistry and in the construction of extended, preorganized molecular structures [48.179]. *Kalkura* et al. reported the crystallization of cholic acid in silica and TMOS gels. Crystals grew with different morphologies, and Liesegang rings were also seen in some cases. The grown crystals were found to be monohydrate in form (Fig. 48.21a,b, [48.54]). Since the bile salts have the ability to form inclusion compounds with a variety of molecules, these studies will be helpful in understanding the role played by cholic acid and its derivatives in cholesterol crystallization in gallbladder.

48.16 Crystallization of Hormones: Progesterone and Testosterone

Steroid hormones are derived from cholesterol, and a very small amount of these exerts potent physiological effects. Progesterone ($C_{21}H_{30}O_2$) is a naturally occurring mammalian hormone. It is known as the hormone of pregnancy because of its importance just prior to and during the gestation period. It is secreted by corpus luteum (material which surrounds the egg). The main functions of progesterone are the following:

1. Preparation of the uterine endometrium for the implantation of fertilized ovum
2. Maintenance of uterus during/after pregnancy
3. Inhibition of spontaneous contraction of the uterus
4. Participation in the development of breast
5. Inhibition of ovulation [48.180].

Progesterone is an important intermediate in biosynthesis of adrenocortical and gonadal steroid hormones. It also forms the basis for oral contraceptive agents.

Testosterone ($C_{19}H_{28}O_2$) belongs to a class of compounds called androgens. Androgens are responsible for the development of the male sex organs. Testosterone is the most active androgen and it is the functioning hormone found in the testes, ovary, and adrenal cortex. This most potent male sex hormone is isolated from testicular tissue and spermatic vein blood. The important function of this group is in the development of masculine sexual characteristics such as deepening of voice, growth of a beard, and distribution of pubic hair. Their structure is similar to that of the female sex hormones, viz. estrogens [48.181]. Crystallization of the above hormones was carried out by *Kalkura* and *Devanarayanan* [48.12–14] by modifying the techniques of crystallization of steroids in gels. Crystals were obtained (Fig. 48.22a,b) when there was a slow diffusion of water into the gel medium containing sex hormones (progesterone and testosterone) in liquid phase. This suggests a possible mechanism for the growth of crystals. Water, while diffusing through the gel, supersaturates the sex hormone which is in a liquid phase in the gel medium, leading to nucleation of the crystals and their subsequent growth. In addition, crystals also appeared in the supernatant solution, due to the slow reverse diffusion of the precipitating solution from the gel medium into the supernatant solution, leading to the supersaturation of the hormones (Fig. 48.23).

Fig. 48.22a,b Crystals of progesterone grown inside (**a**) silica gel, (**b**) crystallized in the supernatant due to reverse diffusion of the precipitating solvent

Fig. 48.23 Crystals of testosterone as grown in silica gel with crystals in the supernatant solution due to the reverse diffusion (*arrow*)

48.17 Pancreatitis

The pancreas in all mammalian species is an important gland located in the upper abdomen behind and below the stomach. The pancreas has both exocrine and endocrine functions. Inflammation of the pancreas is known as pancreatitis. Chronic pancreatitis is a continuing inflammatory disease of the pancreas characterized

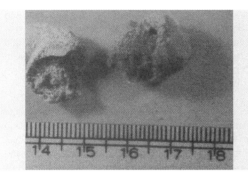

Fig. 48.24 Surgically removed pancreatic stones

by irreversible morphological changes that typically cause pain and permanent loss of pancreatic function. Initially, pain dominates, soon followed by the clinical complication of calcification, leading to stone formation (pancreatic calculi), in the main duct (Fig. 48.24). Biological fluids are generally supersaturated with respect to calcium salts such as oxalates in urine, phosphates in saliva, and carbonates in bile or pancreatic juice. It might become harmful if continuous crystal growth is allowed, leading to the formation of stones. Inhi-

Fig. 48.25 Scanning electron micrograph of a bipyramidal calcite crystal

bition of stones is mainly done by protein inhibitors. In the pancreas, the stone formation is inhibited by a 144-amino-acid glycoprotein (15.5 kDa). However, the role of the pancreatic stone protein (PSP), called lithostathine, is controversial and is expected to contribute to the stone formation.

Studies on the composition of pancreatic stones using x-ray diffraction methods revealed calcium carbonate as the primary constituent [48.182, 183]. In addition, traces of nickel and fatty acids, organic matrix with desquamated epithelium, fibrin mucoid substances, and protein have also been reported [48.182]. Recent studies on pancreatic stones using x-ray diffraction, Fourier-transform infrared (FTIR) spectroscopy, and electron paramagnetic resonance (EPR) by *Narasimhulu* et al. [48.184] support the view that lithostathine has a role in the formation of pancreatic stones.

48.17.1 Calcium Carbonate

Calcium carbonate ($CaCO_3$) exists in three polymorphic forms: aragonite, vaterite, and calcite [48.185]. Calcite occurs as a biomineral and happens to be the major constituent of pancreatic calculi [48.186] and also a constituent of gallbladder and urinary stones [48.156, 159].

Gel method of crystallization has been used extensively to study calcium carbonate crystals to elucidate the biomineralization process. Double-diffusion technique using silica gel has been employed to study the effect of divalent cations Ba^{2+}, Sr^{2+}, Co^{2+}, and Mn^{2+}, in small concentrations (50, 200, and 600 ppm) [48.82]. Furthermore, calcite has also been crystallized using double-diffusion technique with polyacrylamide hydrogels incorporating carboxylate groups [48.93]. *Imai* et al. have crystallized porous aragonite crystals in silica gel [48.94]. *Ramachandran* et al. [48.91] have crystallized $CaCO_3$ in silica gel (Fig. 48.25) and studied the thermal parameters by photoacoustic method.

48.18 Conclusions

The in vivo processes leading to formation and growth of crystals in biological fluids are influenced by various factors which are very complex and not yet fully understood. Gel is the most suitable medium to study biomineralization because of its viscous nature, providing simulation of synovial fluid, cartilage, and other biological fluids where crystallization occurs. An ideal in vitro crystallization technique which mimics or simulates exactly the in vivo mineralization processes is yet to be developed. As described in this chapter, crystallization in gels is an ideal technique to study the pathological biomineralization in vitro, as this method provides a fairly simple technique which closely resembles crystallization in vivo. Furthermore, it can be extended to understand problems concerned with biological, medical, and geological aspects of biominerals.

Almost all major and minor constituents of pathological deposits which are responsible for diseases, such as atherosclerosis, gallstones, urinary stones, etc., can be crystallized in various gel media. Gel method of crystallization is also an ideal method to study the effect on crystallization of electric and magnetic fields, various drugs, herbal extracts, and the epitaxic relationship between the constituents in order to understand the anti- and pronucleating factors, which will help in finding factors that help in dissolution of pathological crystal deposits under physiological conditions. Once stones are formed in our body system, they rarely disappear and always grow in size. The factors responsible for this are not clearly understood. Hence, suitable solvents and inhibitors could be applied to crystals in gel medium to determine the conditions under which already formed crystals can be dissolved, modified or inhibited from the viewpoint of devising new treatments and means to control crystal deposition diseases.

Intense research is going on into the pathogenesis and etiology of crystal deposition diseases. Medical treatment of crystal deposition diseases and prevention from further attacks are still at the experimental stage. Extensive research work is still needed to understand the factors that influence the formation and growth of various crystals in biological fluids.

References

48.1 H.K. Henisch: *Crystal Growth in Gels* (The Pennsylvania State Univ. Press, University Park 1973)

48.2 R.E. Liesegang: Über einige Eigenschaften von Gallerten, Naturwiss. Wochenschr. **11**, 353–362 (1896), in German

48.3 H.K. Henisch: *Crystals in Gels and Liesegang Rings* (Cambridge Univ. Press, Cambridge 1988)

48.4 S.K. Arora: Advances in gel growth: A review, Prog. Cryst. Growth Charact. **4**, 345–378 (1981)

48.5 A.R. Patel, A. Venkateswara Rao: Crystal growth in gel media, Bull. Mater. Sci. **4**, 527–548 (1982)

48.6 F. Lefaucheux, M.C. Robert: Crystal growth in gels. In: *Handbook of Crystal Growth*, ed. by D.T.J. Hurle (Elsevier Science, Amsterdam 1994)

48.7 B. Brezina, M. Havrankova: Growth of KH_2PO_4 single crystals in gel, Mater. Res. Bull. **6**, 537–543 (1971)

48.8 E. Banks, R. Chinanelli, F. Pintchovsky: The growth of some alkaline earth orthophosphates in gelatin gels, J. Cryst. Growth **18**, 185–190 (1973)

48.9 B. Brezina, M. Havrankova, K. Dusek: The growth of $PbHPO_4$ and $Pb_4(NO_3)_2(PO_4)_2 \cdot 2H_2O$ in gels, J. Cryst. Growth **34**, 248–252 (1976)

48.10 S. Narayana Kalkura, V.K. Vaidyan, M. Kanakavel, P. Ramasamy: Crystallization of uric acid, J. Cryst. Growth **132**, 617–620 (1993)

48.11 E.K. Girija, S. Narayana Kalkura, P. Ramasamy: Crystallization of cystine, J. Mater. Sci. Mater. Med. **6**, 617–619 (1995)

48.12 S. Narayana Kalkura, S. Devanarayanan: Crystallization of steroids in gels, J. Cryst. Growth **110**, 265–269 (1991)

48.13 S. Narayana Kalkura, S. Devanarayanan: Crystal growth of steroids in silica gel: Testosterone, J. Cryst. Growth **94**, 810–813 (1989)

48.14 S. Narayana Kalkura, S. Devanarayanan: Growth of progesterone crystals in silica gel and their characterization, J. Mater. Sci. Lett. **7**, 827–829 (1988)

48.15 M.T. George, V.K. Vaidyan: A new electrolytic method to grow silver dendrites and single crystals in gel, J. Cryst. Growth **53**, 300–304 (1981)

48.16 M. Muzikar, V. Komanicky, W.R. Fawcett: A detailed study of gold single crystal growth in a silica gel, J. Cryst. Growth **290**, 615–620 (2006)

48.17 M. Muzikar, P. Polkov, J.C. Fettinger, W.R. Fawcett: Electrochemical growth of platinum particle and platinum-containing crystals in silica gel, Cryst. Growth Des. **6**, 1956–1960 (2006)

48.18 K.V. Saban, T. Jini, G. Varghese: Influence of magnetic field on the growth and properties of calcium tartrate crystals, J. Magn. Magn. Mater. **265**, 296–304 (2003)

48.19 M.H. Rahimkutty, K.R. Babu, K.S. Pillai, M.R.S. Kumar, C.M.K. Nair: Thermal behavoiur of strontium tartrate single crystals grown in gel, Bull. Mater. Sci. **24**, 249–252 (2001)

48.20 N.M. Sundaram, M. Ashok, S. Narayana Kalkura: Observation of cholesterol nucleation in magnetic field, Acta Cryst. D **58**, 1711–1714 (2002)

48.21 N.M. Sundaram: Investigations on the cholesterol, cholesteryl acetate, hydroxyaptite and lysozyme crystallization and the influence of magnetic field on the nucleation process. Ph.D. Thesis (Anna University, Chennai 2004)

48.22 A.R. Patel, A.V. Rao: Gel growth and perfection of orthorhombic potassium perchlorate single crystals, J. Cryst. Growth **47**, 213–218 (1979)

48.23 A.R. Patel, A.V. Rao: Modified gel technique to grow single crystals of $KClO_4$, Indian J. Pure Appl. Phys. **16**, 544–545 (1978)

48.24 B. Sis, T. Canda, Ö. Harmancioglu: Liesegang rings in breast tissue: An unusual component of a for-

48.25 T. Antal, M. Droz, J. Magnin, Z. Racz, M. Zrinyl: Derivation of the Matalon–Packter law for Liesegang patterns, J. Chem. Phys. **21**, 9479–9486 (1998)

48.26 Z. Racz: Formation of Liesegang patterns, Physica A **274**, 50–59 (1999)

48.27 D.S. Chernavskii, A.A. Polezhaev, S.C. Müller: A model of pattern formation by precipitation, Physica D **54**, 160–170 (1991)

48.28 J.H.E. Cartwright, J.M. Garcia-Ruiz, A.I. Villacampa: Pattern formation in crystal growth, Comput. Phys. Commun. **21**, 411–413 (1999)

48.29 G. Venzl, J. Ross: Comments on pattern formation in precipitation process, J. Chem. Phys. **77**, 1308–1313 (1982)

48.30 J. George, G. Varghese: Migrating triplet precipitation bands of calcium phosphates in gelatinous matrix, J. Mater. Sci. **40**, 5557–5559 (2005)

48.31 T. Terada, S. Yamabi, H. Imai: Formation process of sheets and helical forms consisting of strontium carbonate fibrous crystals with silicate, J. Cryst. Growth **253**, 435–444 (2003)

48.32 R.V. Suganthi, E.K. Girija, S. Narayana Kalkura, H.K. Varma, A. Rajaram: Self-assembled right handed helical ribbon of hydroxyapatite, J. Mater. Sci. Mater. Med. (2008) DOI 10.1007/s10856-008-3495-1

48.33 J.M. Garcia-Ruiz, D. Rondon, A. Garcia-Romero, F. Otalora: Role of gravity in the formation of Liesegang patterns, J. Phys. Chem. **100**, 8854–8860 (1996)

48.34 W. Ostwald: Besprechung (Rezension) der Arbeit von Liesegangs "A-Linien", Z. J. Phys. Chem. **23**, 365 (1897), in German

48.35 J. George, G. Varghese: Formation of periodic precipitation patterns: A moving boundary problem, Chem. Phys. Lett. **362**, 8–12 (2002)

48.36 J. George, G. Varghese: Liesegang patterns: Estimation of diffusion coefficient and a plausible justification for colloid explanation, Colloid Polym. Sci. **280**, 1131–1136 (2002)

48.37 J. George, S. Nair, G. Varghese: Role of colloid dynamics in the formation of Liesegang rings in multi-component systems, J. Mater. Sci. Lett. **39**, 311–331 (2004)

48.38 J. George, I. Paul, P.A. Varughese, G. Varghese: Rhythmic pattern formation in gels and Matalon-Packter law: A fresh perspective, Pramana J. Phys. **60**, 1259–1271 (2003)

48.39 F. Izsák, I. Lagzi: A new universal law for the Liesegang pattern formation, J. Chem. Phys. **122**, 184707-1–184707-6 (2005)

48.40 B.A. Grzybowski, K.J.M. Bishop, C.J. Campbell, M. Fiałkowski, S.K. Smoukov: Micro and nanotechnology via reaction diffusion, Soft Matter **1**, 114–128 (2005)

48.41 A. Volford, F. Izsák, M. Ripszám, I. Lagzi: Systematic front distortion and presence of consecutive fronts in a precipitation system, J. Phys. Chem. B **110**, 4535–4537 (2006)

48.42 F. Lefaucheux, Y. Bernard, C. Vennin, M.C. Robert: Gel growth of 2-amino-5-nitropyridinium dihydrogen phosphate crystals followed by holographic interferometry, J. Cryst. Growth **165**, 90–97 (1996)

48.43 G. Sajeevkumar, R. Raveendran, B.S. Remadevi, A.V. Vaidyan: Growth feature of ammonium hydrogen D-tartrate single crystals, Bull. Mater. Sci. **27**, 323–325 (2004)

48.44 S.G. Bhat, S.M. Dharmaprakash: Crystal growth and characterization of antimony thiourea bromide, J. Cryst. Growth **181**, 390–394 (1997)

48.45 R.I. Petrova, R. Patel, J.A. Swift: Habit modification of asparagine monohydrate crystals by growth in hydrogel media, Cryst. Growth Des. **6**, 2709–2715 (2006)

48.46 P.V. Dalal, K.B. Saraf: Growth and study of barium oxalate single crystals in agar gel, Bull. Mater. Sci. **29**, 421–425 (2006)

48.47 M.C. Etter, D.A. John, B.S. Donahue: Growth and characterization of small molecule organic crystals, J. Cryst. Growth **76**, 645–655 (1986)

48.48 O.M. Yaghi, G. Li, H. Li: Crystal growth of extended solids by nonaqueous gel diffusion, Chem. Mater. **9**, 1074–1076 (1997)

48.49 S.K. Arora, A.J. Kothari, R.G. Patel, K.M. Chauhan, B.N. Chudasama: Optical absorption in gel grown cadmium tartrate single crystals, J. Phys. Conf. Ser. **28**, 48–52 (2006)

48.50 T. Jini, K.V. Saban, G. Varghese: Thermal and infrared studies of calcium malate crystals grown in diffusion limited medium, Cryst. Res. Technol. **40**, 1155–1159 (2005)

48.51 S.R. Suthar, M.J. Joshi: Growth and characterization of Mn^{2+} doped calcium L-tartrate crystals, Cryst. Res. Technol. **41**, 664–670 (2006)

48.52 X.S. Shajan, C. Mahadevan: On the growth of calcium tartrate tetrahydrate single crystals, Bull. Mater. Sci. **27**, 327–331 (2004)

48.53 M.V. John, M.A. Ittyachen: Growth and characterization of cerium lanthanum oxalate crystals grown in hydro-silica gel, Cryst. Res. Technol. **36**, 141–146 (2001)

48.54 S. Narayana Kalkura, M. Kanakavel, P. Ramasamy: Crystallization of an organic intercalation compound: cholic acid, Cryst. Res. Technol. **32**, 569–575 (1997)

48.55 A. Elizebeth, V. Thomas, G. Jose, G. Jose, N.V. Unnikrishnan, C. Joseph, M.A. Ittyachen: Studies on the growth and optical characterization of dyspro-

48.56 K. Ambujam, S. Selvakumar, D. Prem Anand, G. Mohamed, P. Sagayaraj: Crystal growth, optical, mechanical and electrical properties of organic NLO material γ-glycine, Cryst. Res. Technol. **41**, 671–677 (2006)

sium gadolinium oxalate single crystals, Cryst. Res. Technol. **39**, 105–110 (2004)

48.57 S.J. Joshi, B.B. Parekh, K.D. Vohra, M.J. Joshi: Growth and characterization of gel grown pure and mixed iron-manganese levo-tartrate crystals, Bull. Mater. Sci. **29**, 307–312 (2006)

48.58 M.J. González-Tejera, S. López-Andrés, M.V. García, M.I. Redondo, J.A.R. Cheda: Single crystal growth of lead (II) n-octa-, n-nona- and n-decanoate, J. Cryst. Growth **152**, 330–333 (1995)

48.59 E. Ramachandran, S. Natarajan: Gel growth and characterization of gel grown DL-methionine, Cryst. Res. Technol. **41**, 411–415 (2006)

48.60 P. Andreazza, F. Lefaucheux, M.C. Robert, D. Josse, J. Zyss: Gel growth of 3- methyl-4-nitropyridine-1-oxide organic crystals: x-ray and nonlinear optics characterization, J. Appl. Phys. **68**, 8–13 (1990)

48.61 R. Mahalakshmi, S.X. Jesuraja, S. Jerome Das: Growth and characterization of L-phenylalanine, Cryst. Res. Technol. **41**, 780–783 (2006)

48.62 S. Selvasekarapandian, K. Vivekanandan, P. Kolandaivel: Vibrational studies of gel grown ferroelectric $RbHC_4H_4O_6$ and $SrC_4H_4O_6 \cdot 4H_2O$ crystals, Cryst. Res. Technol. **34**, 873–880 (1999)

48.63 S.K. Arora, V. Patel, A. Kothari, B. Amin: Gel growth and preliminary characterization of strontium tartrate trihydrate, Cryst. Growth Des. **4**, 343–349 (2004)

48.64 T. Jini, K.V. Saban, G. Varghese: The growth, spectral and thermal properties of the coordination compound crystal-strontium malate, Cryst. Res. Technol. **41**, 250–254 (2006)

48.65 M. Wang, X.Y. Liu, C. Sun, N. Ming, P. Bennema, W.J.P. Van Enckevort: Periodic roughening transitions in diffusion-limited growth, Europhys. Lett. **41**, 61–66 (1998)

48.66 Y. Oaki, H. Imai: Experimental demonstration for the morphological evolution of crystals grown in gel media, Cryst. Growth Des. **2**, 711–716 (2003)

48.67 S.J. Shitole, K.B. Saraf: Growth and study of some gel grown group II single crystals of iodate, Bull. Mater. Sci. **24**, 461–468 (2001)

48.68 S.J. Shitole, K.B. Saraf: Growth, structural and microtopographical studies of calcium iodate monohydrate crystals grown in silica gel, Cryst. Res. Technol. **37**, 440–445 (2002)

48.69 C.M. Pina, L. Fernández-Díaz, J.M. Astilleros: Nucleation and growth of scheelite in a diffusing-reacting system, Cryst. Res. Technol. **35**, 1015–1022 (2000)

48.70 M. Gu, D.X. Wang, Y.T. Huang, R. Zhang: Growing CuI crystals with decomplexation method modified by concentration programming, Cryst. Res. Technol. **39**, 1104–1107 (2004)

48.71 M. Gu, Y.F. Li, X.L. Liu, D.X. Wang, R.K. Xu, G.W. Li, X.P.O. Yang: The influence of CuI·HI complex distribution on CuI crystal growth with decomplexation method in silica gel, J. Cryst. Growth **292**, 74–77 (2006)

48.72 H. Kusumoto, T. Kaito, S. Yanagiya, A. Mori, T. Inoue: Growth of single crystals of $PbBr_2$ in silica gel, J. Cryst. Growth **277**, 536–540 (2005)

48.73 C.M. Pina, L.F. Diaz, M. Prieto: Topotaxy relationships in the transformation phosgenite-cerussite, J. Cryst. Growth **158**, 340–345 (1996)

48.74 K. Mayer, D. Wörmann: Influence of a polycrystalline precipitation zone on the growth of single crystals using a gel method: Growth of single crystals of $PbHPO_4$, J. Cryst. Growth **169**, 317–324 (1996)

48.75 D.S. Bhavsar, K.B. Saraf, T. Seth: Studies on growth and microstructure of lead iodide single crystals, Cryst. Res. Technol. **37**, 225–230 (2002)

48.76 C.M. Pina, L.F. Diaz, M. Prieto: Crystallization of β''-$LiNH_4SO_4$ and $(NH_4)_2SO_4$ in gels: Growth morphology and epitaxy phenomena, J. Cryst. Growth **177**, 102–110 (1997)

48.77 C.M. Pina, F.L. Diaz, J. Lopez-Gareia, M. Prieto: Growth of β-$LiNaSO_4$ and $Li_2SO_4 \cdot H_2O$: Epitaxy and intergrowth phenomena, J. Cryst. Growth **148**, 283–288 (1995)

48.78 R. Kanagadurai, R. Sankar, G. Sivanesan, S. Srinivasan, R. Jayavel: Growth and properties of ferroelectric potassium ferrocyanide trihydrate single crystals, Cryst. Res. Technol. **41**, 853–858 (2006)

48.79 R.I. Petrova, J.A. Swift: Habit changes of sodium bromate crystals grown from gel media, Cryst. Growth Des. **2**, 573–578 (2002)

48.80 R.I. Petrova, J.A. Swift: Growth of crystal enantiomers in hydrogels, J. Am. Chem. Soc. **126**, 1168–1173 (2004)

48.81 J.M. Garcia-Ruiz: On the formation of induced morphology crystal aggregates, J. Cryst. Growth **73**, 251–262 (1985)

48.82 L.F. Diaz, J.M. Astilleros, C.M. Pina: The morphology of calcite crystals grown in a porous medium doped with divalent cation, Chem. Geol. **225**, 314–321 (2006)

48.83 J.P. Reyes-Grayeda, D. Jauregui-Zuniga, N. Batina, M. Salmon-Salazar, A. Moreno: Experimental simulations of the biomineralization phenomena in avian eggshells using $BaCO_3$ aggregates grown inside an alkaline silica matrix, J. Cryst. Growth **234**, 227–236 (2002)

48.84 M. Prieto, F. Diaz, L. Andres: Supersaturation evolution and first precipitate location in crystal growth in gels: application to barium and strontium carbonates, J. Cryst. Growth **98**, 447–460 (1989)

48.85 J.M. Garcia-Ruiz: Crystal growth in gels as a laboratory analogous of the natural crystallization, Estud. Geol. **38**, 209–225 (1982)

48.86 S. Mann, J. Webb, R.J.P. Williams (Eds.): *Biomineralization* (VCH, New York 1989)

48.87 P. Dieppe, P. Calvert: *Crystals and Joint Disease* (Chapman Hall, New York 1983)

48.88 H. Iwata, Y. Abe, S. Nishio, A. Wakatsuki, K. Ochi, M. Takeuchi: Crystal-matrix interrelations in brushite and uric acid calculi, J. Urol. **135**, 397–401 (1986)

48.89 W. Achilles, R. Freitag, B. Kiss, H. Riedmiller: Quantification of crystal growth of calcium oxalate in gel and its modification by urinary constituents in a new flow model of crystallization, J. Urol. **154**, 1552–1556 (1995)

48.90 V.I. Katkova, V.I. Rakin: Bacterial genesis of calcite, J. Cryst. Growth **142**, 271–274 (1994)

48.91 E. Ramachandran, P. Raji, K. Ramachandran, S. Natarajan: Photoacoustic studies of the thermal properties of calcium carbonate – The major constituent of pancreatic calculi, Cryst. Res. Technol. **41**, 64–67 (2006)

48.92 O. Grassmann, R.B. Neder, A. Putnis, P. Lobmann: Biomimetic control of crystal assembly by growth in an organic hydrogel network, Am. Mineral. **88**, 647–652 (2003)

48.93 O. Grassmann, P. Lobmann: Biomimetic nucleation and growth of $CaCO_3$ in hydrogels incorporating carboxylate groups, Biomaterials **25**, 277–282 (2004)

48.94 H. Imai, T. Terada, T. Miura, S. Yamabi: Self-oriented formation of porous aragonite with silicate, J. Cryst. Growth **244**, 200–205 (2002)

48.95 E.K. Girija, S.C. Latha, S. Narayana Kalkura, C. Subramanian, P. Ramasamy: Crystallization and microhardness of calcium oxalate monohydrate, Mater. Chem. Phys. **52**, 253–257 (1998)

48.96 T. Jung, W.S. Kim, C.K. Choi: Crystal structure and morphology control of calcium oxalate using biopolymeric additives in crystallization, J. Cryst. Growth **279**, 154–162 (2005)

48.97 E.V. Petrova, N.V. Gvozdev, L.N. Rashkovich: Growth and dissolution of calcium oxalate monohydrate (COM) crystals, J. Optoelectron. Adv. Mater. **6**, 261–268 (2004)

48.98 N. Srinivasan, S. Natarajan: Growth of some urinary crystals and studies on inhibitors and promoters. I. Standardization of parameters for crystal growth and characterization of crystals, Indian J. Phys. **70A**, 563–568 (1996)

48.99 S. Natarajan, E. Ramachandran, D.B. Suja: Growth of some urinary crystals and studies on inhibitors and promoters. II. X-ray studies and inhibitory or promotory role of some substances, Cryst. Res. Technol. **32**, 553–559 (1997)

48.100 E. Ramachandran, S. Natarajan: XRD, thermal and FTIR studies on gel-grown gypsum crystals, Indian J. Phys. **79**, 77–80 (2005)

48.101 P. Kumareson, S. Devanarayanan: Growth of crystals of calcium sulphate in silica gel in presence of additive and their characterization, Cryst. Res. Technol. **22**, 1453–1458 (1987)

48.102 A. Elizabeth, C. Joseph, M.A. Ittyachen: Growth and micro-topographical studies of gel grown cholesterol crystals, Bull. Mater. Sci. **24**, 431–434 (2001)

48.103 C.R. Loomis, G.G. Shipley, D.M. Small: The phase behavior of hydrated cholesterol, J. Lipid Res. **20**, 525–535 (1979)

48.104 E. Ramachandran, S. Natarajan: Crystal growth of some urinary stone constituents: III. In-vitro crystallization of L-cystine and its characterization, Cryst. Res. Technol. **39**, 308–312 (2004)

48.105 T.K. Anee, N. Meenakshi Sundaram, D. Arivuoli, P. Ramasamy, S. Narayana Kalkura: Influence of an organic and an inorganic additive on the crystallization of dicalcium phosphate dihydrate, J. Cryst. Growth **285**, 380–387 (2005)

48.106 V.S. Joshi, M.J. Joshi: FTIR spectroscopic, thermal and growth morphological studies of calcium hydrogen phosphate dihydrate crystals, Cryst. Res. Technol. **38**, 817–821 (2003)

48.107 G.R. Sivakumar, E.K. Grija, S. Narayana Kalkura, C. Subramanian: Crystallization of calcium phosphates: Brushite and monetite, Cryst. Res. Technol. **33**, 197–205 (1998)

48.108 T. Irusan, S.N. Kalkura, D. Arivuoli, P. Ramasamy: Dendritic structures of brushite in silica gel, J. Cryst. Growth **130**, 217–220 (1993)

48.109 R.H. Plovnick: Crystallization of brushite from EDTA-chelated Ca in agar gels, J. Cryst. Growth **141**, 22–26 (1991)

48.110 T.K. Anee, M. Palanichamy, M. Ashok, N. Meenakshi Sundaram, S. Narayana Kalkura: Influence of iron and temperature on the crystallization of calcium phosphates at the physiological pH, Mater. Lett. **58**, 478–482 (2004)

48.111 E. Ramachandran, S. Natarajan: Crystal growth of some urinary stone constituents: II. In-vitro crystallization of hippuric acid, Cryst. Res. Technol. **37**, 274–1279 (2002)

48.112 M. Ashok, N. Meenakshi Sundaram, S. Narayana Kalkura: Crystallization of hydroxyapatite at physiological temperature, Mater. Lett. **57**, 2066–2070 (2003)

48.113 E.K. Girija, Y. Yokogawa, F. Nagata: Bone like apatite formation on collagen fibrils by biomimetic method, Chem. Lett. **31**, 702–703 (2002)

48.114 G.R. Sivakumar: Investigations on crystallization and characterization of biominerals: Calcium and magnesium phosphates. Ph.D. Thesis (Anna University, Chennai 2000)

48.115 S. Narayana Kalkura, E.K. Girija, M. Kanakavel, P. Ramasamy: In-vitro crystallization of spherulites of monosodium urate monohydrate, J. Mater. Sci. Mater. Med. **6**, 577–580 (1995)

48.116 M. Iijima, Y. Moriwaki: Lengthwise and oriented growth of octacalcium phosphate crystal in polyacrylamide gel in a model system of tooth enamel apatite formation, J. Cryst. Growth **194**, 125–132 (1998)

48.117 E. Ramachandran, S. Natarajan: Crystal growth of some urinary stone constituents: I. In-vitro crystallization of L-tyrosine and its characterization, Cryst. Res. Technol. **37**, 1160–1164 (2002)

48.118 W.G. Robertson, M. Peacock: The pattern of urinary stone disease in Leeds and in the United Kingdom in relation to animal protein intake during the period 1960–1980, Urol. Int. **37**, 394–399 (1982)

48.119 R.L. Ryall: The scientific basis of calcium oxalate urolithiasis. Predilection and precipitation, promotion and proscription, World J. Urol. **11**, 59–65 (1993)

48.120 X. Guan, R. Tang, G.H. Nancollas: The potential calcification of OCP on intraocular lenses, J. Biomed. Mater. Res. **71A**, 488–496 (2004)

48.121 S.R. Qui, A. Wierzbicki, C.A. Orme, A.M. Cody, J.R. Hoyer, G.H. Nancollas, J.J. De Yoreo: Molecular modulation of calcium oxalate crystallization by osteopontin and citrate, Proc. Natl. Acad. Sci. **101**, 1811–1815 (2004)

48.122 R. Tang, L. Wang, C.A. Orme, T. Bonstein, P. Bush, G.H. Nancollas: Dissolution at the nanoscale: Self-preservation of biominerals, Angew. Chem. Int. Ed. **43**, 2697–2701 (2004)

48.123 P. Spirnak, M.I. Resnick: Urinary stones. In: *Smith's General Urology*, ed. by E.A. Tangho, J.W. McAninch (Prentice-Hall International, San Francisco 1990)

48.124 A.D. Seftel, M.I. Resnick: Metabolic evaluation of urolithiasis, Urol. Clin. North Am. **17**, 159–170 (1990)

48.125 A. Randal: The origin and growth of renal calculi, Ann. Surg. **105**, 1009–1027 (1937)

48.126 L. Anderson, J.R. McDonald: The origin, frequency, and significance of microscopic calculi in the kidney, Surg. Gynecol. Obstet. **82**, 275–282 (1946)

48.127 A.J. Butt, E.A. Hauser: The importance of protective urinary colloids in the prevention and treatment of kidney stones, Science **115**, 308–310 (1952)

48.128 C.W. Vermeulon, E.S. Lyon: Mechanisms of genesis and growth of calculi, Am. J. Med. **45**, 684–692 (1968)

48.129 F. Grases, A. Costa-Bauza, M. Kroupa: Studies on calcium oxalate monohydrate crystallization: Influence of inhibitors, Urol. Res. **22**, 39–43 (1992)

48.130 A. Millan, F. Grases, O. Söhnel, I. Krivankova: Semi-batch precipitation of calcium oxalate monohydrate, Cryst. Res. Technol. **27**, 31–39 (1992)

48.131 J. Hofbauer, I. Steffen, K. Hobarth, G. Vujicic, H. Schwetz, G. Reich, O. Zechner: Trace elements and urinary stone formation: New aspect of the pathological mechanism of urinary stone formation, J. Urol. **145**, 93–96 (1991)

48.132 I. Melnick, R.R. Landes, A.A. Hoffman, J.F. Burch: Magnesium therapy for recurring calcium oxalate urinary calculi, J. Urol. **105**, 119–122 (1971)

48.133 M. Ashok, S. Narayana Kalkura, V. Vijayan, P. Magudapathy, K.G.M. Nair: Investigations of the elemental concentration of kidney stones by PIXE analysis, Int. J. PIXE **11**, 21–25 (2001)

48.134 W. Achilles: In Vitro Crystallization systems for the study of urinary stone formation, World J. Urol. **15**, 244–251 (1997)

48.135 Y.M.F. Marickar, P. Koshy: Scanning electron microscopic study of effect of various agents on urinary crystal morphology, Scanning Microsc. **1**, 571–577 (1987)

48.136 A.M. Freitas, N. Schor, M.A. Boim: The effect of *Phyllanthus niruri* on urinary inhibitors of calcium oxalate crystallization and other factors associated with renal stone formation, BJU Int. **89**, 829–834 (2002)

48.137 V.S. Joshi, B.B. Parekh, M.J. Joshi, A.B. Vaidya: Herbal extracts of *Tribulus terrestris* and *Bergenia ligulata* inhibit growth of calcium oxalate monohydrate crystals in vitro, J. Cryst. Growth **275**, e1403–e1408 (2005)

48.138 K.C. Joseph, B.B. Parekh, M.J. Joshi: Inhibition of growth of urinary type calcium hydrogen phosphate dihydrate crystals by tartaric acid and tamarind, Curr. Sci. **88**, 1232–1238 (2005)

48.139 S.R. Khan, P.A. Glenton: Deposition of calcium phosphates and calcium oxalate crystals in the kidney, J. Urol. **153**, 811–817 (1995)

48.140 P.G. Werness, J.H. Bergert, L.H. Smith: Crystalluria, J. Cryst. Growth **53**, 166–181 (1981)

48.141 W.H. Boyce, J.S. King: Crystal-matrix interrelation in calculi, J. Urol. **3**, 351–365 (1959)

48.142 E.K. Girija: Investigations on biological crystals and analyses and epidemiological studies of urinary calculi. Ph.D. Thesis (Anna University, Chennai 1998)

48.143 D. Hirsch, R. Azoury, S. Sarig: DSC, x-ray and NMR properties of cholesterol crystals, Clin. Chim. Acta **174**, 65–82 (1988)

48.144 S. Narayana Kalkura, T.K. Anee, M. Ashok, C. Betzel: Investigations on the synthesis and crystallization of hydroxyapatite at low temperature, J. Biomed. Eng. **14**, 581–592 (2004)

48.145 S. Lazic: Microcrystalline hydroxyapatite formation from alkaline solutions, J. Cryst. Growth **147**, 147–154 (1995)

48.146 G. Morales, J.T. Burgues, R.R. Clemente: Crystal size distribution of hydroxyapatite precipitated in a MSMPR reactor, Cryst. Res. Technol. **36**, 1065–1074 (2001)

48.147 A.L. Braybrook, B.R. Heywood, R.A. Jackson, K. Pitt: Parallel computational and experimental studies

of the morphological modification of calcium carbonate by cobalt, J. Cryst. Growth **243**, 336–344 (2002)
48.148 D.B. Leusmann, H. Niggemann, S. Roth, H.V. Ahlen: Recurrence rate and severity of urinary calculi, Scan. J. Urol. Nephron **29**, 279–283 (1995)
48.149 R.J. Riese, K. Sakhaee: Uric acid nephrolithiasis: pathogenesis and treatment, J. Urol. **148**, 765–771 (1992)
48.150 D.J. Sutor, S. Scheidt: Identification standards for human urinary calculus components, using crystallographic methods, Br. J. Urol. **40**, 22–28 (1968)
48.151 F.L. Coe, A. Evan, E. Worcester: Kidney stone disease, J. Clin. Invest. **115**, 2598–2608 (2005)
48.152 C.Y.C. Pak, O. Waters, L.M. Arnold, K.H. Cox, C.D. Barilla: Mechanism of calcium urolithiasis among patients with hyperuricosuria: Supersaturation of urine with respect to monosodium urate, J. Clin. Invest. **59**, 426–431 (1977)
48.153 R. Boistelle, C. Rinaudo: Phase transition and epitaxies between hydrated orthorhombic and anhydrous monoclinic uric acid crystals, J. Cryst. Growth **53**, 1–9 (1981)
48.154 C. Rinaudo, R. Boistelle: The occurrence of uric acids and the growth morphology of the anhydrous monoclinic modification: $C_5H_4N_4O_3$, J. Cryst. Growth **49**, 569–576 (1980)
48.155 J.E.Z. Caner, J.L. Decker: Recurrent acute arthritis in chronic renal failure treated by periodic dialysis, Am. J. Med. **36**, 571–582 (1964)
48.156 J.F. Fiechtner, P.A. Simkin: Urate spherulites in gouty synovia, J. Am. Med. Assoc. **245**, 1533–1536 (1981)
48.157 N.W. McGill, P.A. Dieppe: The effect of biological crystals and human serum on the rate of formation of crystals of monosodium urate monohydrate in vitro, Br. J. Rheum. **30**, 107–111 (1991)
48.158 K.R. Murray, D.K. Granner, P.A. Mayers, V. Rodwell, O.K. Graner: *Lange Medical Book: Harper's Biochemistry* (Prentice Hall, New Jersey 1988)
48.159 D.S. Sutor, S.E. Wooley: The structure and formation of urinary calculi: oriented crystal growth, Br. J. Urol. **44**, 532–536 (1972)
48.160 M. Daudon, P. Jungers: Drug-induced renal calculi: epidemiology, prevention and management, Drugs **64**, 245–275 (2004)
48.161 R. Vani: Invitro crystallization and characterization of an amino acid: Cystine. M.Phil. Dissertation (Anna University, Chennai 1999)
48.162 B.L. Oser: *Hawk's Physiological Chemistry*, 14th edn. (Tata McGraw-Hill, New Delhi 1976)
48.163 E. Ramachandran: Investigations on certain chemical constituents of urinary stones and some amino acids. Ph.D. Thesis (Madurai Kamaraj University, Madurai 2006)
48.164 D.J. Sutor: Crystal growth in bile, Prog. Cryst. Growth Charact. **4**, 47–58 (1981)
48.165 D.M. Small, G.G. Shipley: Physical-chemical basis of lipid deposition in atherosclerosis, Science **185**, 222–229 (1974)
48.166 P. Libby, M. Aikawa, U. Schönbeck: Cholesterol and atherosclerosis, Biochim. Biophys. Acta **1529**, 299–300 (2000)
48.167 W.H. Admirand, D.M. Small: The physicochemical basis of cholesterol gallstone formation in man, J. Clin. Invest. **47**, 1043–1052 (1968)
48.168 R.P. Mensink, E.H.M. Temme, J. Plat: Dietary fats and coronary heart disease. In: *Food Lipids. Chemistry, Nutrition and Biotechnology*, ed. by C.C. Akoh, D.B. Min (Marcel Dekker, New York 1998) pp. 507–535
48.169 R.T. Holzbach: Factors influencing cholesterol nucleation in bile, Hepatology **4**, 173S–176S (1984)
48.170 G.H. Nancollas: Crystallization in bile, Hepatology **4**, 169S–172S (1984)
48.171 S.M. Strasberg: The pathogenesis of cholesterol gallstones – A review, J. Gastrointest. Surg. **2**, 109–125 (1998)
48.172 M.C. Frincu, S.D. Fleming, A.L. Rohl, J.A. Swift: The epitaxial growth of cholesterol crystals from bile solutions on calcite substrates, J. Am. Chem. Soc. **126**, 7915–7924 (2004)
48.173 A.J. Reginato, B. Kurnik: Calcium oxalate and other crystals associated with kidney diseases and arthritis, Semin. Arthritis Rheum. **18**, 198–224 (1989)
48.174 B.M. Craven: Pseudosymmetry in cholesterol monohydrate, Acta Cryst. B **35**, 1123–1128 (1979)
48.175 N.T. Saraswathi, F.D. Gnanam: Effect of medicinal plants on the crystallization of cholesterol, J. Cryst. Growth **179**, 611–617 (1997)
48.176 M. Ashok, S. Narayana Kalkura, V.J. Kennedy, A. Markwitz, V. Jayanthi, K.G.M. Nair, V. Vijayan: Trace element analysis of South Indian gallstones by PIXE, Int. J. PIXE **12**, 137–144 (2002)
48.177 M. Ashok, T.R. Rautray, P.K. Nayak, V. Vijayan, V. Jayanthi, S. Narayana Kalkura: Energy dispersive x-ray fluorescence analyses of gallstones, J. Radioanal. Nucl. Chem. **25**, 333–335 (2003)
48.178 M. Ashok, D.N. Reddy, V. Jayanthi, S.N. Kalkura, V. Vijayan, S. Gokulkrishnan, K.G.M. Nair: Regional differences in constituents of gallstones, Trop. Gastroenterol. **26**, 73–75 (2005)
48.179 R.P. Bonar-Law, A.P. Davis: Cholic acid as an architectural component in biomimetic/molecular recognition chemistry: synthesis of the first "cholaphanes", Tetrahedron **49**, 9855 (1993)
48.180 J.T. Velardo: *The Endocrinology of Reproduction* (Oxford Univ. Press, New York 1958)
48.181 L. Fieser, M. Fieser: *Steroids* (Reinhold, New York 1959)

48.182 J. Geevarghese: *Calcific Pancreatitis: Causes and Mechanisms in the Tropics Compared to the Subtropics* (J. Varghese, Bombay 1976)

48.183 A.C. Schulz, P.B. Moore, P.J. Geevarghese, C.S. Pitchumoni: X-ray diffraction studies of pancreatic calculi associated with nutritional pancreatitis, Dig. Dis. Sci. **31**, 476–480 (1986)

48.184 K.V. Narasimhulu, N.O. Gopal, J. Lakshmana Rao, N. Vijayalakshmi, S. Natarajan, R. Surendran, V. Mohan: Structural studies of the biomineralized species of calcified pancreatic stones in patients suffering from chronic pancreatitis, Biophys. Chem. **114**, 137–147 (2005)

48.185 S.R. Kamhi: On the structure of vaterite $CaCO_3$, Acta Cryst. **16**, 770–772 (1963)

48.186 H.G. Beger, A.L. Warshaw, D.L. Carr-Locke, J.P. Neoptolemos, C. Russell, M.G. Sarr: *The Pancreas* (Blackwell Scientific, London 1998)

49. Crystal Growth and Ion Exchange in Titanium Silicates

Aaron J. Celestian, John B. Parise, Abraham Clearfield

In situ experiments, whether carried out in-house or at synchrotron sources, can provide valuable information on the nucleation and subsequent growth of crystals and on the mechanism of growth as well as mechanisms of phase changes and ion-exchange phenomena. This chapter describes the types of x-ray detectors, in situ cells, and detectors used in such studies. The procedures are illustrated by a study of the preparation of a tunnel-structured sodium titanium silicate, the partially niobium framework phase, and the mechanism of ion exchange as revealed by time-resolved x-ray data.

- 49.1 X-Ray Methods 1637
  - 49.1.1 X-Rays and Diffraction Theory 1638
  - 49.1.2 Neutron Diffraction Theory 1640
- 49.2 Equipment for Time-Resolved Experiments 1642
  - 49.2.1 In-House X-Ray Sources 1642
  - 49.2.2 Synchrotron Radiation Sources 1642
- 49.3 Detectors 1642
  - 49.3.1 Image Plates 1642
  - 49.3.2 Charge-Coupled Devices 1643
  - 49.3.3 Position-Sensitive Detectors (PSD) 1643
  - 49.3.4 Energy-Dispersive Detectors 1643
  - 49.3.5 Silicon Strip Detector 1644
  - 49.3.6 Other Considerations 1644
- 49.4 Software 1644
- 49.5 Types of In Situ Cells 1645
  - 49.5.1 SECeRTS Cell 1646
  - 49.5.2 Polyimide Environmental Cell 1647
  - 49.5.3 High-Pressure Cells 1647
  - 49.5.4 Hydrothermal Steel Autoclave-Type Cell 1647
  - 49.5.5 Neutron Diffraction Cell.............. 1648
- 49.6 In-Situ Studies of Titanium Silicates (Na-TS) with Sitinakite Topology 1649
  - 49.6.1 Introduction to the Problem 1649
  - 49.6.2 Synthesis and Structure of Sodium Titanium Silicate (Na-TS) 1649
  - 49.6.3 Synthesis Problems and In Situ Hydrothermal Study ... 1650
  - 49.6.4 Ion Exchange of Cs^+ into Na-TS.... 1652
  - 49.6.5 Cesium Ion Exchange into H-TS 1654
  - 49.6.6 Sodium Niobium Titanosilicate (Nb-TS) 1655
  - 49.6.7 In Situ Synthesis of Na-NbTS 1655
  - 49.6.8 In Situ Ion Exchange of Cesium Ion Exchange in Na-NbTS 1656
  - 49.6.9 Cesium Ion Exchange into H-NbTS 1656
- 49.7 Discussion of In Situ Studies 1658
  - 49.7.1 Synthesis of Na-TS and Na-NbTS .. 1658
  - 49.7.2 Exchange Mechanisms 1659
- 49.8 Summary 1660
- References 1660

49.1 X-Ray Methods

Hydrothermal techniques for the synthesis of new materials and for crystal growth have been used extensively since the 1950s, and this has increased substantially over the last decade. The use of in situ studies in many cases is desirable since the data collection and analysis methods do not disrupt the chemical reaction or process. In the ex situ hydrothermal process, especially when new materials are the goal, the experimentalist usually has control over the time/temperature/pressure conditions and reactant ratios of the synthesis. What is missing in these ex situ processes is knowledge of what takes place as the reaction proceeds. It is possible that the phase(s) formed at elevated temperatures may revert to a more stable phase on cooling. Furthermore, several phase transformations, or metastable intermediates, may occur during the synthesis procedure that would otherwise go undetected in a typical *cook-and-look* experiment. Nucleation and crystal growth, mechanisms of phase changes, and ion-exchange processes are readily studied using in situ x-ray diffraction as the major experimental tool [49.1–3]. If the reaction is relatively slow and the resolution required is moderate, the in situ study can be carried out in an in-house x-ray facility or at a neutron facility. New detectors and more powerful x-ray and neutron sources are enabling new experimental techniques that allow routine data collection from in situ experiments. Compared with conventional sealed-tube x-ray sources, synchrotron radiation is 10^4–10^{12} times brighter. At the advanced photon source (APS), Argonne National Laboratory and the newly renovated facility at Brookhaven National Laboratory, National Synchrotron Light Source (NSLS) x-ray powder patterns can be recorded in seconds and with excellent resolution. In what follows, we will describe the types of facilities and provide several examples of how in situ studies are performed at synchrotron and neutron user facilities.

There are compromises involved in the collection and interpretation of time-resolved data. The data quality is affected chiefly by the poorer signal-to-noise discrimination, compared with data collected over longer time frames and high-quality ex situ data. However, by combining data collected in a *static* manner, on materials representing the beginning and end or the reaction pathway, for example, with time-resolved data, a more complete picture of the mechanism emerges; for example structural models of the kinetics of zeolite synthesis [49.4–6], ion exchange in their various cation-exchanged forms [49.5, 7, 8], in their dehydrated and hydrated states [49.9, 10], and with and without sorbents, are important to rationalizing their mode of operation. Monitoring these structural changes in situ and as a function of time allows the mechanism of transformation to be followed. Several case studies involving this approach are given below.

The quality and types of data required to uncover mechanisms depends on the information being sought. The optimization of synthetic conditions only requires identification of the phases and the pressure–temperature–composition conditions over which they are stable, and many of these can be carried out in the laboratory setting, particularly since the wider availability of area detectors. Powder diffraction data suitable for Rietveld refinement require:

1. Access to brighter x-ray beams at second- and third-generation synchrotron storage rings
2. Versatile high-pressure/temperature and hydrothermal cells designs.

49.1.1 X-Rays and Diffraction Theory

A crystal may be defined as a solid composed of atoms, ions or molecules arranged in a pattern that is periodic in three-dimensional space. The smallest repeating pattern containing all elements and symmetry operations is termed the *unit cell*, and is constructed by three non-coplanar vectors $\boldsymbol{a}, \boldsymbol{b}, \boldsymbol{c}$, where a, b, and c are the axial lengths of the unit cell, and three angles α, β, γ, where α, β, and γ are the angles between the axes. It is convenient to focus on the geometry of the periodic array. The crystal is then represented as a three-dimensional array of points, each of which has identical surroundings. There exist 14 such lattices, the Bravais lattices, that describe the geometry of crystals. The lattices are infinite in extent and it is possible to construct many sets of parallel planes that pass through the points. The perpendicular distance between any set of such planes is known as the d-spacing.

When a beam of monochromatic x-rays is passed through a crystal, diffraction occurs because the wavelength of the x-rays and the distances between atoms, or the sets of planes, are of the same order of magnitude. Diffraction only occurs when the waves being diffracted constructively interfere, and therefore not all d-spacing diffraction occurs at the same time. The Bragg equation defines the conditions for diffraction to occur

$$\sin\theta = \frac{n\lambda}{2d}, \qquad (49.1)$$

where λ is the x-ray wavelength, θ is the angle of diffraction, and n is an integer, which for our purposes may be set to unity. From knowledge of the d-spacings it is possible to obtain the values of the unit cell constants. The orientation of a plane in space relative to an axial system may be given in terms of the intercepts of the plane with respect to the axes, or the vector from the origin of the axial system that is perpendicular to the plane. This latter distance is d and the intercepts are designated by $hk\ell$, where $hk\ell$, are integers that are the reciprocals of the intercepts; for example, if the plane cuts the a-axis at $\frac{1}{2}a$ then

$$h = \frac{a}{\frac{1}{2}a} = 2.$$

Similarly, k and ℓ are the reciprocals of the intercepts along the b- and c-axis, respectively. There are six axial systems by which unit cells and lattices are described, or seven if we separate trigonal from hexagonal. For each axial system there is an equation relating d with $hk\ell$. For the cubic system there is only one unknown as all the unit cell dimensions are equal and the angles are all 90°, then

$$d = \frac{a}{[h^2 + k^2 + \ell^2]^{1/2}}. \tag{49.2}$$

In single-crystal studies it is possible to align each set of parallel planes at their respective Bragg angle to the x-ray beam and record all the d-spacings in three-dimensional space. For in situ synthesis studies this method would be time consuming, and most starting materials used in hydrothermal crystallization experiments are not single crystals, but rather solutions, gels or powders. It is advantageous to use the powder diffraction method to record a large number of d-spacings in one or two dimensions quickly.

The powder method is predicated on the fact that the powder will have an equal number of crystallites in all possible orientations to the x-ray beam, and all diffraction from the sets of d-spacings will be generated simultaneously [49.12]. The angle between the incident beam and the diffracted beam is always 2θ (Fig. 49.1). The detector travels about a circle of fixed radius with the sample at the center of the circle. The x-ray powder diffraction pattern (XRPD) is recorded as the diffracted intensities as a function of 2θ, and is a one-dimensional pattern. For complex structures or mixtures, the diffracted intensities may overlap, and extracting the intensities of overlapping features becomes increasingly difficult with increasing number of crystalline phases. Therefore the resolving power of the diffractometer becomes an important feature in planning in situ studies. Resolution afforded by synchrotron-based diffraction is much greater than that of in-house powder diffraction techniques. The breadth of an x-ray diffracted peak is measured as the width of the peak at half-height, designated the full-width at half-maximum (FWHM). For medium-resolution powder diffractometer at a synchrotron storage ring the FWHM is $\approx 0.01°$, as compared with $0.1-0.5°$ for the K_α doublet of Cu radiation from an in-house powder diffractometer. Another point to consider is that x-ray diffraction features generally broaden with increasing 2θ values, so that overlap of peaks at high 2θ values also increases.

X-ray scattering arises from the interaction of electrons with the electromagnetic field of the collimated and coherent x-ray beam. As an element's atomic number increase, so does the total number electrons around the atom, and as the number of electrons of the atom

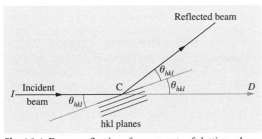

Fig. 49.1 Bragg reflection from a set of lattice planes showing that the angle between the incident beam and the reflected beam is always 2θ

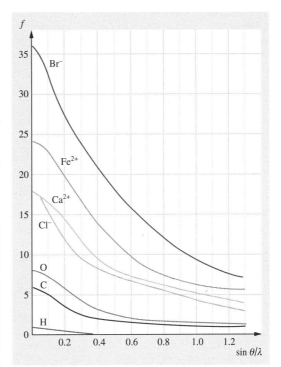

Fig. 49.2 Variation of the atomic scattering factor as a function of $\sin\theta/\lambda$ (after [49.11])

increases, so does the total x-ray scattering power. As a consequence, distinguishing between the scattering power of near neighbors in the periodic table, such as Al and Si, is often difficult because their electron configurations are similar. Also, the atomic scattering factors are angle dependent and fall off with increasing $\sin\theta$, as shown in Fig. 49.2 [49.11]. This dependence decreases the intensities of high-angle reflections. All these factors of x-ray diffraction require the use of high-intensity, well-collimated, narrow-beam radiation available at synchrotron sources. Nevertheless, in-house studies have the advantage of unlimited access where preliminary experimentation data collection may be obtained. The results from in-house data are valuable for determining experimental design, temperature ranges for synthesis, and synthesis times to optimize time spent at remote beam facilities, where access to the most desirable and well-conditioned x-ray and neutron beams is limited.

49.1.2 Neutron Diffraction Theory

Neutrons primarily interact with nuclei of atoms, leading to scattering and diffraction phenomena. According to the de Broglie equation, a beam of particles of mass M and velocity v generates a plane wave of length

$$\lambda = \frac{h}{Mv}, \tag{49.3}$$

where λ is the generated wavelength, v is the velocity, M is the mass of neutron, and h is Planck's constant. At a velocity of 4 km/s,

$$\lambda = \frac{6.625 \times 10^{-27}\,\mathrm{g\,cm^2/s}}{1.67 \times 10^{-24}\,\mathrm{g} \times 4 \times 10^5\,\mathrm{cm/s}}$$
$$= 0.992 \times 10^{-8}\,\mathrm{cm}\,.$$

This simple calculation shows that it is possible to have neutrons whose associated wavelengths are in the range to diffract from crystals or to be scattered by amorphous materials. The characteristics of neutrons provide certain advantages not obtainable with x-rays:

1. The coherent neutron scattering length (b) analogous to the x-ray atomic scattering factors for elements, does not increase with atomic number but fluctuates from element to element, and can be zero [49.13]. Thus, scattering from light elements may be as intense as that from heavy elements.
2. As a consequence of the nature of atomic scattering lengths the differences in scattering of neighboring elements in the periodic may differ greatly, making it easy to distinguish one from the other. Isotopes also have different scattering lengths, making it possible to distinguish H from D, for example.
3. Because the nuclei are point scatterers, the value of the scattering factor (more properly, the scattering length factor) does not change with increasing values of θ. As a result, the intensity of diffracted radiation does not decrease with increased θ as is the case for x-rays.
4. Many materials exhibit low absorption values for neutrons. This fact allows the use of thicker walled vessels for high-temperature/pressure studies. Further elements (vanadium), or alloys (Ti-Zr) with $b = 0$, are exceedingly useful as null-scattering sample containers. There is no equivalent to this application for x-ray studies, where even amorphous glass adds considerable parasitic scattering to the powder diffraction pattern.
5. Magnetic behavior arises from the presence of unpaired electrons in atoms. Because neutrons possess a magnetic dipole moment, they interact with unpaired electrons. This gives rise to an additional scattering effect that results in the appearance of weak peaks in the diffraction pattern. These extra peaks can be indexed as a superlattice of the x-ray lattice and reveals the magnetic ordering of the unpaired electrons. Thus neutron diffraction can be utilized to study ferro-, ferri-, and antiferromagnetic phenomena not readily accessible to x-ray methodology.

Neutron beams can be obtained from nuclear (steady-state) reactors, where they are typically monochromated by means of curved germanium or silicon crystals (Fig. 49.3).

A second method of neutron generation is through a process termed *spallation*. A heavy-metal target, such as uranium, mercury or tungsten, is bombarded with 450 MeV protons in short uniform bursts which separate the neutrons from the target nucleus. The released neutrons are at high kinetic energy levels and must be thermalized, moderated or cooled (slowed) to be in the 0.2–5.0 Å wavelength range for suitable diffraction studies. This is accomplished by bringing epithermal neutrons into thermal equilibrium with hydrogen-rich moderators such as polyethylene and liquid methane that reduce the velocity of the neutrons. Hydrogen, being about the same mass as the neutron, is most efficient for this process. Because of the high flux of this neutron beam and its range of neutron energies, it is advantageous to use the energy-dispersive method (Fig. 49.3). The Spallation Neutron Source at Oak Ridge National Laboratory uses a time-of-flight methodology. The sample is held at the center of a large ring (source-to-sample distance is 20 m) with banks of detectors held at specific angles to the incident beam, ±30°, ±90°, and ±120°. Use of multiple detector banks increases the total collected intensities, which is necessary because the neutron beam seldom interacts with the sample. In the time-of-flight method, the time for a diffracted beam to reach the detector is measured. Combining the Bragg equation (49.1) in the form of d and substituting for λ the value given by the de Broglie equation yields

$$d = \frac{n\lambda}{2\sin\theta} = \frac{ht}{2ML\cdot\sin\theta}, \quad (49.4)$$

where $v = L/t$. Here we utilize the definition of velocity as length L divided by time. Because h, O, M, and L are fixed quantities the d-spacing is proportional to time. The larger the d-spacing, the longer the time for the diffracted radiation to reach the detector. Thus the powder pattern plot of I versus time inverts the normal order of the powder pattern. The smallest d-spacings are recorded at short times and the largest d-spacing is the last peak in the pattern.

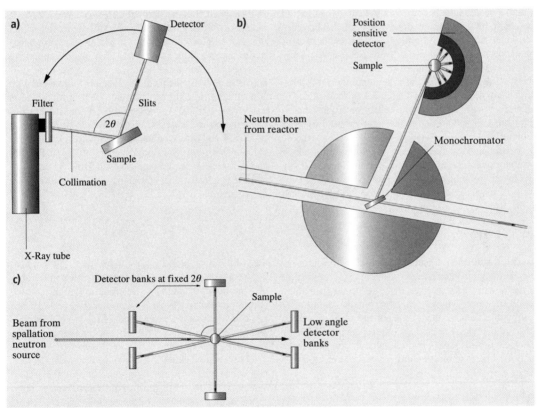

Fig. 49.3a–c Comparison of typical geometries for (**a**) a laboratory-based x-ray source, and for powder diffractometers found at (**b**) reactor and (**c**) spallation neutron sources. In (**a**) the incident x-ray beam of characteristic radiation is selected by the use of a filter (e.g., Ni filter for Cu-K_α radiation) while at a steady-state (reactor) source a single-crystal monochromator selects the desired, narrow, wavelength range from the Maxwellian distribution of neutron energies emerging from the reactor. The scattering from the sample is detected in a position-sensitive detector. In (**a**), at a spallation (pulsed) neutron source the energies of the neutrons scattered at fixed 2θ are determined according to their time of flight along the flight path relative to the origin of the neutron pulse. This is analogous to the use of energy-dispersive x-ray techniques (see text)

49.2 Equipment for Time-Resolved Experiments

49.2.1 In-House X-Ray Sources

For time-resolved studies the experimenter needs to record a large number of powder diffraction patterns rapidly to capture time-dependent chemical reactions, and therefore a high-intensity source of x-rays is required. A standard sealed-tube x-ray tube may provide a power of 2 kW or 40 mA at 50 kV. Special sealed tubes may be rated as high as 3 kW. The power limit is controlled by the rate at which heat can be conducted away from the anode, and x-ray intensities are directly proportional to the power. About 98% of the energy of a sealed tube is converted to heat, which would soon melt the target unless otherwise cooled. Therefore a constant stream of cold water flows across the back of the anode to conduct the heat away. Sealed-tube x-ray intensities may also be increased by the use of optical fibers. Each fiber consists of a large number of hollow glass capillaries. These capillaries act as waveguides in which the x-rays undergo total reflection from the capillary surfaces and are led to the sample with little loss of intensity.

For more powerful in-house units a rotating-anode generator may be used. In a rotating-anode x-ray generator, the anode is rotated rapidly to always supply a relatively cool metal target for the bombarding electrons. The anode is connected to a shaft that rotates through vacuum-tight seals in the tube housing. Such units come in two power models: 12 and 18 kW. Some of the new models of rotating anodes approach the x-ray flux observed at second-generation synchrotrons.

49.2.2 Synchrotron Radiation Sources

Synchrotron radiation [49.14] is produced by the acceleration of electrons moving at near the speed of light. The charged particles move in circular orbits within an evacuated chamber termed a *storage ring*. Magnetic fields are used to alter the trajectory of the particles, and this acceleration causes them to radiate energy tangential to the ring. This radiation is then made available to the experimenter via a beamline pipe containing suitable optical components under vacuum. Compared with conventional sealed-tube sources, synchrotron radiation has a flux 10^4–10^{12} times brighter. Other advantages of synchrotron radiation are the broad spectral range produced and small radiation divergence, which have the advantage of wavelength selection (typically 0.4 and 1.5 Å at a bending magnet), narrow beam collimation (typically 0.01–0.6 mm), and very low beam divergence.

49.3 Detectors

To obtain a diffraction pattern suitable for structure refinement from in situ diffraction studies, a large portion of the powder pattern must to be recorded in a short period of time. Typically monochromatic radiation is used with a detector such as an image plate, position-sensitive detectors, or charge-coupled device (CCD) or using energy-dispersive radiation and a multichannel analyzer. Only the most common types of detectors are discussed below.

49.3.1 Image Plates

Image plates (IP) consist of a thick layer of x-ray-sensitive phosphor on an optically transparent backing. When the x-rays strike the plate, the phosphor grains ionize. The released electrons are trapped at F-centers, which are point defects of the phosphor solid. The plate is then scanned with a small-diameter laser beam that liberates the trapped electrons, which recombine with the ions from which they were liberated. This transition produces light that is collected in a photomultiplier that amplifies the signal. The amount of light emitted is proportional to the x-ray intensity. Any residual image can be erased by exposing the film to a light source and reused, and the IP is usually read and erased online during the experiment.

The main advantage of using an IP for data collection is the size. The IP represents one of the largest detectors for the collection of diffracted x-ray intensities, with sizes up to and exceeding 34.5 cm. Using a large detector allows data to be collected to approximately 0.8 Å depending on wavelength choice and sample–detector distance. Another advantage of using this type of area detector is that the entire Debye–Scherrer ring can be collected. During in situ crystallization studies, the material may not form in abundance to produce smooth diffraction rings. These rings will often appear spotty and incomplete contributing to the problems associated with small cells that produce small quantities of crystallites. An area detector can overcome this problem as it collects data in two-dimensional space as opposed to point detectors and position-sensitive detectors. The disadvantage of IP detectors is the readout time. The IP must be digitally scanned and then erased before the next exposure can begin. The process of reading and erasing can take up to 2.5 min, during which time no data is being collected from the experiment. Although this *gap* in data collection appears to be problematic, the advantages of obtaining high-resolution diffraction data in one exposure outweigh those disadvantages.

49.3.2 Charge-Coupled Devices

A charge-coupled device (CCD) is simply a semiconductor chip where one side of the chip is subdivided into rectangular sections (pixels) that are approximately 10 μm on edge. Between the x-rays and the CCD are a phosphor and a photodiode that convert the incoming light into electrons. Depending on where the light strikes the phosphor, the electrons build in number and are trapped in the rectangular pixel on the CCD semiconductor surface. After a complete exposure, the CCD is then read using analogue-to-digital conversion, which counts the amount of electrons on each pixel, where the number of electrons is proportional to the original light intensity. In this process, it is possible to have rapid data collection and readout time with little delay between exposures.

However, there are caveats to using a CCD for x-ray diffraction studies. For example, if a pixel fills with electrons before the exposure is over, those excess electrons will *spill* into the surrounding wells. This can cause problems when trying to read the detector after x-ray exposure, and may be visible on the processed image as a vertical, or horizontal, streak across the entire image. Another problem is that most CCD chips are quite small, typical ranging in size from 512×512 to 2048×2048 pixels. Therefore, such small chips are made usable by having a large phosphor front plate, ≈ 15 cm in size, and a fiber-optic taper that reduces the image down to the CCD chip, which is ≈ 2 cm in size. Naturally, information will be lost during the image size reduction, and care must also be taken to remove the distortion in the fiber-optic taper prior to data reduction. The construction of larger CCD chips will eventually lead to the removal of the fiber-optic taper and thus produce a distortion-free raw image.

49.3.3 Position-Sensitive Detectors (PSD)

A PSD is a proportional counter that is position sensitive. The proportional counter consists of a tube filled with a noble gas such as xenon. A thin tungsten wire runs down the center of the tube and is positively charged. A thin window of low-absorbing glass allows the x-ray photons to enter the tube and ionize the xenon, which releases a cloud of electrons. The released electrons are drawn to the positively charged wire, giving rise to a charge pulse. The size of the pulse is proportional to the energy of the incident photon, which allows x-rays of different wavelengths to be distinguished. In a PSD, the wire can be long and curved to coincide with the diffractometer circle. The PSD is placed so as to intercept any x-rays diffracted by the sample within the angular range of the wire. The x-ray beam strikes the wire at particular 2θ values dictated by diffraction from the powder sample. The time of travel taken for the pulse to reach the end of the tungsten wire fixes the position or 2θ value. A multichannel analyzer is able to sort out all the times and amplitudes into digital form and the output is that of a conventional powder pattern.

PSDs are available in a range of sizes starting from $\approx 4°$ to $120°$ in 2θ. All the reflections within this range are recorded simultaneously. By choosing a smaller wavelength it is possible to record the entire usable range of diffraction data in a single exposure using a $30°$ or $60°$ PSD. The advantage of a PSD is their readout time, which far exceeds that of the IP, and allows near-continuous diffraction patterns to be collected without interruption. Most PSDs are linear, thus only one-dimensional diffraction data can be obtained, which may result in diffracted intensities being missed because of incomplete Debye–Scherrer rings from non-ideal powder samples.

49.3.4 Energy-Dispersive Detectors

At a synchrotron source, a broad range of x-ray energies can be produced. The peak of the energy distribution depends on the ring characteristics, including the energy of the charged particles within the ring and the strength of the magnets used to bend particles around the ring. A beam of well-defined energy is produced for monochromatic studies by intercepting this *white* radiation with a single-crystal monochromator. Alternatively, if the powder sample is held at a specific angle and the incident beam contains a broad spectrum of wavelengths, a particular set of planes of fixed d-spacing will diffract only that wavelength that satisfies Bragg's law for that value of d. Thus, all the different wavelengths which satisfy this condition for all the planes in the sample will enter the detector and be sorted out by the multichannel analyzer. The entire pattern may be recorded in less than 1 min. The main advantage of using white radiation is the intensity of the beam. The high flux, and hard x-rays, have a higher penetration depth and can pass steel environmental cells. One disadvantage of the energy-dispersive detector is that its peak resolution is poorer compared with monochromatic beams.

49.3.5 Silicon Strip Detector

A new detector, offered by Brucker Corporation, contains 192 strips of silicon which act as 192 individual detectors. This results in an almost 200 times increase in intensity. This detector together with fiber optics may open many opportunities for in situ studies in-house.

49.3.6 Other Considerations

There are several disadvantages when area detectors are used without collimation, some of which are due to the nondiscriminating nature of both imaging plates and charge-coupled devices. The synchrotron beam inevitably excites fluorescence within the hutch and care must be taken to shield these devices from stray radiation. The difficulty in designing slits for these devices also decreases the signal-to-noise discrimination since scattering from sample containers, or environmental chambers, often contaminates the pattern. Although this can be eliminated using subtraction [49.15, 16], another possibility when powder averaging is not a problem and angular resolution can be relaxed ($\Gamma \approx 0.03°$), is to use an energy-discriminating PSD fitted with a slit. While most commercial PSDs operate in the so-called *streaming mode*, proportional counting and energy discrimination are possible with these devices.

The highest possible time resolution is afforded by energy-dispersive diffraction. Since the whole pattern is recorded at once, this resolution is determined by the brightness of the beam and the readout time of the multichannel analyzer. Quantitative interpretation of crystal structure is hampered by systematic errors such as energy-dependent absorption corrections, absorption edges, and definition of the intensity versus energy curve. Reliable results have been obtained, however, and the software and method to enable structure determination using the Rietveld method are now well established [49.16, 17]. The energy-dispersive x-ray technique does have distinct advantages. The experiment allows for straightforward collimation and data collection at fixed angles, which makes it easier to discriminate parasitic scattering from sample containers.

In general, for phase identification and determination of unit cell parameters, energy-dispersive x-ray diffraction data are suitable and offer distinct advantages in terms of spatial and time resolution. In those cases where accurate determination of structural parameters is the objective, monochromatic data are preferred. Many beamlines are capable of changing between these two modes of operation, and a description of such a setup is given in two recent reviews [49.18, 19].

49.4 Software

Collection of either x-ray or neutron diffraction data with ever-increasing time resolution, as is envisioned at third-generation synchrotrons, will inevitably lead to data glut. While visual inspection, to identify the appearance of one phase or the disappearance of another, may be sufficient in some studies, an unbiased method which can be automated to provide some real-time feedback on the course of a reaction is desirable. Such a method might provide information on systematic errors as well as the appearance of new phases and the disappearance of others. It should also provide visual queues to allow the choice of a manageable number (3–5) of diffraction patterns to analyze out of the hundreds collected. One example in situ XRD investigation, ion exchange in porous titanium silicates, is discussed below and employs iterative target transform factor analysis (ITTFA) [49.20], an unbiased mathematical treatment of the diffraction data that looks for changes as a function of time. From this processing, kinetic information and clues as to which patterns to use first for Rietveld refinement were obtained. Problems and questions such as which patterns to explore, whether measurable changes are occurring, and determining if the reaction has completed are vital for efficiently managing the ever-increasing quantities of data. Also general structure analysis software (GSAS), Fullprof+winplotr, and Powerd3D are freeware that are capable of handling large sets of one-dimensional diffraction data. Since new software is continuously appearing on the Internet, the authors cannot give a detailed software overview. However, the CCP14 website contains a vast collection of software that is useful in interpreting time-resolved data [49.21].

49.5 Types of In Situ Cells

Fig. 49.4 Beamline geometry suitable for time-resolved synchrotron powder diffractometry in monochromatic mode. A double-crystal monochromator (M) is used to select the incident energy that is focused on the sample, capillary or flat plate holder (SAC), and recorded on a position-sensitive detector (PSD), image plate (IP) or charge-coupled device (CCD). An incident-beam monitor (IC) is needed to normalize the data to the same relative intensity scale because the synchrotron beam decays with time (after [49.22]) ▶

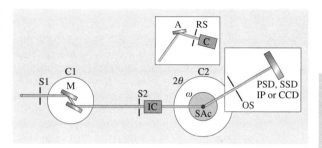

A schematic diagram of the beamline geometry suitable for time-resolved synchrotron diffractometry is shown in Fig. 49.4 [49.22]. For in situ studies where structural studies are to be performed monochromatic radiation is preferred. The beam passes through an incident beam monitor ion chamber (IC) to record the intensity of the incident beam during the entire run. Monitoring the incident beam is necessary because the synchrotron beam decays with time, and therefore the total intensity of the diffraction pattern also decreases with time. If a synchrotron runs in the so-called top-off injection mode, where electrons are continuously supplied to the storage ring, an IC is still required because the storage ring is usually not topped off to a constant value. The monitored data is then used to normalize the data to the same intensity scale.

Our own interests are twofold: to grow crystals from gels that possess ion-exchange properties and then to study the mechanism of ion exchange. The crystal growth is done to temperatures of 150–250 °C but the ion exchange may be carried out at room temperature and at $\approx 60\,°C$. Two different reaction cells are used for these purposes. The crystal growth experiments are carried out hydrothermally, requiring elevated pressures. A quartz capillary reaction cell introduced by *Norby* [49.23] is shown in Fig. 49.5. The capillary is closed at one end (A) and is held in place with a Swagelock fitting (B) mounted on a modified goniometer head. Heating can be effected by heating plates placed above and below the capillary tube so as not to obstruct the path of the x-ray beam, or by using an air blower heater. A small amount of gel to be crystallized is placed into the capillary and fixed into the Swagelock fitting, and a pressure of N_2 is applied through the fitting. The detector is an imaging plate of which only a portion is exposed to the x-rays. This is done by placing two lead shields (D) over the IP, exposing a portion of the plate to the x-rays. At the end of each cycle the IP film is moved to expose a fresh surface. A cycle consisted of

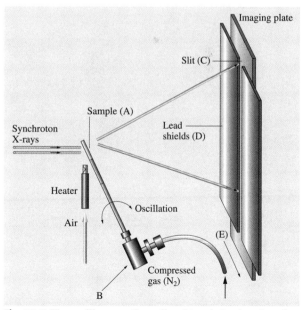

Fig. 49.5 The capillary reaction cell and translating imaging-plate detection system as used at the NSLS

a 60 s exposure and erasure, and exposing a new IP surface required 1.5 min for data processing or 2.5 min per cycle. Recent improvements in the storage ring at the NSLS Brookhaven National Laboratory and detection system have now reduced the time per cycle to seconds. This cell has been widely used by many investigators [49.3].

49.5.1 SECeRTS Cell

A somewhat different type of cell was used for the ion-exchange studies, and is also easily adaptable to synthesis studies. The design of a small environmental cell for real-time studies (SECeRTS) [49.24] is shown in Fig. 49.6. A quartz, or sapphire, capillary (0.3–1 mm OD) tube (A), open at both ends is mounted in a Swagelock tee (B) using a Vespel ferrule. A pressure can be applied, through the connected tube (C), to the surface of the reaction mixture in the capillary. For titration synthesis studies, injection takes place through a 0.3 mm quartz capillary (A) that goes through the tee and into the 1 mm capillary. This 0.3 mm capillary is mounted between a Swagelock elbow and tee with Vespel ferrules, all of which are mounted on a modified goniometer head (E). Injection under pressure through the elbow (A) is possible via a gas-chromatography syringe (not shown) mounted on an aluminum holder. A screw connected to the piston of the syringe ensures pressurization. By turning the screw, the piston is depressed and a controlled volume can be injected into the 1 mm capillary through the 0.3 mm capillary. Alternatively, the position of the sample capillary (A) can be plugged, and the assembly can be extended to expose the injection capillary location in the x-ray beam. Ports D and C then become the supply and exhaust lines, respectively, for flow-through ion-exchange experiments or solid/gas-phase reactions.

It is important to note that completely sealed capillaries pose a significant risk to experimentalists and equipment. The pressure generated during hydrothermal reactions is difficult to control in small-diameter capillaries, and they could burst at any time during the heating of the cell. Having an open end where an overpressure of inert gas is applied is a safety feature to

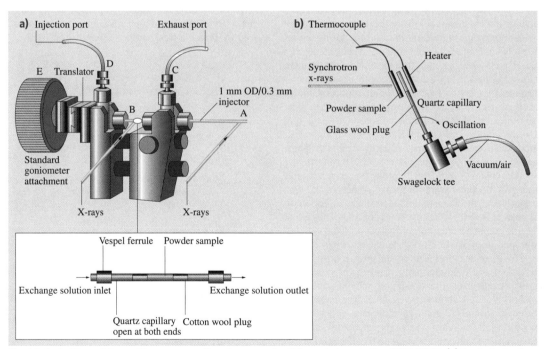

Fig. 49.6a,b View of the capillary-type small environmental cell for real-time studies (SECeRTS). (**a**) In situ hydrothermal titration cell and (**b**) simplified quartz or sapphire hydrothermal cell (courtesy of John Parise and coworkers, Suny Stony Brook)

prevent the cell from exploding while still maintaining hydrothermal conditions.

49.5.2 Polyimide Environmental Cell

The polyimide environmental cell (PEC) was designed to alleviate the problems and safety issues associated with quartz capillaries and gas-driven solutions. Initially designed for flow-through experiments [49.26], the PEC is easily adaptable to in situ synthesis experiments by simply sealing one end and applying an overpressure of nitrogen to the other. Polyimide tubing was chosen to replace the quartz/sapphire capillaries based on multiple criteria. The tubing is flexible, thin walled, and x-ray transparent (within the radiation range of $0.1-1.54\,\text{Å}$) in the wide-angle region. Controlled solution flow through a larger-inner-diameter tube is easier to maintain and reproduce. Tube breakage is no longer a problem during sample loading or during the experiment because of the flexibility of polyimide. Tube transparency in the wide-angle region ($2\theta > 2°$) means only x-ray scattering from the sample and solution contributes, with no parasitic scattering from the polyimide cell, to the diffraction patterns in that region. Thin walls (typically < 0.01 mm for a 0.8 mm OD tube) allows more sample in the beam and increases peak-to-background ratios.

49.5.3 High-Pressure Cells

High-pressure cells commonly in use can often be adapted with slight modification for use in in situ studies at a synchrotron. For example, the large-volume high-pressure device originally designed for diamond growth (DIA) has been successfully interfaced to a synchrotron at the Photon Factory, Japan [49.13]. Similar installations were constructed at the NSLS Brookhaven, the advanced photon source (APS) Argonne, and at Deutsches Elektronen Synchrotron (DESY) in Germany. They are utilized in the study of materials at high pressure and temperature using mainly energy-dispersive diffraction. In some cases the use of monochromatic radiation together with the IP was more effective. An example is the partitioning of Fe in a mixture of olivine and the β-phase $(\text{Fe},\text{Mg})_2\text{SiO}_4$ [49.16] using an IP and monochromatic radiation resulted in considerably better time resolution, and the mechanism of the transformation from the olivine structure to the spinel structure was observed [49.27].

Other sample cells, such as furnaces and diamond-anvil cells, are also easily transferred. Similarly, steel hydrothermal reaction vessels can be transferred from the laboratory to the synchrotron.

49.5.4 Hydrothermal Steel Autoclave-Type Cell

Evans et al. [49.25] introduced a hydrothermal autoclave reaction vessel (Fig. 49.7) that has been extensively used in many studies [49.28]. The cell consists of a modified Parr reaction vessel (25 ml) with a section of the wall milled down to a thickness of 0.4 mm. This thinner portion of the wall permits transmission of white x-radiation and has a maximum operating temperature of $230\,°\text{C}$. Higher operating temperature is possible with thicker-walled cells. However they require the focusing optics available at a third-generation source [49.29]. Attached to the top of the cell is a head consisting of a pressure transducer, a safety relief valve, and an injunction reservoir (2 ml). This reservoir consists of a remotely placed gate valve that permits injection of a second solution into the cell at

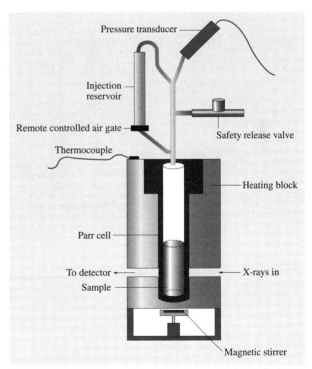

Fig. 49.7 The hydrothermal autoclave-type reaction cell due to *Evans* et al. [49.25]. Walls are milled down to allow x-ray transmission while permitting a maximum operating temperature of $300\,°\text{C}$

controlled time and temperature conditions. Temperature control of the cell is done via a resistance-heated aluminum block and is monitored with K-type thermocouples in contact with the outside of the reaction vessel.

49.5.5 Neutron Diffraction Cell

For neutron diffraction studies, a vanadium (V) tube is used instead of polyimide, quartz or sapphire. The coherent neutron scattering length of vanadium is small, $b = -0.5$ fm, compared with that of oxygen, which is 5.8 fm, making it an ideal sample container since it contributes primarily to the incoherent background and not to the Bragg diffraction in a powder pattern (Fig. 49.3b). Also, the V-tube is much larger in diameter (6.5 mm) to allow more powder to be used (Fig. 49.8). The low absorption cross-section of neutrons allows the use of larger samples, which perhaps will be more representative of the situation when using titanium silicate (TS) (see below) in the field.

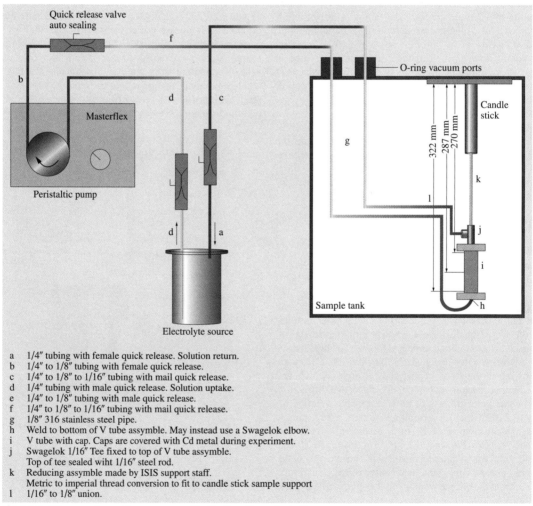

a   1/4″ tubing with female quick release. Solution return.
b   1/4″ to 1/8″ tubing with female quick release.
c   1/4″ to 1/8″ to 1/16″ tubing with mail quick release.
d   1/4″ tubing with male quick release. Solution uptake.
e   1/4″ to 1/8″ tubing with male quick release.
f   1/4″ to 1/8″ to 1/16″ tubing with mail quick release.
g   1/8″ 316 stainless steel pipe.
h   Weld to bottom of V tube assymble. May instead use a Swagelok elbow.
i   V tube with cap. Caps are covered with Cd metal during experiment.
j   Swagelok 1/16″ Tee fixed to top of V tube assymble.
    Top of tee sealed wiht 1/16″ steel rod.
k   Reducing assymble made by ISIS support staff.
    Metric to imperial thread conversion to fit to candle stick sample support
l   1/16″ to 1/8″ union.

Fig. 49.8 Diagram of vanadium can assembly for flow-through studies

49.6 In-Situ Studies of Titanium Silicates (Na-TS) with Sitinakite Topology

49.6.1 Introduction to the Problem

Large quantities of nuclear waste were generated as a byproduct of the US nuclear weapons programs. The waste is held in large, underground steel tanks at both the Hanford and Savannah River sites. A search for efficient, cost-effective methods of removing cesium, strontium, and actinides from the waste solutions is an ongoing project of the Department of Energy. A sodium titanium silicate (Na-TS) was found to be highly selective for Cs^+ and Sr^{2+} in moderately alkaline solutions. When the Na-TS material was immersed in the highly alkaline wastes its selectivity for Cs^+ was low, but in a partially Nb-substituted form (Na-NbTS) the material was highly effective at removing Cs^+ [49.30]. Several problems are attendant to the use of these titanium silicates for separation of high-level waste.

The nuclear waste is a highly basic solution containing 1–3 M NaOH. It also has high sodium ion concentration, 5–7 M, and contains a variety of other ionic species. The ions that are the highest γ-emitters are the highly radioactive ^{137}Cs and ^{90}Sr, with 30 years half-lives. In addition, small amounts of Pu and Np also need to be removed from the mass of waste. This high-level waste (HLW) would be sealed in a special glass and stored in steel tanks below ground. The remaining low-level waste (LLW) would be taken up in a grout and stored aboveground.

The questions are: what are the structural and/or chemical properties of sodium titanium silicate that display such high uptake of Cs^+ in neutral to mildly basic solution, but not in the waste solutions? Also, what effect does the substitution of Nb for some Ti have on the increased selectivity toward Cs^+? In short, what is the origin of the high ion selectivity in these compounds? In addition, the hydrothermal syntheses by which these titanium silicates are prepared are often accompanied by the appearance of impurity phases. What are these impurities, and how can they be eliminated from the synthesis [49.5, 6]?

49.6.2 Synthesis and Structure of Sodium Titanium Silicate (Na-TS)

The ideal composition of this titanosilicate is $Na_2Ti_2O_3SiO_4 \cdot 2H_2O$. However, hydrolysis occurs in water to yield a composition of $Na_{1.64}H_{0.36}Ti_2O_3SiO_4 \cdot 1.8H_2O$ [49.31]. The crystal structure of the sodium ion phase and the partially exchanged Cs^+ phase were determined from powder x-ray studies [49.31].

The crystals are tetragonal, with $a = 7.8082(2)$ Å, $c = 11.9735(4)$ Å, space group $P4_2/mcm$, and $Z = 4$. The titanium atoms occur in clusters of four grouped about a 4_2 axis, two up and two down, rotated by $90°$. Each titanium is octahedrally coordinated, sharing edges in such a way that an inner core or four oxygens and four Ti atoms form a distorted cubane-like structure (Fig. 49.9).

These cubane-type structures are bridged to each other through silicate groups along the a- and b-axis directions. The titanium–oxygen clusters are 7.81 Å apart in both the a- and b-axis directions, with the Si atoms at $c = \frac{1}{4}, \frac{3}{4}$. The c-axis is ≈ 12 Å long, which is twice the distance from the center of one cubane-like cluster to its neighbor in the c-axis direction. Two views of the framework are shown in Figs. 49.10 and 49.11. The net result of this framework arrangement is that one-dimensional tunnels are formed along the c-axis di-

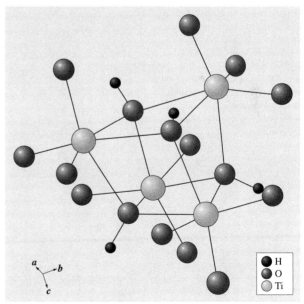

Fig. 49.9 A portion of the titanosilicate structure showing the cluster of four titanium–oxygen octahedra sharing edges to form the cubane-like Ti_4O_4 group. The oxygens within the cubane group are each bonded to a proton making them hydroxo groups. The Ti atoms are *light brown*, oxygens are *dark brown*, and hydrogens are *black*

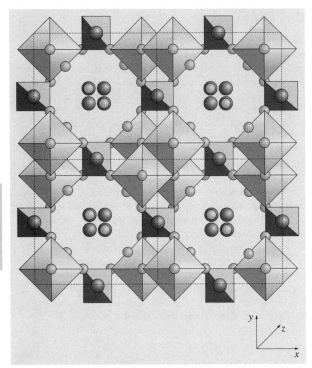

Fig. 49.10 Top view (down the c-axis) of sodium titanium silicate showing the clusters of four TiO_6 octahedra bridged by silicate groups. The tunnels are filled with Na^+ and water molecules. The Na^+ on top of the tetrahedra symbolizes the Na^+ ions sandwiched between silicate groups within the framework. Refer to Fig. 49.11 for color legend

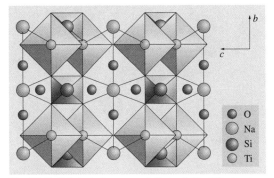

Fig. 49.11 Polyhedral representation of Na-TS structure as viewed down the a-axis, showing the hexagonal-shaped framework cavities in which half the Na^+ reside. The water molecules reside within the 8MR tunnels

rection. Perpendicular to the tunnels are vacancies in the faces, or four sides, of the tunnels. These cavities are large enough to enclose sodium ions. Four silicate oxygens bond to the sodium ion at a distance of 2.414(5) Å (Fig. 49.11). The sodium ion coordination is completed by bonding to two water molecules in the tunnels at a Na–O bond distance of 2.765(1) Å. Half the sodium ions are thus accounted for in the framework sites as there are sodiums in each face at $c = 0, \frac{1}{2}$ over one c-axis cell length for a total of four out of the eight required per unit cell. The remaining sodium ions reside within the tunnels along with the water molecules. As a historical note, these compounds were originally misnamed as crystalline silico titanate (CST). However, they are silicates, so their preferred nomenclature is as a titanium silicate M-TS, where M is the exchangeable ion.

The Na–O bond distances within the tunnels are longer than the sum of the ionic radii (2.42 Å) [49.32] at 2.76(1) Å. This bond distance measurement was made with only 64% of the sodium ion sites occupied. The deficiency of sodium arises from hydrolysis during washing so that the actual formula was $Na_{1.64}H_{0.36}Ti_2O_3(SiO_4) \cdot 1.8H_2O$. Because of the deficiency of Na^+, the sodium ion positions were found to be disordered with partial occupancy by water molecules. It is possible to obtain the fully occupied sodium phase by not washing the product of the hydrothermal reaction with water, or washing with strong $NaNO_3$ solution.

On exchanging Cs^+ for Na^+ by a flow-through procedure in near-neutral CsCl solution, a composition of $Na_{1.49}Cs_{0.2}H_{0.31}Ti_2O_3SiO_4 \cdot H_2O$ was obtained [49.31]. The Cs^+ occupied two positions within the large tunnel. Site one (Cs1) is located at the center of the eight-membered ring (8MR) window at $\frac{1}{4}c, \frac{3}{4}c$, and Cs2 is located at approximately $0.13c$ and $0.63c$ off from the 8MR window. Cs1 has eightfold bonding coordination to eight oxygens of the framework, four above and four below the Cs, with bond lengths of 3.18 Å. Cs2 has fourfold bonding coordination to the framework oxygens and twofold coordination to the interstitial water with bond lengths of 3.06 and 2.95 Å, respectively. The bulk of the Cs^+ was in site 1 and about 20% in site 2.

49.6.3 Synthesis Problems and In Situ Hydrothermal Study

A problem arose in the actual sol–gel synthesis of the Na-TS crystals. Often an impurity began to form that on further investigation was determined to be a sodium titanium silicate of composition Na_2TiSiO_5 referred to as sodium titanium oxide silicate (STOS). In ex situ hydrothermal studies it was found that the amount of STOS in the mixture increased with the alkalinity of the starting gel. In contrast, the conditions for favorable Na-TS formation required a lower Na_2O content as well as lower pH. Based on this ex situ study we prepared two gels, one designed to yield pure STOS and the other to yield pure Na-TS [49.6]. The gel composition for Na-TS had the composition $1.0TiO_2 : 1.98SiO_2 : 6.77Na_2O : 218H_2O$. In situ experiments were carried out at the Brookhaven synchrotron. The source of Ti was titanium isopropoxide, to which was added NaOH and silica dissolved in NaOH. The apparatus used is that shown in Fig. 49.5, and the results are provided in Fig. 49.12. Figure 49.12 shows a three-dimensional plot of the x-ray diffraction pattern as a function of time during the TS gel heating. As can be seen from the figure, the process starts with the formation of a phase having a broad peak at about 9.5–10 Å. This phase was identified as sodium nonatitanate (SNT) $Na_4Ti_9O_{20}$, a semicrystalline compound that is highly selective for Sr^{2+} in alkaline solution [49.33, 34]. It starts forming at an early stage of reaction, as confirmed by the collection of x-ray powder diffraction patterns of the dried starting gel. The intensity of the (001) SNT reflection does not change as the reaction progresses until the growth of the TS phase begins. The process of transformation of the SNT phase to the TS phase started after 1 h of constant heating at 220 °C, with rapid decrease of the intensity of the SNT peaks and in-growth of the TS peaks (Fig. 49.13). The whole process of transformation lasted approximately 45 min. During the period following the transformation to TS, no significant changes occurred, except a minor increase in the TS peak intensities. No other phases were observed in this experiment.

Interestingly the highly alkaline base gel in ex situ experiments also crystallized as the SNT phase first, as shown in Fig. 49.14. After about 12 h at 200 °C it is converted to the highly crystalline STOS phase in an additional 15 h. Thus, it appears that in gels of different composition and alkalinity the SNT phase forms initially. Then the sodium silicate that remains converts the SNT phase to the Na-TS phase at low sodium content and moderate alkalinity, while at higher sodium content and higher alkalinity the STOS phase prevails. Several avenues of synthesis are now open. The SNT phase is

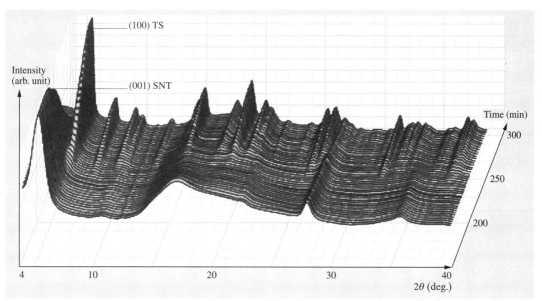

Fig. 49.12 Time-resolved x-ray powder diffraction spectra of SNT-TS transformation for the gel with composition $1.0TiO_2 : 1.98SiO_2 : 6.77Na_2O : 218H_2O$. Powder patterns are collected in 2.5 min intervals

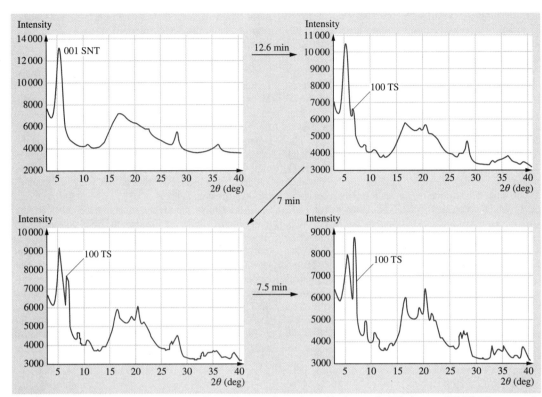

Fig. 49.13 XRD patterns of SNT (*upper right*) and its transformation with time to the TS phase at 220 °C. The pattern in the *upper right-hand corner* is that of pure SNT

readily prepared from a mixture of titanium isopropoxide and NaOH. By adding the correct balance of silica dissolved in NaOH it is possible to avoid the presence of the STOS impurity. Furthermore, by proper control of the reaction it may be possible to prepare a mixture of Na-TS and SNT, such as to remove Cs^+ and Sr^{2+} in a simple one-step process.

49.6.4 Ion Exchange of Cs^+ into Na-TS

The ion-exchange experiments were carried out at the X7B beamline of the National Synchrotron Light Source. The open-capillary SECReTS cell was used (Fig. 49.6). Exchange of Cs^+ from a 1 mM solution of CsCl was used for the Na-TS phase and a 10 μM solution for the exchange with the protonated H-TS phase. The lower-concentration solution was necessary to slow the rate of exchange to fully observe the changes occurring in the H-TS phase. Diffracted intensities were recorded on a MAR345 imaging plate. Each diffraction pattern was collected for 1 min with a 1.5 min lag time to read and erase the IP.

As the Cs^+ was exchanged for Na^+ in Na-TS, the unit cell volume increased from 728.8 to 732.4 Å^3 [49.26]. The unit cell parameters a and b showed a continuous increase from 7.8060(1) to 7.8435(1) Å and the c-axis decreased from 11.9599(2) to 11.9054(4) Å. The ion exchange occurred in two steps. The first step, from minutes 0 to 245, involved occupancy of Cs^+ in the Cs2 site to a fractional occupancy of 0.116(5). Simultaneous with the increased occupancy of site Cs2, the water site OW2 began to decrease in occupancy (Fig. 49.15). Figure 49.15 also shows a concomitant loss of Na^+ as the Cs^+ was taken up. The second step of the ion exchange from minutes 245 to 375 was initiated when site OW2 had zero occupancy (minute 245) and involved the rapid filling of site Cs1 (Fig. 49.15). During minutes 245 to 275 the occupancy of Cs2 increases and reaches a maximum occupancy of 0.136(7). The maximum occupancy for

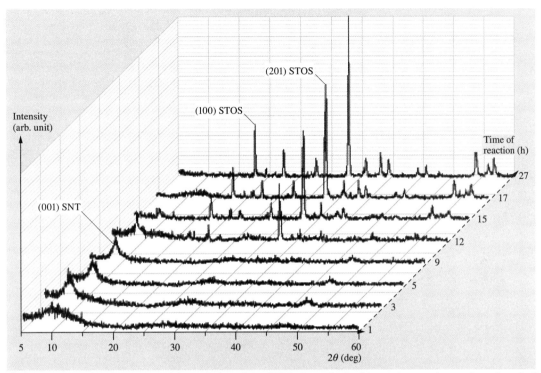

Fig. 49.14 Dynamic XRD spectra of the evolution of the STOS phase obtained in ex situ experiments from the gel of composition $1.0\text{TiO}_2 : 1.98\text{SiO}_2 : 10.53\text{Na}_2\text{O} : 218\text{H}_2\text{O}$, $T = 200\,°\text{C}$

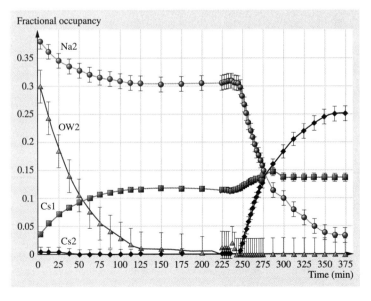

Fig. 49.15 Results of fractional occupancy refinements for Na2, OW2, Cs2, and Cs1 during the Cs^+ ion exchange into Na-TS

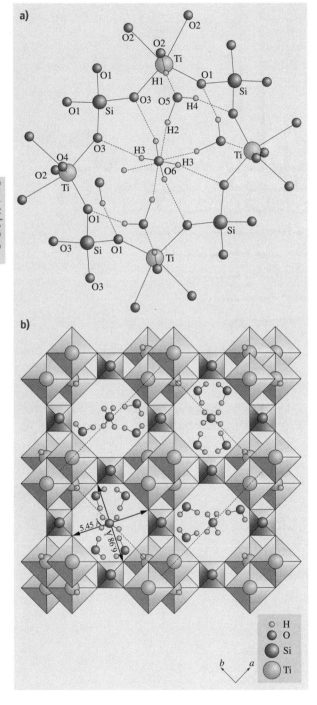

Fig. 49.16a,b Ball-and-stick representation of a portion of the proton phase $H_2Ti_2O_3(SiO_4)_4 \cdot 1.5H_2O$ as viewed down the c-axis, showing the hydrogen bonding scheme between the acid proton and the water molecules O5 and water–water and water–framework oxygens (**a**) and a polyhedral representation showing the elliptical nature of the tunnel along the c-axis direction and the new unit cell (*bold line*) along the diagonals of the Na-TS unit cell ◂

the Cs1 site was 0.25. These values yield a formula of $Na_{1.06}Cs_{0.26}H_{0.68}Ti_2O_3SiO_4 \cdot H_2O$. The proton content was due to hydrolysis and the amount was calculated to balance the charge. Almost all the Na^+ in the tunnel was lost to hydrolysis and exchange for Cs^+. The maximum amount, or fraction, of Cs^+ that can be exchanged into the TS phase is 0.5. This has been shown by titration with CsOH [49.35, 36] and from ex situ structural studies. The Cs^+ is too large to occupy the Na1 framework sites and for the same reason can only occupy half the tunnel sites.

49.6.5 Cesium Ion Exchange into H-TS

There is a major difference between Cs^+ ion-exchange pathways in Na-TS and H-TS. Replacement of the sodium ions by protons is accompanied by a space group change from $P4_2/mcm$ to $P4_2/mbc$ and unit cell dimensions [49.35, 37] of $a, b = 11.039(5)$ Å, $c = 11.880(5)$ Å. The protons are covalently bonded to the cubane oxygens and hydrogen bond to water molecules in the tunnel. The reason for the change is that the a, b axes transition from 7.806 Å to the larger value (actually the ab square diagonal), and the symmetrical eight-membered ring perimeter of the tunnel is distorted into an elliptical shape (Fig. 49.16). The ratio of the length to the width of the tunnel in now 1.2. A very fast exchange of Cs^+ into the Cs2 site (six-coordinate site) occurs in the first 10 min of exchange, as shown in Fig. 49.17 [49.5]. The structure then reverts back to the $P4_2/mcm$ space group, restoring the symmetrical tunnel shape. At this point the Cs1 site begins to fill up, occurring simultaneously with the space group transformation. The diffraction patterns contained both phases to the end of the experiment. After 20 min, both sites continued to fill until a final composition of $Cs_{0.36}H_{1.64}Ti_2O_3SiO_4$. Since complete filling with Cs^+ would have the Cs fractional content of 0.5, the exchange was 72% complete. This value is quite high considering how dilute the exchange solution was initially.

49.6.6 Sodium Niobium Titanosilicate (Nb–TS)

Niobium forms a solid solution with the sodium titanium silicate Na-TS up to about 25% replacement of Ti by Nb. This phase, of ideal composition $Na_{1.5}Ti_{1.5}Nb_{0.5}O_3SiO_4 \cdot 2H_2O$, has a higher selectivity for Cs^+ in highly basic solutions containing higher levels of Na^+ than does the non-niobium-containing phase. As a first step in determining why this phase possesses such a high selectivity for Cs^+, the crystal structure of the cesium ion phase was determined [49.38]. The presence of Nb^{5+} is statistically distributed over the titanium sites and reduces the sodium ion occupancy in the tunnel by half. However, in this phase considerable hydrolysis occurs, so that the tunnel contains almost no Na^+ but only water [49.38]. This lack of competing ions in the tunnel changes the coordination number of Cs^+ in the tunnel to 12. That is, there are eight water molecules per unit cell in the tunnel and two Cs^+. With no Na^+ in the tunnel all the water molecules bond to Cs^+ at a bond distance of 3.13 Å, compared with 3.26 Å for the eight framework oxygens. This increase in coordination number apparently greatly decreases the free energy of exchange as Cs^+ changes from a weak coordination in the aqueous phase to very high coordination within the tunnel.

49.6.7 In Situ Synthesis of Na-NbTS

The preparation of the niobium Na-TS phase is often plagued with an impurity that is probably a polymeric Keggin ion, referred to as Na-IPX. A gel was prepared from Nb_2O_5, titanium isopropoxide, silicon tetraethoxide, and 3.3 M NaOH. The ratio of constituents was $1.0TiO_2:0.167Nb_2O_5:1.33SiO_2:6.9Na_2O:228H_2O$. This gel was heated in a Teflon-lined pressure vessel at 200 °C for 4 days. The resultant Na-NbTS was highly crystalline but contained a small amount of Na-IPX. A similar gel was prepared for the in situ study at the X7B beamline of the National Synchrotron Light Source, Brookhaven National Laboratory. The wavelength used was 0.9223 Å and an external N_2 gas pressure of 250 psi was applied to the sapphire capillary. A sapphire capillary was used because silica is solubilized at the high-pH conditions of the starting gel and would weaken the integrity of the capillary. The gel was heated to 210 °C in 2 min and x-ray diffraction patterns were recorded every 2.5 min.

The initial pattern of the gel exhibited two broad peaks centered at 17.6° and 24.5° in 2θ. The results

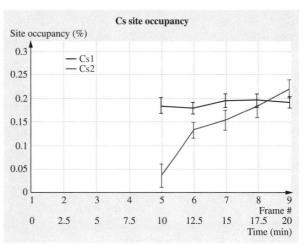

Fig. 49.17 In situ time-resolved x-ray diffraction showing the onset of Cs^+ for H^+ ion exchange to be a site-selective process. The Cs2 position first filled to $\approx 20\%$. After Cs2 filled, Cs1 began to fill the channels along the [001] direction and completed at $\approx 20\%$. The mechanism of exchange is seen as an opening of the [001] channels as Cs2 bonds to the Ti − O − Si polyhedra linkages, allowing Cs1 to fill the center of the eight-membered ring

are shown in Fig. 49.18. Between minutes 0 and 65, the broad features began to diminish. Simultaneously a third broad feature developed at $\approx 5.1° 2\theta$, associated with the crystalline ordering of the Na-IPX phase. From minutes 65 to 75, the reflections from the impurity phase began to increase in intensity, with the strongest peak arising from the broad feature at $\approx 5.1°$ (Fig. 49.18). From minute 75 to the end of the experiment, the impurity phase ceased to increase in intensity as the Na-NbTS phase began to crystallize. The Na-NbTS phase continued to grow for 25 min, and after 100 min no further changes in the diffraction pattern were observed.

Unit cell refinements of Na-TS during crystallization showed an initial increase in the a- and b-axes and a decrease in the c-axis in a brief span of 5 min from minute 75 to 80 (Fig. 49.19). After the initial anomalous expansion, the unit cell volume and unit cell parameters increased until the end of the experiment. The unit cell volume increased from 743 to 745 Å^3 from minute 75 to the completion of the run. By slight adjustments of the gel composition it was possible to achieve phase purity by elimination of Na-IPX formation, as illustrated by this in situ study.

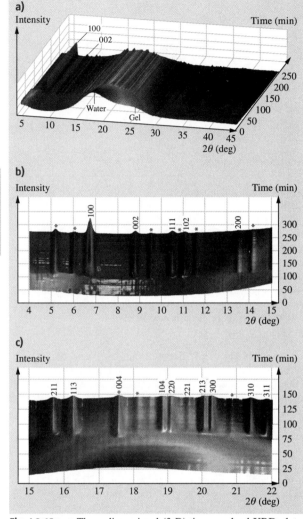

Fig. 49.18a–c Three-dimensional (3-D) time-resolved XRD plots. (**a**) The entire range of the acquired diffraction pattern, showing the main Na-NbCTS, water, and gel signals contributing to the pattern. (**b**) Zoomed plot showing the main Na-NbCTS phase index and peaks from the impurity phases (marked). (**c**) Zoomed plot of higher-angle diffraction data. Asterisk denotes the presence of an unidentified phase

49.6.8 In Situ Ion Exchange of Cesium Ion Exchange in Na-NbTS

The time-resolved x-ray patterns collected during Cs^+/Na^+ exchange [49.26] are shown in Fig. 49.20. Refinements of the cell parameters during the exchange process indicated that no structural transformation had occurred during exchange. The space group remained $P4_2/mcm$ and the eight-membered ring (8MR) of the tunnel remained circular. The changes in unit cell dimensions as exchange proceeded are provided in Fig. 49.21. The a- and b-axes increased from 7.8470(5) Å to 7.8535(5) Å while the c-axis decreased from 11.980(1) Å to 11.963(1) Å. However, after an initial increase in unit cell volume due to the more rapid increase in the a, b parameters, the unit cell volume was very nearly the same at the end of the Cs^+ uptake as for the original Na-NbTS. This exchange process was interpreted as continuous filling of sites Cs1 and Cs2 with little structural distortion, which contrasts with the exchange studies in the Na-TS and H-TS forms. The break in the curves of Fig. 49.21 was due to filling of the storage ring when no x-rays were available.

49.6.9 Cesium Ion Exchange into H-NbTS

A major difference between the Na-NbTS phase and the Na-TS phase is that no space group change occurred when the Na^+ was replaced by H^+. The Ti/Nb fractional occupancies refined to 0.77(3) and 0.23(3), respectively, and were fixed to a ratio of 0.75/0.25 based on previous compositional studies [49.38, 39]. No Na^+ was detected by microprobe analysis. Thus, the formula, both by elemental analysis and x-ray refinement, was fixed at $H_{1.5}Nb_{0.5}Ti_{1.5}SiO_7 \cdot 2H_2O$. The proton parameters could not be determined by the x-ray refinement, but it was assumed that they are covalently bonded to the cubane oxygen sites in a similar fashion as seen in the H-TS structure. The water molecules were also distributed in crystallographic sites analogous to Na-TS. Water (OW1) is located near the walls of the 8MR and the intersection of the 6MR, i.e., those cavities where the Na^+ in the framework is located. Water (OW2) is located near the center of the channel along the c-axis, but not on the 4_2-screw special position. This offset from the screw axis results in the partial occupancy of the OW2 site where its maximal allowed occupancy is 1/2 due to positional disorder.

No structural change was observed in the exchange reaction since all peaks in the diffraction patterns could be indexed on the basis of a single unit cell (Fig. 49.22). The unit cell dimensions initially increased to minute 70, then showed a continuous volume decrease as the a- and b-axes decreased from 7.837(1) to 7.828(1) Å and the c-axis decreased from 11.921(2) to 11.912(2) Å (Fig. 49.23). Rietveld refinement of the

Fig. 49.19 Results of the unit cell refinements of Na-NbTS during crystallization. Initial cell lengths were $a, b = 7.844(2)$ Å, $c = 12.073(4)$ Å and final cell lengths were $a, b = 7.857(2)$ Å, $c = 12.067(4)$ Å ▶

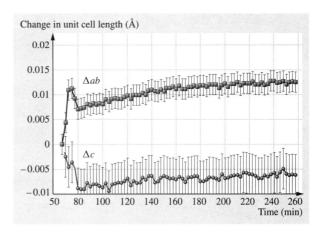

occupancies of the sites populated by Cs$^+$ and water indicated that ion exchange occurred in two distinct steps (Fig. 49.24). The first occurred from minutes 0 to 197.5 and involved a gradual increase in Cs$^+$ uptake from 0 to 0.115(6) fractional occupancy at site 1. After the initial Cs$^+$ uptake, at minute 197.5, H$_2$O at site OW2 decreased in fractional occupancy from 0.51 to 0.24(3) within 1 min, amounting to slightly more than 0.5 molH$_2$O. From minute 200 to the end of the experiment, the fractional occupancy of H$_2$O at site OW2 remained constant at 0.24(3). The second step of the exchange begins at minute 197.5, during which the Cs$^+$ uptake increased to 0.339(7) fractional occupancy at the end of the experiment (700 min). The final formula as obtained from the x-ray data was $H_{1.33}Cs_{0.17}Nb_{0.5}Ti_{1.5}SiO_7 \cdot 1.5H_2O$.

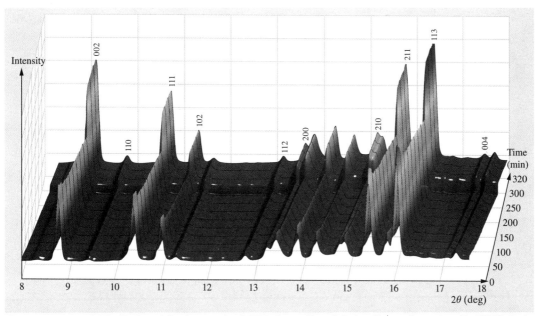

Fig. 49.20 Zoomed 3-D display of the time-resolved XRD data collected during Cs$^+$ exchange into Na-NbTS. Peaks from the Na-NbTS phase are indexed. Diffraction patterns are highlighted (*black*) every 25 min. Optimization of beam path flight tube resulted in an increase of overall intensity near minute 275

49.7 Discussion of In Situ Studies

49.7.1 Synthesis of Na–TS and Na–NbTS

Both the Na-TS and Na-NbTS are formed in a very narrow window of gel composition and alkalinity. An excess of silica relative to titanium is necessary to form Na-TS, generally a 2:1 ratio. The critical factor in crystallizing a pure TS compound is the amount and concentration of NaOH in the initial gel. At high molar ratios and concentrations of NaOH, substantial amounts of STOS are obtained. At a concentration of 3.3 M and a ratio of $6.67Na_2O : 1.0TiO_2$ a pure Na-TS phase is obtained. However, even with higher ratios and concentrations of NaOH, sodium nonatitanate is found to form first and then is converted to STOS, Na_2TiSiO_5, a synthetic analogue of the mineral natisite. Since the nonatitanate has a high affinity for Sr^{2+} in alkaline solution [49.40, 41], it may be possible to provide a combination of Na-TS and the nonatitanate to remove much of the high-level waste, ^{137}Cs and ^{90}Sr, from weapons-type nuclear waste with a combination of Na-TS and SNT. Time-resolved synthesis experiments and optimized synthesis procedures can be used to tailor the Na-TS : SNT ratios for a specified removal capacity

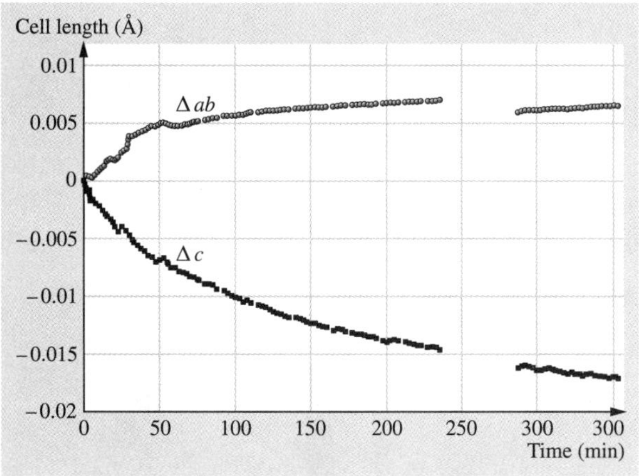

Fig. 49.21 Results of the unit cell refinements for data collected during Cs^+ exchange into Na-NbTS. Unit cell lengths changed from $a, b = 7.8470(5)$ Å, $c = 11.980(1)$ Å at minute 0 to $a, b = 7.8535(5)$ Å, $c = 11.963(1)$ Å at minute 355.5. The *gap* in data is when no x-ray beam was available. *Error bars* are smaller than plot points ◄

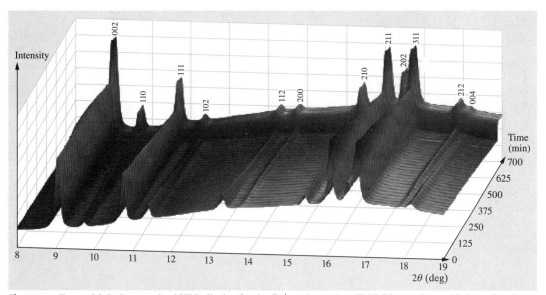

Fig. 49.22 Zoomed 3-D time-resolved XRD display for the Cs^+ exchange into H-NbTS experiment. No impurity phases were present in the diffraction patterns. Diffraction patterns are highlighted (*black*) every 25 min. Optimization of beam path flight tube resulted in an increase of overall intensity near minute 625

of ^{137}Cs and ^{90}Sr, respectively, without waste of ion-exchange product.

Addition of Nb to the Na-TS reactant mix suppresses the formation of SNT but tends to form Na-IPX [49.42] as an impurity phase. A systematic study is required to determine which conditions give rise to the impurity phase to prevent its occurrence. The broad feature at $2\theta = 24.5°$ in the in situ study of Na-NbTS may result from scattering by amorphous solids that are formed by hydrolysis of the alkoxides during heating to the temperature of the reactant mix. In the case of the gels prepared for synthesis of Na-TS, H_2O_2 was added so all the oxides dissolved in basic solution. The lack of solubility of TiO_2 in alkaline solution may have been responsible for the absence of any crystallization in the first 60 min of the in situ study for Na-NbTS.

49.7.2 Exchange Mechanisms

The results of Cs^+ exchange into H-NbTS indicated that the Cs1 site was the only site filled by Cs^+. The uptake of Cs^+ in this exchanger was slow relative to exchange in H-TS. Only 0.115 fractional occupancy was achieved in the first 200 min of exchange into H-NbTS. This represents just 0.23 Cs^+ on average per unit cell, compared with an exchange capacity of two Cs^+ per unit cell. There are four water molecules per unit cell, and half of them are lost when Cs^+ has an occupancy of 0.23. After this occupancy, site Cs1 continues to increase in occupancy. This is a significantly different uptake pathway then any other TS, or NbTS, phase where the filling of one Cs site resulted in structural transition allowing a second Cs site occupancy. One might expect that, as the Cs^+ enters the tunnel, a proton coordinately bonded to a cubane oxygen would form a hydronium ion and leave the unit cell. Thus, for each Cs^+ exchanged, a water molecule as a hydronium ion would leave the solid phase. This evidently does not occur in the H-NbTS phase. Rather it might be speculated that the proton follows a hopping mechanism, traversing from lattice oxygen to water to lattice oxygen. As the Cs^+ diffuses into a unit cell the protons leave that unit cell to maintain charge balance. At the surface, the proton hops to a water molecule on the interface of the unit cell, and then leaves the solid phase.

The exchange of Cs^+ into the H-TS and H-NbTS involves different mechanisms. In H-TS there is a rapid uptake of Cs^+ into the Cs2 site. Only when sufficient cesium is present in this site does the 8MR change from its elliptical shape, with length-to-width ratio $L : S = 1.53$, to the circular geometry with $L : S = 1$. Thereafter

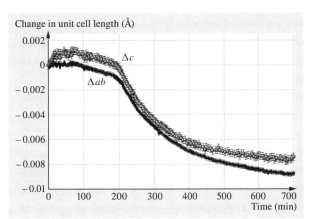

Fig. 49.23 Changes in unit cell parameters of the H-NbTS structure as Cs^+ exchanged for H^+. The two-step process is seen between minutes 0–200 and 200–700. Unit cell lengths at minute 0 were $a, b = 7.837(1)$ Å, $c = 11.919(3)$ Å, and at minute 700 were $a, b = 7.828(1)$ Å, $c = 11.912(3)$ Å

continued diffusion takes place with the filling of the Cs1 site while the occupancy of the Cs2 site remains static.

The results of the time-resolved x-ray diffraction study of Cs^+ exchange into Na-TS illustrated a crystallographic-site-dependent process. Cesium first occupies site Cs2, located outside of the 8MR, dur-

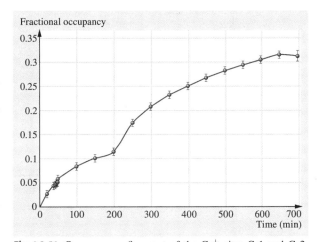

Fig. 49.24 Occupancy refinement of the Cs^+ sites Cs1 and Cs2. Cs2 is not shown because the refinements were 0.00(1) occupancy. The two-step process of Cs occupancy corresponds with the unit cell trends (Fig. 49.22). Cs^+ fractional occupancy at minute 200 was 0.115(6) and at minute 700 was 0.339(7)

ing which time water occupancy at site OW2 decreased rapidly. The process of loading site Cs2 occurred within 250 min, and was much slower in comparison with Cs^+ exchange into H-TS. When OW2 occupancy had reached zero, Cs^+ exchanged into site Cs1 located at the center of the 8MR. The reduction in water occupancy in the center of the 8MR during Cs^+ exchange for Na^+ may be caused by spatial restrictions due to the incorporation of the large Cs^+ ionic radius at site Cs2, and indirectly forces water to migrate out of the channels. This suggests that direct replacement of Cs^+ for Na^+ was favored, and the disruption of the water–hydroxyl bond network in the center of the 8MR was not favored. Once the OW2 occupancy reached zero, the Cs1 site became available and dominated the exchange process.

Although structure refinements of Cs^+ exchange into Na-NbCST could not be preformed, a possible scenario of the exchange process may be inferred in conjunction with previous static exchange studies [49.34] and these time-resolved studies. Starting with the unit cell parameters (Fig. 49.18) as a basis for the argument, the ion exchange may have proceeded in two distinct steps. The first step was Cs^+ uptake into site Cs1, where no H_2O occupied the center of the circular 8MR. This was indicated by a nonlinear increase in the a- and b-axes lengths and a decrease in the c-axis length. Incorporation of Cs^+ at site Cs1 would have continued for at least 235 min along the continuous unit cell lengths trajectories. Between minutes 235 and 287 no x-ray beam was available, but a decrease in the a and b cell lengths occurred, while the c-axis maintained a decreasing trajectory. The decrease in the a and b cell lengths would be the onset of the second step, and at this point, site Cs^+ would have started to fill site Cs2. The availability of site Cs2 would have been initiated by a decrease in OW2 occupancy. Similar exchange mechanisms were observed in the case of Cs^+ into Na-TS, where the Cs2 site filled first to force OW2 out of the structure, and once OW2 achieved a fractional occupancy of ≈ 0.1, Cs^+ began to fill site Cs1.

49.8 Summary

An understanding of the time-dependent nature of chemical reactions and processes is essential for the quantitative description of synthesis processes and ion-exchange mechanisms. Current time-resolved diffraction techniques do not have enough resolution to be used as the sole method for modeling complex crystallographic relationships. With current technologies and techniques, in situ diffraction studies must also be complemented by ex situ diffraction studies of the end-member phases. The kinetics of material synthesis observed here can be described as a series of compound precursor phases that eventually lead to the desired product as a result of changing gel compositions from precipitating phases. Only using in situ techniques can these cursory phases be described in the four-dimensional space of time, temperature, pressure, and composition. In addition, the intermediate precursor phases may themselves be important materials that would normally go unnoticed during ex situ synthesis. In any case, the crystallographic and chemical make-up of the cursory phases will elucidate the mechanisms of crystal growth.

In situ ion-exchange studies illustrate the complex mobility of ion and water mobility within the voids of crystalline phase, and the time and structure dependency of the ion-exchange pathway. These types of in situ time-resolved diffraction studies can be extremely useful in understanding the selectivity and fundamental mechanisms of the ion-exchange process in open-framework materials.

While we have concentrated on the latest results from our own in situ studies, it should be recognized that many other techniques are used as in situ time-resolved procedures, particularly for crystal growth studies. A representative few are listed [49.43–48] to illustrate the extent and power of in situ methods.

References

49.1 P. Norby, C. Cahill, C. Koleda, J.B. Parise: A reaction cell for in situ studies of hydrothermal titration, Mater. Sci. Forum **31**, 481–483 (1998)

49.2 P. Norby: In-situ time resolved synchrotron powder diffraction studies of syntheses and chemical reactions, Mater. Sci. Forum **228–231**, 147–152 (1996)

49.3 J.B. Parise, C.L. Cahill, Y. Lee: Dynamic powder crystallography with synchrotron x-ray sources, Can. Mineral. **38**, 777–800 (2000)

49.4 C.L. Cahill, Y.H. Ko, J.C. Hanson, K.M. Tan, J.B. Parise: Structure of microporous QUI-MnGS-1 and in situ studies of its formation using time-resolved synchrotron x-ray powder diffraction, Chem. Mater. **10**, 1453 (1998)

49.5 A.J. Celestian, D.G. Medvedev, A. Tripathi, J.B. Parise, A. Clearfield: Optimizing synthesis of $Na_2Ti_2SiO_7 \cdot 2H_2O$ (Na-CST) and ion exchange pathways for $Cs_{0.4}H_{1.6}Ti_2SiO_7 \cdot H_2O$ (Cs-CST) determined from in-situ synchrotron x-ray powder diffraction, Nucl. Instrum. Methods Phys. Res. **238**, 61–69 (2005)

49.6 D.G. Medvedev, A. Tripathi, A. Clearfield, A.J. Celestian, J.B. Parise, J. Hanson: Crystallization of sodium titanium silicate with sitinakite topology: Evolution from the sodium nonatitanate phase, Chem. Mater. **16**, 3659–3666 (2004)

49.7 A.J. Celestian, J.B. Parise, C. Goodell, A. Tripathi, J. Hanson: Time-resolved diffraction studies of ion exchange: K^+ and Na^+ exchange into (Al,Ge) gismondine $Na_{24}Al_{24}Ge_{24}O_{96} \cdot 40H_2O$ and $K_8Al_8Ge_8O_{32} \cdot 8H_2O$, Chem. Mater. **16**, 2244–2254 (2004)

49.8 Y. Lee, B.A. Reisner, J.C. Hanson, G.A. Jones, J.B. Parise, D.R. Corbin, B.H. Toby, A. Freitag, J.Z. Larese: New insight into cation relocations within the pores of zeolite rho: In situ synchrotron x-ray and neutron powder diffraction studies of Pb- and Cd-exchanged rho, Phys. Chem. B **105**(30), 7188–7199 (2001)

49.9 J.B. Parise, T.E. Gier, D.R. Corbin, D.E. Cox: Structural changes occurring upon dehydration of zeolite-rho – A study using neutron powder diffraction and distance-least-squares structural modeling, J. Phys. Chem. **88**, 1635–1640 (1984)

49.10 Y. Lee, T. Vogt, J.A. Hriljac, J.B. Parise, G. Artioli: Pressure-induced volume expansion of zeolites in the natrolite family, J. Am. Chem. Soc. **124**, 5466–5475 (2002)

49.11 J.P. Glusker, K.N. Trueblood: *Crystal Structure Analysis: A Primer*, 2nd edn. (Oxford Univ. Press, New York 1985)

49.12 B.D. Cullity, S.R. Stock: *Elements of X-ray Diffraction*, 3rd edn. (Prentice Hall, Upper Saddle River 2001), Chap. 2

49.13 G.E. Bacon: *Neutron Diffraction*, 1st edn. (Clarendon, London 1962)

49.14 P. Coppens (Ed.): *Syncrotron Radiation Crystallography* (Academic, London 1992)

49.15 J.H. Chen, R. Li, J.B. Parise, D.J. Weidner: Pressure-induced ordering in $(Ni,Mg)_2SiO_4$ olivine, Am. Mineral. **81**, 1519 (1996)

49.16 J.H. Chen, D.J. Weidner: X-ray diffraction study of iron partitioning between alpha and gamma phases of the $(Mg,Fe)_2SiO_4$ system, Physica A **239**, 78–86 (1997)

49.17 Y.S. Zhao, R.B. VonDreele, T.J. Shankland, D.J. Weidner, J.Z. Zhang, Y.B. Wang, T. Gasparik: Thermoelastic equation of state of jadeite $NaAlSi_2O_6$: An energy-dispersive Rietveld refinement study of low symmetry and multiple phases diffraction, Geophys. Res. Lett. **24**, 5–8 (1997)

49.18 J.B. Parise, J. Chen: Studies of crystalline solids at high pressure and temperature using the DIA multi-anvil apparatus, Eur. J. Solid State Inorg. Chem. **34**, 809–821 (1997)

49.19 J.B. Parise, D.J. Weidner, J. Chen, R.C. Liebermann, G. Chen: In situ studies of the properties of materials under high-pressure and temperature conditions using multi-anvil apparatus and synchrotron x-rays, Annu. Rev. Mater. Sci. **28**, 349–374 (1998)

49.20 X. Liang, J.E. Andrews, J.A. Haseth: Resolution of mixture components by target transformation factor analysis and determinant analysis for the selection of targets, Anal. Chem. **68**, 378–385 (1996)

49.21 CCP14: Collaborative Computational Project No. 14, http://www.ccp14.ac.uk/index.html (last accessed September 18, 2009)

49.22 D.E. Cox, B.H. Toby, M.M. Eddy: Acquisition of powder diffraction data with synchrotron radiation, Aust. J. Phys. **41**, 117–131 (1988)

49.23 P. Norby: Hydrothermal conversion of zeolites: an in-situ synchrotron x-ray powder diffraction study, J. Am. Chem. Soc. **119**, 5215–5221 (1997)

49.24 P. Norby, C. Cahill, C. Koleda, J.B. Parise: A reaction cell for in situ studies of hydrothermal titration, J. Appl. Crystallogr. **31**, 481–483 (1998)

49.25 J.S.O. Evans, R.J. Francis, D. O'Hare, S.J. Price, S.M. Clark, J. Gordon, A. Nield, C.C. Tang: An apparatus for the study of the kinetics and mechanism of hydrothermal reactions by in-situ energy dispersive x-ray diffraction, Rev. Sci. Instrum. **66**, 2442–2445 (1994)

49.26 A.J. Celestian: Time-resolved structural characterization of microporous silicotitanates and niobium silicotitanates. Ph.D. Thesis (Stony Brook Univ., Stony Brook 2006)

49.27 J. Chen, D.J. Weidner, M.T. Vaughan, R. Li, J.B. Parise, C.C. Koleda, K.J. Baldwin: Time-resolved diffraction measurements with an imaging plate at high pressure and temperature, Rev. High Press Sci. Technol. **7**, 272–274 (1998)

49.28 R.J. Francis, D. O'Hare: The kinetics and mechanisms of the crystallization of microporous materials, J. Chem. Soc. Dalton Trans., 3133–3148 (1998)

49.29 S. Shaw, S.M. Clark, C.M.B. Henderson: Hydrothermal formation of the calcium silicate hydrates, tobermorite ($Ca_5Si_6O_{16}(OH)_2 \cdot 4H_2O$) and xonotlite

49.30 I.A. Bray, K.J. Carson, R.J. Ellorich: *Initial Evaluation of Sandia National Laboratory Prepared Crystalline Silico-Titanates for Recovery of Cesium* (Pacific Northwest National Laboratory, Richland 1993), WA. PNL-8847

49.31 D.M. Poojary, R.A. Cahill, A. Clearfield: Synthesis, crystal-structures, and ion-exchange properties of a novel porous titanosilicate, Chem. Mater. **6**, 2364–2368 (1994)

49.32 C.T. Prewitt, R.D. Shannon: Use of radii as an aid to understanding the crystal chemistry of high pressure phases, Trans. Am. Crystallogr. Assoc. **5**, 51–60 (1969)

49.33 A. Clearfield, J. Lehto: Preparation, structure, and ion-exchange properties of $Na_4Ti_9O_{20} \cdot xH_2O$, J. Solid State Chem. **73**, 98–106 (1988)

49.34 J. Lehto, A. Clearfield: The ion-exchange of strontium on sodium titanate $Na_4Ti_9O_{20} \cdot xH_2O$, J. Radioanal. Nucl. Chem. **118**, 1–13 (1987)

49.35 D.M. Poojary, A.I. Bortun, L.N. Bortun, A. Clearfield: Structural studies on the ion-exchanged phases of a porous titanosilicate, $Na_2Ti_2O_3SiO_4 \cdot 2H_2O$, Inorg. Chem. **35**, 6131–6139 (1996)

49.36 A. Clearfield: Structure and ion exchange properties of tunnel type titanium silicates, Solid State Sci. **3**, 103–112 (2001)

49.37 P. Pertierra, M.A. Salvado, S. Garcia-Granda, A.I. Bortun, A. Clearfield: Neutron powder diffraction study of $Ti_2(OH)_2OSiO_4 \cdot 1.5H_2O$, Inorg. Chem. **38**, 2563–2566 (1999)

49.38 A. Tripathi, D.G. Medvedev, M. Nyman, A. Clearfield: Selectivity for Cs and Sr in Nb-substituted titanosilicate with sitinikite topology, J. Solid State Chem. **175**, 72–83 (2003)

49.39 V. Luca, J.V. Hanna, M.E. Smith, M. James, D.R.G. Mitchell, J.R. Bartlett: Nb-substitution and Cs^+ ion-exchange in the titanosilicate sitinakite, Microporous Mesoporous Mater. **55**, 1–13 (2002)

49.40 P. Sylvester, E.A. Behrens, G.M. Graziano, A. Clearfield: An assessment of inorganic ion-exchange materials for the removal of strontium from simulated Hanford tank wastes, Sep. Sci. Technol. **34**, 1981–1992 (1999)

49.41 E.A. Behrens, P. Sylvester, A. Clearfield: An assessment of a sodium nonatitanate and pharmacosiderite-type ion exchangers for strontium and cesium removal from DOE waste simulants, Environ. Sci. Technol. **32**, 101–107 (1998)

49.42 M. Nyman, F. Bonhomme, T.M. Alam, M.A. Rodriguez, B.R. Cherry, J.L. Krumhansl, T.M. Nenoff, A.M. Sattler: A general synthetic procedure for heteropolyniobates, Science **297**, 996–998 (2002)

49.43 S. Hermes, T. Witte, T. Hikov, D. Zacher, S. Bahnmüller, G. Langstein, K. Huber, R.A. Fischer: Trapping metal-organic framework nanocrystals: An in-situ time-resolved light scattering study on the crystal growth of MOF-5 in solution, J. Am. Chem. Soc. **129**, 5324–5325 (2007)

49.44 G. Cao, M.J. Shah: In situ monitoring of zeolite crystallization by electrical conductivity measurement: New insight into zeolite crystallization mechanism, Microporous Mesoporous Mater. **101**, 19–23 (2007)

49.45 W. Fan, M. Ogura, G. Sanker, T. Okubo: In-situ small-angle and wide-angle x-ray scattering investigation on nucleatron and crystal growth of nanosized zeolite A, Chem. Mater. **19**, 1906–1917 (2007)

49.46 H. Lee, S.W. Yoon, E.J. Kim, J. Park: In-situ growth of copper sulfide nanocrystals on multiwalled carbon nanotubes and their application as novel solar cell and amperometric glucose sensor materials, Nano Lett. **7**(3), 778–784 (2007)

49.47 J. Schörmann, S. Potthast, D.J. As, K. Lischka: In-situ growth regime characterization of cubic GaN using reflection high energy electron diffraction, Appl. Phys. Lett. **90**, 041918/1–041918/3 (2007)

49.48 C.-H. Cheng, G. Juttu, S.F. Mitchell, D.F. Shantz: Synthesis, characterization and growth rates of aluminum- and Ge,Al-substituted silicalite-1 materials grown from clear solutions, J. Phys. Chem. B **110**(45), 22488–22495 (2006)

50. Single-Crystal Scintillation Materials

Martin Nikl, Anna Vedda, Valentin V. Laguta

Scintillation materials are employed to detect x-ray and γ-ray photons or accelerated particles. Wide-bandgap semiconductor or insulator materials with a high degree of structural perfection are suitable for this purpose. They must accomplish fast and efficient transformation of incoming high-energy photon/particles to a number of electron–hole pairs collected in the conduction and valence bands, respectively, and their radiative recombination at suitable luminescence centers in the material. Generated ultraviolet (UV) or visible light can then be detected at high sensitivity by conventional solid-state semiconductor- or photomultiplier-based photodetectors, which are an indispensable part of scintillation detectors.

An insight into this field will be provided for a wider scientific audience and at the same time we will point out some current hot topics. After reviewing the historical issues and fundamental physical processes of the $x(\gamma)$-to-visible light transformation occurring in scintillators, practically important material parameters, characteristics, and related measurement principles will be summarized. An overview of selected modern single-crystal and optical ceramic materials will be given. Particular attention will be paid to the relation between the manufacturing technology used and the occurrence of material defects and imperfections. The study and understanding of related trapping states in the forbidden gap and their role in the energy transfer and storage processes in the material will be shown to be of paramount importance for material optimization. Correlated experiments of time-resolved luminescence spectroscopy, wavelength-resolved thermally stimulated luminescence, and electron paramagnetic resonance offer a powerful tool for this purpose. Future prospects and directions for activity in the field will be briefly mentioned as well.

50.1	**Background**	1663
	50.1.1 Historical Aspects	1664
	50.1.2 Fundamentals	1664
	50.1.3 Material Characteristics	1665
	50.1.4 Characterization Methods	1668
50.2	**Scintillation Materials**	1670
	50.2.1 Lead Tungstate ($PbWO_4$) Single Crystals	1670
	50.2.2 Aluminum Perovskite $XAlO_3$:Ce ($X = Y$, Lu, Y/Lu)-Based Scintillators	1673
	50.2.3 Aluminum Garnet $X_3Al_5O_{12}$:Ce ($X = Y$, Lu, Y/Lu)-Based Scintillators	1677
	50.2.4 Ce-Doped Silicate Single Crystals	1681
	50.2.5 Ce-Doped Rare-Earth Halide Single Crystals	1684
	50.2.6 Optical Ceramics and Microstructured Materials	1687
50.3	**Future Prospects**	1689
50.4	**Conclusions**	1691
	References	1691

50.1 Background

In the introductory part of this chapter, the history of scintillator material development will be briefly reviewed. Fundamental aspects of $x(\gamma)$-ray transformation to ultraviolet/visible photons will be explained and practical scintillation parameters will be listed, which will aid in the quantitative evaluation of candidate materials. Preparation techniques, occurrence of defects, and their role will be discussed. Finally, experimental techniques suitable for correlated characterization of physical phenomena in scintillators will be described.

50.1.1 Historical Aspects

In November 1895 Wilhelm Conrad Röntgen noticed the glow of a barium platinocyanide screen placed next to his operating discharge tube, thus discovering a new invisible and penetrating radiation [50.1, 2], named *x-rays* in English or *Röntgen radiation* in some other languages. For x-ray registration, simple photographic film was found to be rather inefficient and the search for materials able to convert x-to-visible started immediately in order to couple with sensitive photographic-film-based detectors. Pupin introduced $CaWO_4$ powder, which was used for this purpose for more than 75 years. Together with ZnS-based powders introduced later by Crookes and Regener, $CaWO_4$ powder is the oldest phosphor material employed for detection of x-rays.

In early 1896, the natural radioactivity (spontaneously released high-energy radiation) of uranium was discovered by *Henri Becquerel* [50.3] and followed by pioneering works of *Marie Curie Sklodowska* and her husband *Pierre Curie*, which led to the discovery of the strongly radioactive elements polonium and radium [50.4, 5]. This kind of radiation is commonly named gamma rays and its energy is higher than the above-mentioned x-rays. Among the various methods for the detection of gamma radiation, scintillator-based detectors have been widely employed. It is also interesting to note that, in 1897, the cathode ray tube was invented by Braun, in which the energy of an accelerated electron beam is converted into visible light (in a process called cathodoluminescence) by a phosphor material. The mechanism of this conversion is quite similar [50.6] to that functioning in the case of x-ray conversion. Thus, at the end of the 19th century, two energetic (photon and particle) radiations were available, together with gamma rays generated by natural radioactive elements. Their enormous application potential was realized especially during World War I, which stimulated the development of phosphor and scintillator materials necessary for their practical exploitation.

The history of single-crystal scintillators begins in the late 1940s with the introduction of NaI:Tl and CsI:Tl scintillators by *Hofstadter* [50.7, 8]. Since that time a number of material systems have been reported, see Fig. 50.1 and [50.9] for an overview. The above two materials and $Bi_4Ge_3O_{12}$ (BGO), introduced by *Weber* and *Monchamp* [50.10], became widespread scintillator materials and are often used as *standard samples* to evaluate newly developed materials. Within the last 15 years there has been considerable activity in this field, triggered mainly by the needs of high-energy physics and advanced imaging applications in science, medicine, and industry. This has led to a considerable number of articles in the literature dealing mostly with Ce^{3+}-doped materials, due to the fast decay time of the 5d–4f radiative transition of the Ce^{3+} center (typically 20–60 ns) and its high quantum efficiency (close to 1) at room temperature (RT). For reviews see [50.11, 12].

Due to the great practical importance and relatively long history of this field, there already exists a large amount of published information on the topic. Let us mention the superb recent survey of luminescent materials by *Blasse* and *Grabmaier* [50.13], the overview of methodology in use in the radiation measurement by *Knoll* [50.14], the monograph devoted to scintillator materials from *Rodnyi* [50.15], and the useful phosphor handbook written by *Shionoya* and *Yen* [50.16]. Numerous featured review papers on phosphor and scintillator materials and their applications exist in the scientific literature as well, for instance [50.17–22].

50.1.2 Fundamentals

Wide-bandgap materials are employed for transformation of the $x(\gamma)$-ray to ultraviolet/visible photons; see the sketch in Fig. 50.1. Consistent phenomenological descriptions of the scintillation conversion process, efficiency criteria, etc., were developed already in the 1970s [50.23] and later further refined [50.24]. Scintillation conversion is a relatively complicated pro-

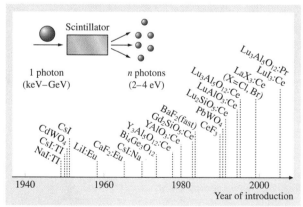

Fig. 50.1 Sketch of the scintillator principle and a timeline of single-crystal scintillators

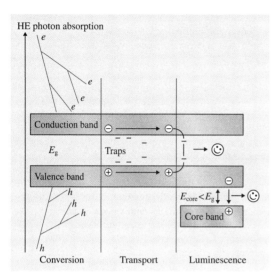

Fig. 50.2 Sketch of scintillator conversion of a high-energy (HE) photon

cess, which can be divided into three consecutive subprocesses: *conversion*, *transport*, and *luminescence* (Fig. 50.2). During the initial conversion a multistep interaction of a high-energy photon with the lattice of the scintillator material occurs through the photoelectric effect, Compton scattering effect, and pair production; for photon energies below 100 keV the first of these is of major importance. Many electron–hole pairs are created and thermalized in the conduction and valence bands, respectively. This first stage is concluded within less than 1 ps. More detailed considerations about the conversion processes have been published in [50.25, 26]. In the transport process electrons and holes (eventually created excitons) migrate through the material, repeated trapping at defects may occur, energy losses are probable due to nonradiative recombinations, etc. Considerable delay in the migration can be introduced due to this charge-carrier recapture at trapping levels in the material forbidden gap. This stage is the least predictable, as point defects, flaws, surfaces, and interfaces can introduce energy levels into the forbidden gap and strongly modify/degrade otherwise high intrinsic scintillation performance. These phenomena are strongly dependent upon manufacturing technology [50.27]. During the final stage, luminescence consists of consecutive trapping of the electron and hole at the luminescent center and their radiative recombination. Nonradiative energy transfer towards the emission center via exciton-based state is also possible. In a particular group of materials the light generation occurs in radiative transitions between the valence and first core bands (sketched in Fig. 50.2); these are so-called cross-luminescence scintillators [50.15]. The latter mechanism enables very fast, even subnanosecond, scintillation response, which is, however, usually accompanied by much slower exciton-related luminescence. This phenomenon is reported in the literature mainly for BaF_2 and other halide single crystals [50.28]. The physics of luminescent centers is usually well understood due to advanced experimental methods available for their selective and time-resolved study; see, e.g., [50.13, 14].

50.1.3 Material Characteristics

As mentioned above, in the early days of x-ray usage, just phosphor powders in the form of thin screens were employed for their registration, coupled together with a photographic film [50.17]. Later, due to the need to additionally detect and monitor higher-energy x- or γ-rays, scintillator materials were introduced in the form of single crystals (Fig. 50.1).

Despite the fact that the underlying physics is identical, scientific communities working on phosphors and scintillators have been partially separated, mainly due to the different demands of related applications and different preparation technologies employed [50.16–18]. Commonly, materials are called phosphors when used in applications using photon-integrating (steady-state) mode detection, while scintillators are employed in the (x- or γ-ray) photon-counting regime. Today, the separation between phosphor (powders) and scintillator (solid state) materials is somewhat diminished as some materials are used in both detection modes, in powder, bulk or other forms, depending on the application [50.21].

Scintillation Parameters

In the case of scintillators, x(γ)-ray photon counting consists of accumulating the generated light arriving soon after the initial conversion stage is accomplished (Fig. 50.2), as the scintillator works as a high-energy photon counter. Strongly delayed light, e.g., due to retrapping processes mentioned above cannot be technically exploited in the counting mode. Also x(γ)-ray photons of different energy should be resolved for some applications. The most important characteristics of scintillation materials for detectors are as follows:

1. Light yield (LY)
2. X(γ)-ray stopping power

3. Scintillation response – decay time
4. Spectral matching between the scintillator emission spectrum and photodetector
5. Chemical stability and radiation resistance
6. Linearity of light response with the incident $x(\gamma)$-ray photon energy – energy resolution.

The overall scintillation efficiency of $x(\gamma)$-ray-to-light conversion is determined by both intrinsic and extrinsic material characteristics. The number of UV/visible photons N_{ph} produced in the scintillation conversion per energy E of incoming $x(\gamma)$-ray photon can be expressed as [50.23–25]

$$N_{ph} = [E/(\beta E_g)]SQ, \quad (50.1)$$

where E_g represents the forbidden gap of the material, S and Q are the quantum efficiencies of the transport and luminescence stages, respectively, and β is a phenomenological parameter which is typically found to be 2–3 for most materials. The relative efficiency can then be obtained as

$$\eta = E_{vis} N_{ph}/E, \quad (50.2)$$

where E_{vis} is the energy of generated UV/vis photons. The most efficient material among the phosphors and scintillators today is ZnS:Ag with $\eta \approx 0.2$. If we consider that the bandgap of the scintillator materials discussed herein is higher than that of ZnS (≈ 3 eV), then even more efficient materials could be found within those with yet narrower bandgap.

The LY of a scintillator is always an inferior value in terms of (50.1) as it represents the fraction of generated visible photons that arrived at the photodetector within a certain time gate after the high-energy photon absorption; typical values of such a time gate are in practice set between 100 ns and 10 μs.

The $x(\gamma)$-ray stopping power (attenuation coefficient) for a material of given thickness depends on its density ρ and its effective atomic number Z_{eff} (for the calculation of Z_{eff}; see, e.g., [50.18]). Considering only interaction through the photoelectric effect, the stopping power is proportional to ρZ_{eff}^{3-4} [50.20].

The kinetics of the light response of a scintillator are governed by the characteristics of the transport and luminescence stages in Fig. 50.2, as they are far slower than the initial conversion. The decay rate of the luminescence center itself is defined by its transition dipole moment from the excited to ground state and can be further enhanced by additional nonradiative quenching or energy transfer processes away from the excited state. Such quenching or energy transfer, however, results in the decrease of parameter Q in (50.1), and the overall conversion efficiency is also decreased. In the most simple case of exponential decay, the emission intensity $I(t)$ is

$$I(t) \approx \exp(-t/\tau), \quad (50.3)$$

where τ is the decay time. While decay times of the parity and/or spin-forbidden transitions for most rare-earth ions are typically of the order of several tens of μs up to ms, in the case of allowed 5d–4f transitions of Ce^{3+} and Pr^{3+} the values scale down to tens of nanoseconds, and fully allowed singlet–singlet transition in organic molecules is still about ten times faster [50.29]. The fastest emission transitions are offered by the radiative decay of Wannier excitons in direct-gap semiconductors, where subnanosecond values have been reported for compounds such as ZnO, CuX, CsPbX$_3$ (X = Cl, Br), PbI$_2$, and HgI$_2$ [50.30]. The emission rate is enabled by the coherent nature of the exciton state spread over a (large) number of elementary cells [50.31]. However, a serious disadvantage of the latter group of materials is the lower binding energy of their excitonic state, which results in partial ionization and consequently quenching of the exciton-related emission at RT. In bulk materials, further efficiency decrease follows from the small Stokes shift, which leads to noticeable reabsorption losses [50.32].

Due to the aforementioned retrapping processes during the transport stage sketched in Fig. 50.2, the light emission response of a material under high-energy excitation is often further complicated by slower nonexponential components. These processes are currently quantified by the afterglow (sometimes called a persistence), which is defined like a residual light intensity at some time (from a few ms to min) after the excitation is cut off [50.27]. The delayed radiative recombination will contribute to slow scintillator response, lower LY values, and worse signal-to-background ratio [50.33]. In some studies, a parameter α has been introduced to provide a comparative measure for the occurrence of delayed radiative recombination within tens to hundreds of μs. Due to the repetitive nature of the measurement, they are reflected in the signal level before the rising edge of the decay curve [50.33, 34], as illustrated in Fig. 50.3. In practical evaluation of both phosphors and scintillators, sometimes simple $1/e$ or $1/10$ decay times are provided, which are defined as the time when the light intensity decreases to $1/e$ or $1/10$ of its initial intensity after a flash excitation, or the so-called mean decay time can be calculated [50.22].

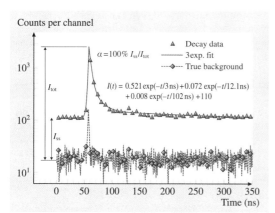
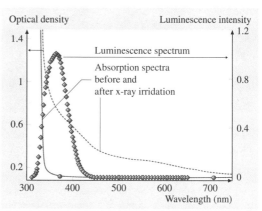

Fig. 50.3 Spectrally unresolved scintillation decay of un-doped PbWO$_4$, fitted by three exponential functions, $I(t)$, convoluted with the instrumental response (*solid line*). The true background level is displayed as well. Background enhancement in the repetitive decay measurement is reflected in I_{SS}, the superslow decay component amplitude which comes from preceding excitations and is evaluated using the coefficient α

Fig. 50.4 Absorption spectra of 2 cm-thick YAlO$_3$:Ce^{3+} single crystal before and after γ-ray irradiation (^{60}Co, 500 Gy dose) at RT. Overlap of the irradiation-induced absorption with luminescence spectrum of the material is demonstrated

Spectral matching between the scintillator emission band and the photodetector spectral sensitivity dependence is an obvious requirement. Classical criteria were defined some time ago as near-UV/blue emission is optimum for a photomultiplier detector, while for a photodiode the green–red spectral region was considered best. In recent years enormous development of semiconductor photodetectors has occurred and the latest generation of back-illuminated charge-coupled devices (CCDs) show enhanced sensitivity down to 200 nm.

Chemical stability concerns mainly the hygroscopicity of materials, which in some cases severely limits their long-time operation in the open air (NaI:Tl, CsI:Na, and LaX$_3$:Ce with X = Cl, Br).

Radiation resistance of materials regards mainly the performance changes and instabilities due to the induced absorption resulting from material irradiation and creation of color centers. It is a matter of concern mainly in bulk scintillation materials [50.22]. In the case of overlap between induced absorption and the emission spectrum, reabsorption losses occur with resulting loss of overall efficiency and LY. An example of this behavior is shown in Fig. 50.4. While this parameter has been considered mainly in research into scintillators for high-energy physics, it should be noted that it has importance also in several medical imaging techniques [50.20] and in the case of industrial flaw detection or synchrotron beam diagnostics applications as well.

The linearity of light response with incident x(γ)-ray photon energy in scintillators can be monitored through the dependence of LY on the incident photon energy. The value for LY is energy dependent, partly due to abrupt changes of the attenuation coefficient around the K and L edges of the elements constituting the compound, but also due to the nonequal conversion efficiency of the photoelectric and Compton scattering effects, which become progressively more important with increasing energy of incoming x(γ)-rays. As a result the energy resolution of a scintillator material is degraded with respect to the intrinsic limits based purely on statistical grounds [50.35].

The methodology of radiation detection is described in [50.14]. To quantify the characteristics and parameters described above, the set of usually used routine methods is described in [50.21].

Preparation Technology and Material Defects

Single-crystal scintillators are readily manufactured using crystal growth from the melt, i. e., by Czochralski, Bridgman or similar techniques. Though such techniques are well established and large crystals can be grown [50.36], point or extended defects are present in these artificial materials and often limit their performance. In recent years, the micropulling down growth technique has become frequently reported in the literature [50.37]. Enabling synthesis of circular or shaped

Fig. 50.5a,b Single crystal with square rod shape grown by the micropulling-down method: (**a**) sapphire single crystals and (**b**) LuAG single crystal (after [50.38])

single-crystal fibers of diameter up to several millimeters (Fig. 50.5) grown from the melt and of very good structural quality, this is a fast and economic research tool in the search for new scintillator materials [50.38] or in producing *device-size* crystal elements without the need for extensive cutting and polishing procedures.

The transport stage of scintillator conversion (sketched in Fig. 50.2) was mentioned as a critical and unpredictable period, during which the material performance can be degraded due to defects and flaws arising in the manufacturing process. In all materials, intrinsic point defects arise (e.g., cation and anion vacancies, interstitial atoms) as a result of general thermodynamic conditions and are further accompanied by extrinsic point (accidental impurities) or extended (dislocations, grain and domain interfaces) defects. In many cases, such defects introduce energy levels in the forbidden gap, which can be involved in the capture of migrating charge carriers. In the process of material optimization it is necessary to diminish or deactivate such defects to increase the material performance to its intrinsic/theoretical limit. In the case of accidental impurities or extended structural defects, the solution consists of material purification and optimization of the technological process to improve material purity and structure. However, in case of intrinsic defects, alternative solutions are pursued such as codoping the material by aliovalent ion(s) that do not participate in the energy transfer and storage processes. Due to the different charge state of the dopant ion (with respect to the original substituted lattice ion) the Coulombic equilibrium of the lattice is changed and this usually induces changes in the concentration of intrinsic point defects. Examples are the optimization of $PbWO_4$ by doping or codoping with trivalent ions at the Pb^{2+} site [50.22], the codoping with Zr^{4+} in $YAlO_3$:Ce [50.39] or the recent discovery of positive influence of Eu^{2+} codoping in CsI:Tl to reduce its afterglow [50.40].

50.1.4 Characterization Methods

To understand the relation between material defects and the occurrence of traps in the material forbidden gap, their involvement in energy capture and storage, and interconnection with the production technology, a set of specific characterization methods must be used in close coordination with the manufacturing technology. Correlated experiments using time-resolved luminescence spectroscopy, wavelength-resolved thermally stimulated luminescence, and electron spin resonance in an extended temperature interval (10–300 K at least) are very powerful tools for this purpose.

Time-resolved UV/visible emission spectroscopy enables detailed understanding of the luminescence center itself, its intracenter transitions, and eventual nonradiative quenching pathways [50.13, 41]. At the same time its interconnection with the host lattice environment via energy transfer processes can be monitored. Selective pulse excitation in the vacuum ultraviolet (VUV)/UV/visible spectral region completed with x-ray or single-photon gamma-ray excitations is used for this purpose.

Thermally stimulated luminescence (TSL) or thermoluminescence allows the study of point defects giving rise to localized energy levels (traps) in a material forbidden gap. A TSL measurement is a two-stage process, which involves at least one electron trap and one hole trap. A highly simplified scheme is proposed in Fig. 50.6; see [50.42, 43] for a systematic description of these phenomena and measurement schemes. During irradiation at a given temperature (RT or cryogenic) free carriers are trapped at electron or hole centers. A subsequent heating cycle allows the carriers in shallow traps to be released and to recombine radiatively with carriers of the opposite sign stably localized at recombination centers. In the example scheme illustrated, electron traps are emptied while hole traps act as re-

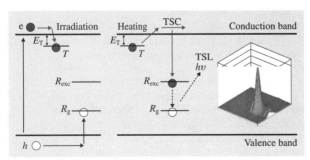

Fig. 50.6 Sketch of the TSL process

combination centers. The emitted light as a function of temperature (glow curve) displays a peak at a characteristic temperature related to the trap structure. If, before recombination, electrons are transferred to the conduction band, a thermally stimulated conductivity (TSC) signal can also be measured. In the simplest first-order kinetics model [50.42, 43] the intensity $I(t)$ of emitted photons is

$$I(T) = sn_0 \exp\left(-\frac{E_T}{k_B T}\right)$$
$$\times \exp\left[-\frac{s}{\beta}\int_{T_0}^{T} \exp\left(-\frac{E_T}{k_B T'}\right) dT'\right], \quad (50.4)$$

where n_0 (cm^{-3}) is the number of traps filled by irradiation, s (s^{-1}) is a *frequency factor*, β is the heating rate, and E_T (eV) is the trap depth. In the same simplified model, if instead of being heated the sample is held at a constant temperature after irradiation, the mean time spent by carriers in the trap before recombination is

$$\tau = \frac{1}{s} \exp\left(\frac{E_T}{k_B T}\right). \quad (50.5)$$

This time governs the phosphorescence which can be spontaneously emitted at constant temperature. According to trap parameters, τ can vary in a very wide interval from ns–μs to hundreds or thousands of years. In suitable cases a detailed analysis of (50.4) or of more general ones by numerical methods [50.42, 43] results in the determination of the trap parameters (E_T, s). The mean time τ at RT can thus be evaluated, allowing direct comparison with slow components in time-resolved scintillation measurements or correlation with LY. Spatial correlation between traps and recombination centers can also sometimes be elucidated. Moreover, more detailed information can be obtained by measuring the wavelength spectrum of the emitted light (see the example in Fig. 50.6). So, whereas temperature resolution can separate different trapping processes, measurement of the TSL emission spectrum can evidence different recombination processes when several centers act simultaneously.

The nature and local structure of the defects, traps, and recombination centers mentioned above can be studied in a detailed manner by electron spin resonance (ESR) and related methods such as electron nuclear double resonance (ENDOR) and optically detected magnetic resonance (ODMR) [50.44, 45]. All of these methods are so-called local methods and exploit the interaction of the magnetic moment of the

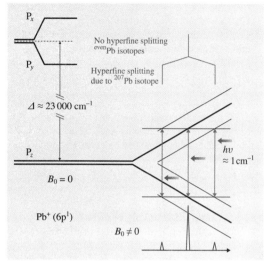

Fig. 50.7 ESR measurement principle

unpaired electron(s) or hole(s) and of the nuclei in a defect with each other and with external magnetic fields.

In a magnetic resonance experiment, the position of the resonance lines reflects both local symmetry and degree of lattice distortion around paramagnetic particle. In addition, the electron–nuclear interaction can give direct evidence for the nuclear (thus also the chemical) species coupled to the paramagnetic particle. This can be illustrated in Fig. 50.7 with an example related to the Pb$^+$ center in a PbWO$_4$ crystal [50.46] where an electron is localized at the Pb$^+$ p$_z$ orbital around an oxygen vacancy. With the application of an external magnetic field B_0 the lowest energy level of Pb$^+$ splits nearly linearly with field strength. The microwave energy is absorbed by the sample when the frequency of the irradiation is in resonance with the energy splitting. For lead ions with nonzero nuclear spin there is an additional splitting of the energy levels (hyperfine splitting) due to the interaction of electron and nuclear magnetic moments, which gives rise to low-intensity satellites around the main resonance. Detection of hyperfine satellites in many cases allows unambiguous determination of the local structure of defects.

In favorable cases one can thus gain the following information from ESR experiments:

1. Symmetry of a defect
2. Concentration of defects

3. Identification of nuclei (atoms) constituting a defect
4. Charge state of paramagnetic ion(s) involved
5. Change of defect charge state under external influences (irradiation, temperature, etc.)
6. Dynamic properties of a defect, e.g., decay or growth of the concentration of defects after the irradiation is switched on (off).

The last item allows the direct determination of the energy of the local electronic level of a defect created by irradiation and thus can be correlated with TSL characteristics.

Further refinement of the electronic structure of absorption or luminescence centers can be obtained using the technique of optically detected ESR via magnetic circular dichroism (MCD) [50.45]. This technique provides a direct correlation between ESR and MCD by the *tagged MCD* method, when one keeps the magnetic field constant on one of the resonance positions and changes the wavelength at which the optical absorption or luminescence is detected. Therefore, ODMR offers the most compelling interrelation between atomic-scale defects and the related optical absorption or luminescence bands.

50.2 Scintillation Materials

Due to the wide variety of applications, about 20 candidate scintillator systems have been systematically studied, their manufacturing industrialized, and most of them also successfully commercialized to be used in scintillation detectors (Fig. 50.1). Parameters of selected materials are given in Table 50.1. In this section, an overview of several material systems introduced and/or intensively researched and optimized within the last 10–15 years will be given. Specifically, manufacturing technology aspects will be mentioned and essential luminescence characteristics will be described together with the materials' scintillation performance. Further emphasis will be placed on the description of point defects, which are typical for each material group, and to their role in the processes of energy transfer and capture in a scintillator. Correlated use of several experimental techniques and well-defined samples is vital to gain necessary insight into the underlying complex physical phenomena at the atomistic level. The optimization of scintillation materials toward their intrinsic limits can then be guided by knowledge gained about the limitations in the scintillator conversion and their relation to the manufacturing technology.

50.2.1 Lead Tungstate (PbWO$_4$) Single Crystals

Single crystals of PbWO$_4$ (PWO) were the subject of increased interest for scintillation detection in the early 1990s because of their potential use in electromag-

Table 50.1 Survey of characteristics of selected single-crystal scintillators (for those based on rare-earth halides, see Table 50.2). Light yield values are spectrally corrected

Crystal	Density (g/cm^3)	Light yield (phot/MeV)	Dominant scint. decay time (ns)	Emission maximum (nm)	$\Delta E/E$ at 662 keV (%)
CsI:Tl	4.51	61 000	800	550	6.6
NaI:Tl	3.67	41 000	230	410	5.6
BaF$_2$ (only crosslumin.)	4.88	1500	0.6–0.8	180–220	7.7
Bi$_4$Ge$_3$O$_{12}$	7.1	8600	300	480	9.0
PbWO$_4$	8.28	300	3	410	30–40
CdWO$_4$	7.9	20 000	5000	495	6.8
YAlO$_3$: Ce	5.6	21 000	20–30	360	4.6
LuAlO$_3$: Ce	8.34	12 000	18	365	8–10
Y$_3$Al$_5$O$_{12}$: Ce	4.56	24 000	90–120	550	7.3
Lu$_3$Al$_5$O$_{12}$: Ce	6.67	12 500	55	530	11
Lu$_3$Al$_5$O$_{12}$: Pr	6.67	19 000	20	308	5–6
Gd$_2$SiO$_5$: Ce	6.7	12 500	60	420	7.8
Lu$_2$SiO$_5$: Ce	7.4	26 000	35	390	7.9

netic calorimeters – the inner parts of huge detectors used in high-energy physics accelerators. The essential luminescence characteristics of PWO had already been studied by the 1970s [50.47–49] and the growth of tungstate and molybdate single-crystal systems was also reported in the literature [50.50]. However, the first systematic scintillation-oriented reports appeared in 1992 [50.51, 52], showing the favorable characteristics of PWO for the above-mentioned applications.

PWO melts congruently at 1123 °C. It shows an exceptionally high density (8.28 g/cm^3) when compared with other scintillators. It exists in several structural modifications, but only the scheelite structure characterized by the tetragonal space group $I4_1/a$ or C_{4h} can be obtained using the high-temperature melt-based methods. Both Czochralski [50.53] and Bridgman [50.54] methods have been widely employed, and industrial size crystals up to about 30 cm length and 7–10 cm diameter were grown [50.55, 56]. The stoichiometry of the melt [50.57, 58], atmosphere of the growth, and postgrowth annealing [50.59] were widely studied to optimize the scintillation performance. Once the growth process was optimized, the doping of PbWO$_4$ by selected trivalent ions was employed as another efficient tool to further increase the figure of merit of this scintillator. See [50.22, 55, 56] for reviews of reported results.

It is well established that the PWO emission spectrum consists of two dominant components. The blue component, peaking at ≈ 420 nm (2.9 eV), was ascribed to the regular lattice center, namely the (WO$_4$)$^{2-}$ group [50.47, 48], and identified later as the self-trapped exciton (STE). At higher temperatures, however, this component originates from various localized exciton states, because the STE is thermally disintegrated around 150 K [50.60, 61]. The green component, peaking at 480–520 nm (≈ 2.5 eV), was ascribed to a defect center (WO$_3$) [50.49]. Moreover, a Mo impurity (i.e., the (MoO$_4$)$^{2-}$ group) was also found to give rise to an emission component at 520 nm [50.62, 63]. Temperature dependencies and temporal characteristics of these two green emissions were studied in detail and compared in a number of different PWO crystals [50.64]. Photoluminescence of PWO is heavily quenched at RT; this is the reason why its scintillation response is dominated by a decay time of a few ns only and its LY is very low (Table 50.1).

The decay kinetics of the blue and green emission components has been reported already in [50.47, 48]. Very slow nonexponential components have been observed in the 10^{-6}–10^{-4} s time scale even under selective excitation within the lowest absorption peak of PWO above 180 K [50.65, 66], or in the scintillation decays excited by the 511 keV photons of ^{22}Na radioisotope [50.34, 67]. The thermally induced decomposition of the STE and excited green emission centers above 150 and 200 K, respectively [50.61, 68], establishes an intrinsic equilibrium between the localized and free electron–hole states at higher temperatures. This explains the coexistence of the first- and second-order decay mechanisms due to the prompt and delayed radiative recombination processes, respectively, observed clearly in the decay at RT [50.66] (Fig. 50.8). Undesirable slowing of the luminescence response was observed especially in PWO samples containing the WO$_3$ green emission centers [50.64].

Because of the mentioned thermal disintegration of the excited emission centers, any shallow trap states in the PWO lattice taking part in carrier capture processes become important. They modify the migration characteristics of free charge carriers through retrapping. Consequently, the speed of their delayed radiation recombination is altered. Monitoring such trap states by TSL measurements was reported some time ago [50.69], and more recent measurements confirm the rather rich variety of such shallow trap states in the PWO lattice [50.66, 70, 71], which are strongly dependent upon synthesis route. Simultaneous use of TSL and ESR experiments was applied efficiently to gain understanding about the nature of traps and details about the energy storage in PWO lattice. Specifically, electron self-trapping at the (WO$_4$)$^{2-}$ complex anion was revealed by ESR [50.72]. The related paramag-

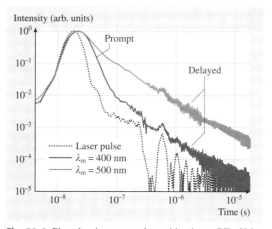

Fig. 50.8 Photoluminescence decay kinetics at RT of Mo-doped (160 ppm) PWO. $\lambda_{\text{ex}} = 308$ nm (XeCl excimer laser)

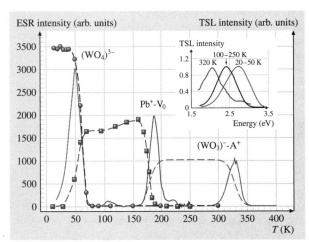

Fig. 50.9 Temperature dependence of ESR (*dashed line*) and TSL (*solid line*) signals related to $(WO_4)^{3-}$, Pb^+-V_O, and $(WO_3)^-$-A^+ electron centers. In the *inset* the TSL emission spectra within the given temperature intervals are shown

netic $(WO_4)^{3-}$ center can survive up to about 50 K and its thermal disintegration is accompanied by a TSL peak (Fig. 50.9) [50.73]. In undoped PbWO$_4$, a fraction of released electrons is stored in deeper Pb^+-V_O traps [50.46, 74, 75], i.e., at the Pb^+ p-orbital around an isolated oxygen vacancy V_O, creating an F^+ center. This center is stable up to about 190 K and during its thermal disintegration again a TSL peak appears (Fig. 50.9). Similar to the situation at ≈ 50 K, a fraction of the released electrons are stored at still deeper traps, namely the oxygen-deficient $(WO_3)^-$-A^+ complex (A^+ is a monovalent impurity or possibly a cation vacancy at a Pb position). Thermal destruction of this center occurs around 320–330 K and is again accompanied by a TSL peak [50.76]. It is worth noting that TSL emission spectra of the peaks at ≈ 50, 190, and 320 K show maxima at different energies (inset of Fig. 50.9), pointing to the existence of three different hole traps. However, no detailed information about them could be obtained since no related hole centers have been detected by ESR thus far.

A significant change in the concentration of TSL-monitored trapping states was observed in PWO samples doped with large and stable trivalent A^{3+} ions ($A^{3+} = La^{3+}$, Y^{3+}, Lu^{3+}, and Gd^{3+}) [50.77–79]. Similar effects were obtained by using any of these dopants. Traps and related TSL peaks above 150 K were practically eliminated. La doping completely removes electron traps associated with oxygen vacancies and stabilizes the self-trapped electron in its vicinity up to about 90–100 K [50.80]. New insight into the effect of La doping was recently gained in the study of doubly doped PbWO$_4$:Mo,La [50.81]. The Mo^{6+} ion is an efficient electron trap and the resulting $(MoO_4)^{3-}$ center is stable up to about 240–250 K [50.82]. Codoping with La resulted in a considerable increase of $(MoO_4)^{3-}$ center concentration, which was detected even without prior irradiation of the samples. Moreover, electrical conductivity above RT appeared and is governed by the same activation energy as that obtained from thermally induced disintegration of the $(MoO_4)^{3-}$ center. From these measurements, it was concluded that La^{3+} at Pb^{2+} sites introduces free electrons in the PbWO$_4$ conduction band, i.e., La^{3+} behaves as a donor impurity in a semiconductor. At the same time, a noticeable decrease of TSL signal related to $(MoO_4)^{3-}$ thermal disintegration around 240–250 K was found in doubly doped PbWO$_4$:Mo,Y [50.83]. To explain the increase of electron centers observed by ESR and the decrease of related TSL signal in trivalent-ion-codoped PbWO$_4$:Mo, the suppression of related hole traps must be considered. Hence, the effect of trivalent-ion doping has been understood, consisting of simultaneous suppression of oxygen vacancies and some of the hole traps. Furthermore, free excess electrons generated in the conduction band can fill electron traps of any kind. Such effects lower the capacity of the PbWO$_4$ lattice to trap and store generated charge carriers after x- or γ-irradiation. Consistently, the faster scintillation and photoluminescence decays are obtained [50.77, 84] and also the transmission [50.77, 84] and especially radiation resistance [50.78, 85] characteristics are noticeably improved (Fig. 50.10). Theoretical calculations established the energy levels of the defects based especially on vacancies and clarified further the effect of the La doping [50.86]. They are consistent with the experimental findings with the exception of the Pb^+-V_O center, which was not found in the calculations.

For heavily A^{3+}-doped samples (above a few hundred molar ppm in the crystal), a decrease of LY was observed; this was explained by the possible creation of new nonradiative recombination traps related to La dimers/small aggregates [50.77]. Surprisingly, a very large concentration of trivalent ions could be accommodated in the PWO structure (even more than 10% of La) without significant loss of transmittance or radiation hardness. Due to extremely low LY, such heavily A^{3+}-doped PWO-based scintillators were proposed as a Cherenkov radiator [50.87]. Coulombic compensation of the excess charge of trivalent ions in such samples was proposed as a self-compensation process in which a part of the La ions can reside at the W site [50.88]. Introduction of an interstitial oxygen was considered by the same authors as well [50.89]. Recent theoretical calculations favor the latter possibility [50.90].

Radiation damage processes and the nature of related defects were of special interest, because the resulting loss of transmission, instability of LY or even induced changes in the scintillation mechanism can strictly limit the applicability of PWO as a scintillator material in the severe radiation environments of planned detectors. Related deep traps must be stable enough at RT so that they are reflected in TSL glow curves above RT [50.91]. An attempt was made to decompose the induced absorption spectrum (inset of Fig. 50.10, curve *a*) into a sum of Gaussians, which should belong to particular color centers [50.91]. Based on correlated decomposition of irradiation-induced, high-temperature-annealing-induced, and as-grown undoped PWO absorption spectra, four bands were identified, peaking at 3.5, 2.9, 2.4, and 1.8 eV. Two components were found to contribute to the 3.5 eV band [50.92]. Ascription of these and other bands, extracted from absorption spectra, was attempted by several research groups [50.90–95]. However, none of these suggestions was supported by more conclusive methods such as ESR. An interesting suggestion about the physical origin of coloration of PbWO$_4$ was recently formulated [50.96], taking into account the variation of tungsten valence in oxygen-deficient clusters, which can be created or destroyed by an appropriate post-growth annealing. In general, deep traps associated with radiation-damage-induced defects are expected to be rather complex, taking into account that the $(WO_3)^-$-A^+ electron trap responsible for the 300–330 K TSL peaks includes an oxygen vacancy and another defect nearby [50.76].

Recently, an effort has been made to increase the LY of PbWO$_4$ [50.97], which could allow its application outside of high-energy physics. Such a possibility is worth attempting since the thermal quenching of PbWO$_4$ emission is not due to intracenter quenching, but rather due to a nonradiative electron–hole recombination over the material bandgap [50.68] or around other defects [50.60]. If such thermally released electrons and holes can be efficiently and quickly radiatively recombined at an additional center, integral scintillation efficiency would rise. Doubly doped crystals (Mo, A^{3+}) were reported, in which the LY value was increased about 3–4 times with respect to undoped material, while scintillation response was kept reasonably fast [50.98–100]. The effect of PbWO$_4$ nonstoichiometry on the LY value was studied as well [50.101]. Recently, another strategy was introduced, namely fluorine doping [50.102]. A common problem in all of these studies is that additional emission centers frequently give rise to additional shallow traps, which delay radiative recombination and keep the LY below $\approx 10\%$ of BGO even if the integral scintillation efficiency could be increased up to a level comparable to BGO, i.e., about 25 times that of undoped PbWO$_4$ [50.103].

50.2.2 Aluminum Perovskite XAlO$_3$:Ce (X = Y, Lu, Y/Lu)-Based Scintillators

Ce doping has provided scintillators with the highest figure of merit in this group of materials so far. Fast and efficient luminescence and scintillation is achieved due to the parity- and spin-allowed 5d–4f transition of Ce^{3+}, which gives rise to the emission band at 360–370 nm (Fig. 50.11). Moreover, it is free from thermally induced nonradiative quenching at least up to 500 K. Luminescence of YAlO$_3$:Ce (YAP:Ce) was reported by *Weber* [50.104] and favorable properties of this material for scintillation applications were described later by *Takeda* et al. [50.105] and *Autrata* et al. [50.106]. A review paper devoted also to this group of scintillation materials has been published recently [50.22, 107].

Yttrium aluminate YAlO$_3$ has an orthorhombically distorted perovskite structure with space group *Pbnm*.

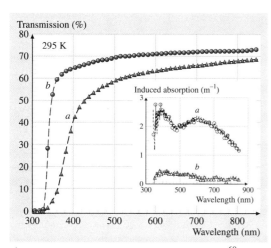

Fig. 50.10 Initial transmission and irradiation (^{60}Co radioisotope, 10 Gy dose) induced absorption of equivalently grown undoped (*a*) and La-doped (*b*) PbWO$_4$ single crystals (after [50.78])

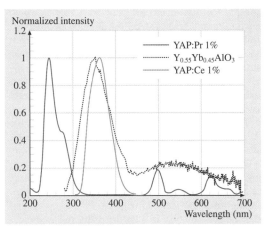

Fig. 50.11 Normalized radioluminescence spectra of doped YAlO$_3$ (see the legend) at RT

Crystals are grown from nonstoichiometric melt at temperatures of 1835–1875 °C [50.108], in which the 1 : 1 melt composition (Y$_2$O$_3$:Al$_2$O$_3$) is stable. In the case of LuAP this temperature region is shifted up by about 80 K. The nearly congruent growth of aluminum perovskites from high-temperature melt is further troubled by twin formation [50.109, 110]. Presence of light scattering centers was also reported [50.111]. The growth process is usually performed in Ar, Ar + H$_2$ or N$_2$ atmospheres or even under vacuum [50.108–112]. Optimized YAP:Ce single crystals with Ce concentration up to 0.6 at.% in the crystal can be grown in boules up to about ⌀ 5 cm × 15 cm in length in CRYTUR Ltd., Turnov, Czech Republic (Fig. 50.12). The boules of

Fig. 50.12 Zr-codoped YAP:Ce single crystal (45 mm diameter and 125 mm length)

LuAP:Ce grown by the Bridgman method [50.113] approach dimensions of about ⌀1.2 cm × 6 cm in length with similar Ce concentrations. Due to the difficulties associated with the growth of pure LuAP single crystals (instability of perovskite phase and frequent appearance of the garnet one), attempts were made to grow mixed Y$_{1-x}$Lu$_x$AP:Ce crystals ($x \leq 0.3$) [50.114]. Mixed Y$_{1-x}$Lu$_x$AP:Ce ($0.15 \leq x \leq 0.8$) crystals were grown also by vertical Bridgman method [50.115]. Mixed (Y/Lu)AP:Ce crystal growths, $x = 0.6$–0.7, using Czochralski method, have been already mastered by a few industrial producers [50.116].

Five bands in the 180–300 nm region were clearly distinguished in the excitation spectrum [50.117] of the 370 nm emission of Ce^{3+}, in good agreement with the expected splitting of the 5d Ce^{3+} excited-state level due to the low Y^{3+} site symmetry. The 370 nm emission band shows a single-exponential fast photoluminescence decay (about 17 ns decay time) and slightly slower scintillation decay governed by decay time between 22 and 38 ns followed by a minor slow component of a few hundred ns [50.118, 119]. The latter component can be explained by delayed recombination of charge carriers at Ce^{3+} centers (similar to the recombination decay components in PbWO$_4$). The scintillation response of pure LuAP:Ce is dominated ($\approx 80\%$ of the overall intensity) by the fast decay component of about 17–19 ns, and the rest of the intensity is released with a decay time of 160–180 ns. In the case of Lu-rich mixed ($x = 0.65$–0.7) crystals, the intensity of the fast component is lowered considerably (40–54%) and accordingly the second slower component of about 190 ns decay time is more intense [50.116]. The origin of this degrading phenomenon has not yet been understood. High LY of YAP:Ce, up to 20 000–22 000 phot/MeV, was reported by several laboratories [50.120] (Table 50.1), decreasing with the admixture of Lu down to ≈ 12 000 phot/MeV in LuAP:Ce. In the process of energy transfer towards the Ce emission centers the excitonic mechanism was evidenced [50.117], but sequential capture of holes and electrons at Ce centers seems to be more important due to low efficiency of lattice exciton creation by hot electrons in the process of initial energy conversion in the YAP and LuAP lattices [50.121].

Doping aluminum perovskites by Pr^{3+} and Yb^{3+} was accomplished as well to obtain scintillators with still faster response with respect to the Ce^{3+}-doped ones; related radioluminescence spectra are shown in Fig. 50.11. The 5d–4f transition of Pr^{3+} provides a ≈ 1.5 eV high energy shifted and faster emission with respect to the Ce^{3+} center in the host matrices with medium–strong crystal field, where the lowest 5d state is shifted below the $^1S_0$ level of Pr^{3+}. Both YAP:Pr [50.122] and LuAP:Pr [50.123] showed at least two times shorter dominant decay time with respect to the Ce-doped hosts. However, the reported LY is 2–4 times lower, so that their figure of merit is significantly degraded. The reason for the latter phenomenon is not clear as the RT quantum efficiency of Pr^{3+} luminescence center in YAP was found to be close to unity [50.124]. Cross-relaxation among Pr^{3+} ions [50.123] and lower efficiency of the energy transfer from the host to Pr^{3+} [50.124] were tentatively proposed as explanations.

Yb^{3+} doping gives rise to the charge transfer luminescence extending over near-UV and visible regions in a YAP host (Fig. 50.11). This luminescence mechanism became of renewed interest, as in the case of Yb^{3+} this kind of radiative transition is parity and spin allowed [50.125, 126] and the luminescence lifetime is typically below 100 ns. However, in most hosts it shows onset of thermal quenching well below RT, so that relatively low radioluminescence intensity and extremely fast, subnanosecond scintillation decay have been reported in $Y_{1-x}Yb_xAlO_3$ [50.127]. The highest radioluminescence intensity is obtained for $x = 0.3$–0.4. Favorably high Yb concentration is allowed due to limited concentration quenching in this system [50.128].

If one compares the results from different laboratories, the Ce^{3+} emission data in the YAP and LuYAP matrices exhibit slightly different shapes and position shifts. Several inequivalent Ce^{3+} positions were found also in ESR measurements [50.129]. The presence of stable Ce^{4+} ions is indicated in [50.130], and Ce^{4+}-related absorption transitions are ascribed to the parasitic absorption overlaying the region of the Ce^{3+} emission. Detailed studies of the temperature dependence of the LY below RT indicated the importance of shallow traps in the process of energy transfer in the YAP matrix; correlations between LY, TSL intensity, and scintillation decay were found and explained by a model based on the participation of trapping state(s) in the processes of electron transfer to trapped holes at Ce^{3+} sites [50.131]. However, no interpretation of the nature of the defects involved was proposed. Deeper electron traps related to TSL peaks at about $50\,°C$, 100–$130\,°C$, $175\,°C$, and $225\,°C$ in YAP:Ce and (Lu/Y)AP:Ce were ascribed to oxygen-vacancy-related defects; a thermally assisted tunneling process with the nearby lying Ce^{4+} hole center could

explain the calculated trap depth and frequency factors of these traps [50.132]. It is worth noting that in an early study [50.133] antisite defects (Y at Al site and vice versa) were considered for the stabilization of F or O^- centers, and their relation to various absorption bands in the 290–760 nm region was hypothetically assumed. The occurrence of color centers is a frequent problem in YAP and there are numerous studies in the literature dealing with this problem, [50.134, and references therein]. The induced absorption after γ-irradiation (radiation damage) of YAP:Ce was also reported [50.135] (Fig. 50.4), and bands peaking at 2.2–2.9 eV, ≈ 3.3 eV, and ≈ 3.89 eV were obtained by decomposition of the induced absorption spectra into Gaussian components. Deviation from stoichiometry, presence of transition-metal and rare-earth impurity ions, and temperature conditions during growth process strongly influence the presence and concentration of defects and color centers in as-grown crystals. Postgrowth annealing is frequently employed to decrease parasitic absorptions of the crystals in UV/visible spectral regions.

Literature information on intrinsic paramagnetic active defects in YAP crystals is rather limited. The autolocalization of holes in the form of O^--bound small polarons was evidenced in YAP some time ago using ESR and optical techniques [50.136]. Recent ESR measurements of lightly (50 ppm) Ce-doped YAP have shown at least four different O^- center configurations caused by site inequivalence of the oxygen ions in the lattice. They are progressively transforming within themselves with increasing temperature as shown in Fig. 50.13. At the same time thermally induced decay

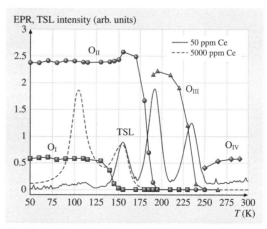

Fig. 50.13 Correlation between EPR intensity of O^- centers and corresponding TSL peaks in YAP crystals

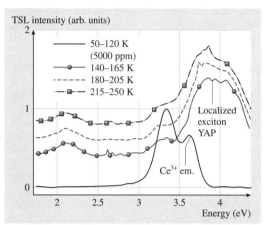

Fig. 50.14 TSL spectra in the temperature range indicated in the legend and related to glow curves in Fig. 50.6. Curves for temperatures above 140 K refer to 50 ppm Ce. Curves are vertically shifted for clarity

of each O^- center is accompanied by a TSL peak. This means that, when a hole is liberated from an O^- center, it is either captured at a deeper hole trap or migrates until a localized electron is met and such electron–hole pair radiatively recombines giving rise to a TSL glow curve peak.

The site of the just mentioned radiative-recombination process was revealed from TSL emission spectra within the glow curve peaks (Fig. 50.14). Dominant glow curve peaks related to O_I-O_{III} annihilation show the emission at 3.5–4.0 eV ascribed to a localized exciton luminescence. Localization of excitons was assumed to be close to F or F^+ centers [50.137]. In undoped YAP, also time-resolved luminescence of F^+ and F centers themselves was reported at 350 nm and 440 nm, respectively [50.138].

The above hypothesis is further confirmed by finding F_A^+ centers in the same sample by ESR measurements [50.139]. They are also found in several configurations with gradually increasing stability within the range 200–300 K, which enables the radiative recombination suggested from TSL. The F_A^+ centers are attributed to the charged oxygen vacancy near the antisite yttrium ion where, however, electron density is essentially shifted to this antisite ion. Such a defect can be described to the Al^{3+}-V_O-Y^{2+}(Al) complex. In a heavily Ce-doped sample, only the self-trapped hole O_I center survives and holes are dominantly trapped at Ce^{3+} ions, as evidenced in TSL spectra (Fig. 50.14). The TSL glow curve is dominated by another electron-trap-related peak at 100–105 K, the origin of which is still unknown. Based on the recent finding in YAG:Ce that the TSL peak at 92 K is related to an electron trap around the Y_{Al} antisite defect [50.140] and taking into account the structural and chemical similarity of such a defect in both aluminum perovskite and garnet structure $((Y_{Al}O_6)^{9-}$ octahedron) one can tentatively suggest a similar interpretation for the 100–105 K peak in YAP:Ce.

Impurity-related centers can come either from the raw materials or due to the particular synthesis technology used. In the former case, mainly iron [50.108] and titanium ions in YAP and furthermore Yb ions in (Lu/Y)AP were evidenced by ESR. In the latter case molybdenum ions were evidenced coming from the crucible used for the growth of bulk crystals [50.108, 112]. Mo^{3+} centers in YAP were described in detail with the help of ESR [50.141]. Being transition-metal elements with easily variable charge state they can be ionized and/or serve as nonradiative recombination centers.

The above-described color centers and impurity ions in YAP crystals give rise to local electronic levels in the forbidden gap, as presented in Fig. 50.15. The Ce^{3+} impurity ion is exploited as a fast and efficient radiative recombination center in YAP:Ce scintillator. In parallel, Mo and Ti impurities serve as ionization centers which produce free electrons and holes under irradiation with energy lower then the bandgap. The position of the Ti^{3+} level is about 4.2 eV above the valence band (VB) edge [50.142]. The Mo^{4+} level is somewhat lower as 330 nm UV irradiation is already enough to excite electrons from the VB to the Mo^{4+} ions, which become Mo^{3+}. The ionization energy of Ce^{3+} ions is at least 5.5 eV [50.143]. Migrating holes can be trapped or self-trapped by the O sublattice in the form of O^--like

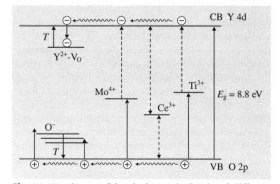

Fig. 50.15 scheme of local electronic levels of different defects in the forbidden gap of YAP

centers. The Y antisite ions give rise to the Y_{Al}^{2+}-V_O defects acting as electron traps.

All the above-mentioned possibilities for hole and electron trapping can deteriorate scintillation performance of the crystal because they compete with or delay the radiative recombination at Ce^{3+} ions. In YAP:Ce, it has been proposed [50.131] that only 25% of the electrons generated by the ionizing photon/particle are transferred promptly to the Ce ion. The rest become trapped in three discrete sites. Consequently, there is considerable room for material optimization, particularly if the existence of these traps can be avoided. Such attempts have already been made by codoping with an aliovalent ion, which can change the Coulombic equilibrium and therefore the overall defect concentrations. Namely, Zr^{4+} codoping was used in YAP:Ce [50.39, 144], which resulted in a dramatic decrease of deep oxygen-vacancy-related electron traps monitored by TSL above RT (Fig. 50.16). It is worth noting that recent atomistic calculations of *Stanek* et al. [50.145] dealt systematically with defect reaction energies in rare-earth aluminum perovskites and the above effect was explained considering the Frenkel recombination of Zr-induced oxygen interstitials with oxygen vacancies. Moreover, it has been shown [50.145, 146] that the relative importance of antisite and Schottky (vacancy) defects varies strongly with the cation size. While Schottky defects are of major importance in YAP, antisite defects might play a major role in LuAP. Such a difference may be of some importance to explain the decreasing LY in LuAP:Ce. Combination of mentioned computer simulations and systematic experiments together with electronic structure calculations can provide a tool for deeper understanding of these scintillation materials and the possible technological approaches leading to their optimization.

50.2.3 Aluminum Garnet $X_3Al_5O_{12}$:Ce (X = Y, Lu, Y/Lu)-Based Scintillators

Single crystals of $Y_3Al_5O_{12}$ (YAG) were among the first oxide materials grown by Czochralski technique during the 1960s [50.147]. Their development was stimulated mainly by the application for solid-state (Nd-doped) lasers, but soon the potential of Ce^{3+}-doped YAG single crystal for fast scintillators was realized as well [50.148]. The first comprehensive description of YAG:Ce scintillator characteristics was reported by *Moszynski* et al. [50.149], who included this material among the high-figure-of-merit oxide scintillators. Isostructural LuAG has a higher density ($6.67\,g/cm^3$) than YAG ($4.56\,g/cm^3$), which is advantageous in the case of hard x- and γ-ray detection (see the definition of stopping power in Sect. 50.1). LuAG:Ce scintillator became of interest relatively recently [50.150, 151]. An early study of the Ce^{3+} and Pr^{3+} photoluminescence decay kinetics in YAG host [50.152] revealed the absence of nonradiative thermal quenching up to about 550 and 250 K, respectively. Consequently, at RT the 5d–4f transition of Ce^{3+} center can be exploited for fast and efficient scintillation in both YAG and LuAG matrices.

The growth of YAG or LuAG crystals by Czochralski technique provides large and high-quality crystals (Fig. 50.17), but also the micropulling down technique has been successful in growing single-crystal rods of comparable structural quality (Fig. 50.5). The former is usually accomplished using either an iridium crucible in an inert atmosphere (N_2) with small addition of oxygen [50.153], or a molybdenum crucible in a reducing (Ar + H_2) atmosphere [50.154]. Isostructural cubic garnet structures of YAG and LuAG form a solid solution and any intermediate mixed composition can be grown. The solidification points of mixed compositions range between 2010 °C (LuAG) and 1930 °C (YAG) [50.153]. The garnet structure appears very flexible for cation substitution; such substitution was used to prepare materials with tailored lattice constants [50.155] and can, in principle, be used for the preparation of highly substituted crystals with homo-

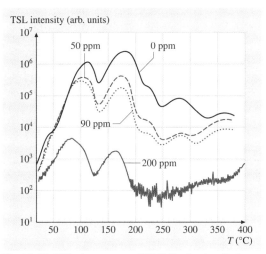

Fig. 50.16 TSL glow curves of YAP:Ce after irradiation at RT. Concentration of Zr in the sample is indicated on the curves, see also [50.39, 144]

geneous composition. However, at the same time this readiness of solid-solution formation in the garnet structure points to a relative ease of compositional defect formation. Ce^{3+} predominantly substitutes for the Y (Lu) cation, but a small fraction can be found at the antisite (Al) position as well. The segregation coefficient of Ce^{3+} in YAG and especially in LuAG host is relatively small (less than 0.1), which is an obstacle to obtaining homogeneous doping profile in large crystals.

Due to high crystal field at the dodecahedral Y (Lu) site, the 5d states of Ce^{3+} are shifted to lower energy in garnet with respect to perovskite host. The lowest $4f$–$5d_1$ absorption and emission bands peak around 450–460 nm and in the green–yellow part of the spectra, respectively (Fig. 50.18). With the help of photoionization threshold and excited-state absorption measurements [50.156] it was established that the lowest $5d_1$ relaxed state is placed about 1.2 eV below the conduction band of YAG, which is enough to completely inhibit undesired ionization of Ce^{3+} relaxed excited state around RT. In a LuAG host, Ce^{3+} emission is high energy shifted by about 0.1 eV with respect to YAG, pointing to a somewhat weaker crystalline field and/or covalent bonding at the Lu site. Consistently, a slightly shorter Ce^{3+} photoluminescence decay time (≈ 55 ns) is obtained at RT in LuAG with respect to YAG (≈ 62 ns) host.

While the steady-state scintillator efficiency (radioluminescence intensity) of both YAG:Ce and LuAG:Ce of high (5–6N) purity can reach up to 700–800% of that of BGO [50.33, 157], their LY is considerably lower. Namely, values up to 300 and 150% of

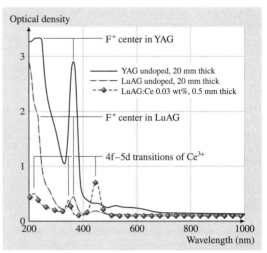

Fig. 50.18 Absorption spectra of YAG and LuAG crystals at RT

BGO were reported for YAG:Ce and LuAG:Ce, respectively [50.120]. This points to a considerable amount of *slow scintillation light* in both materials. Very slow components are reflected in the *increased background* before the rising edge of the LuAG:Ce scintillation decay, which can be quantified by the already mentioned α coefficient (Fig. 50.3). It is worth noting that the *background* level is systematically higher in LuAG:Ce than in YAG:Ce of comparable doping level, quality, and purity [50.33] in agreement with the trend in their LY values. The scintillation decay of LuAG:Ce can be approximated by a two-exponential fit with decay times of about 55–60 ns and 300–1000 ns. A recent study of LuAG:Ce [50.158] has shown that the consecutive capture of a hole and an electron at Ce^{3+} centers is the dominant energy transfer pathway in the scintillation mechanism. More than 65% of the light in the LuAG:Ce scintillation decay is released in the above-mentioned slower decay component, which appears due to delayed radiative recombination at Ce^{3+} ions. The occurrence of slow decay components over an extended time scale suggests the existence of several traps with different thermal depths, which participate in charge-carrier trapping processes.

Scintillation decay characteristics of Pr^{3+}-doped YAG were reported for the first time in 1992 [50.122] and became of renewed interest recently [50.159] when, despite partially quenched emission of the Pr^{3+} center itself, spectrally uncorrected LY values of about 300% with respect to BGO and leading to a scintillation decay

Fig. 50.17 LuAG:Ce crystal, ⌀ 20×30 mm

component below 20 ns were measured in high-quality single crystals. Similar to Ce^{3+}, the $5d_1$–4f emission of Pr^{3+} at the dodecahedral Y site is low energy shifted with respect to perovskite host and shows the double peak band at $318 + 375$ nm. Pr^{3+}-doped LuAG scintillator characteristics were reported only very recently [50.160]. The $5d_1$–4f emission of Pr^{3+} is high energy shifted by about 0.12 eV and the onset of thermal quenching is shifted to noticeably higher temperatures (above 350 K) with respect to YAG:Pr. At RT, the photoluminescence and leading scintillation decay times are very similar (20 ns). The integral of the radioluminescence spectra in LuAG:Pr is 1.4–1.5 times higher than in LuAG:Ce, suggesting higher intrinsic scintillation efficiency. A spectrally uncorrected LY obtained in optimized Czochralski-grown crystal exceeds 300% that of BGO [50.161]. Another advantage consists in the negligible intensity of slower 4f–4f emission lines in radioluminescence spectra in the 450–700 nm range for any Pr concentration, contrary to the situation found in YAG, YAP or LuAP matrices [50.122, 159, 160].

Among the point defects and extrinsic impurities studied in aluminum garnets it is worth mentioning the so-called *antisite* defects, F and F^+ centers (two and one electron in an oxygen vacancy, respectively), and iron impurity ions. Furthermore, recently in LuAG single crystals, the presence of Yb^{3+} was reported. Due to the variable charge state of iron ($Fe^{2+} \leftrightarrow Fe^{3+}$) it can easily participate in charge-carrier capture and give rise to a competing energy loss pathway. At higher concentrations (above 0.1%) Fe^{3+} is evidenced by a characteristic absorption doublet at $407 + 415$ nm and a luminescence band around 800 nm [50.162]. At trace Fe concentrations, the lattice distortions due to embedding Fe^{3+} at the Al site were studied in detail by ESR. Presence of Fe^{3+} in the YAG lattice is also manifested through the charge transfer transition resulting in a wide absorption band at 255 nm [50.163] (Fig. 50.18). Also a Yb^{3+} ion can introduce additional nonradiative losses in LuAG:Ce scintillator as its charge transfer luminescence (similar in shape and about 0.14 eV high energy shifted with respect to that of YAP:Yb; Fig. 50.11) is heavily quenched at RT [50.164]. Fe and Yb ions come from the Al_2O_3 and Lu_2O_3 raw materials, and can be effectively detected by ESR [50.163, 165]. The paramagnetic Ce^{3+} ion in YAG was studied via ESR as early as the 1960s [50.166]. However, only very recently has the occurrence of Ce^{3+} ions at Al sites and possible space correlation between the regular Ce_{Lu} and Lu_{Al} antisite defect been reported in LuAG host [50.167, 168].

Oxygen-vacancy-based color centers are typical defects in aluminum garnets, because of the inert or even reducing atmosphere used during crystal growth. Such atmosphere is necessary to suppress the oxidation of the Ce^{3+} ion, which can decrease scintillator performance. Oxygen vacancies form deep electron traps; they can also effectively lower the scintillation efficiency of Ce-doped aluminum garnets. F^+ centers in YAG host show absorption bands at 235 and 300 nm (Fig. 50.18) and under excitation within these bands, F^+ luminescence was reported at ≈ 400 nm. Evaluation of polarization of this luminescence led to the proposal that F^+ centers might be perturbed by an adjacent Y_{Al} antisite defect [50.169]. Absorption bands at 240 and 200 nm were ascribed to F centers; excitation within these bands yields luminescence band at 460 nm [50.170]. The most studied defect in $A_3B_5O_{12}$ garnet structure (where A and B are are trivalent cations), however, is the antisite defect created by an A cation occupying a B site and vice versa (Fig. 50.19). Such defects arise due to the presence of two equally charged cations and high growth temperature, which induces thermodynamical lattice disorder. In Ce (Pr)-doped aluminum garnets the A_{Al} (A = Y, Lu) one is of special importance (Fig. 50.19) as it gives rise to: (i) slower emission centers in the UV region, peaking at RT within 300–350 nm, which create an unwanted competitive de-excitation pathway in addition to Ce^{3+} (Pr^{3+}); (ii) shallow electron traps, which effectively delay the radiative recombination at the fast emission center and strongly degrade scintillator timing characteristics and LY value. These defects were detected by the existence of satellite 4f–4f emission lines of RE^{3+} ions (Er^{3+}, Nd^{3+}) revealing an octahedral (rather then dodecahedral) emission site symmetry [50.171, 172]. Systematic study of characteristics of the mentioned UV emission has been published recently using time-resolved emission spectroscopy under synchrotron excitation [50.173, 174]. Direct comparison of the bulk single crystals with single-crystalline films grown by liquid-phase epitaxy (LPE) was made. Such films are grown at substantially lower temperature ($\approx 1000\,°C$) and therefore are free of such defects. Moreover, it was shown in aluminum garnet structure that doping by trivalent ions of strongly different radius with respect to original cations, either at the dodecahedral or octahedral sites, induces new UV emission bands, i.e., an effect analogous to process (i). La or Sc doping are some of the most studied cases [50.175, 176]. Despite the easy doping and high overall scintillator efficiency achieved in the latter case, long decay times of this lumines-

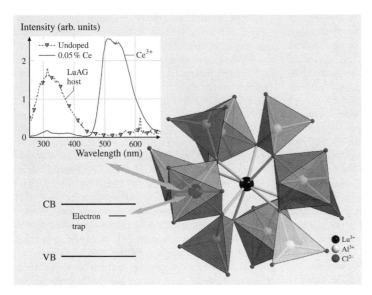

Fig. 50.19 The Lu_{Al} antisite defect in the LuAG structure. The resulting electron trap in the material forbidden gap is sketched on the *left*. The emission band within the range 300–350 nm due to the antisite defect and its competition with that of the Ce^{3+} center can be derived from radioluminescence spectra at RT (*upper left*). Emission lines around 312 nm and 615 nm in the undoped sample are due to Gd^{3+} and Eu^{3+} accidental impurities, respectively

cence (600–800 ns at RT) prevent practical application of such materials as fast scintillators.

It is worth noting, however, that in optimized YAG:Ce and LuAG:Ce bulk single crystals the residual UV emission contributes only a few percent to the total scintillation intensity, and no substantial energy transfer from the UV to Ce^{3+} band was found [50.158], so that the slowing of the scintillation response mentioned above is mainly due to the mechanism (ii), i.e., electron trapping at the antisite defect-related shallow traps before their radiative recombination at Ce^{3+} centers. In YAG:Ce, the TSL peak at 92 K was ascribed to such a trap, taking advantage of the comparison between a bulk single crystal and optical ceramics [50.140, 178]. In LuAG:Ce a comparison was made between a bulk single crystal and LPE-grown films [50.177]; the TSL structure within the 120–200 K interval was ascribed to such a trap (Fig. 50.20). Strong support for this ascription was obtained from TSL measurements of LuAG:Pr [50.179], where the glow curve below 250 K is found to be closely similar to that of LuAG:Ce, while TSL spectra are different, featuring the Pr^{3+} and Ce^{3+} emission spectra, respectively. Using the partial cleaning method, the thermal depth of the 92 K peak in YAG:Ce was calculated to be about 0.18 eV, while the triple TSL structure with peaks at 147, 169, and 187 K in LuAG:Ce was characterized by the energy values of 0.29, 0.40, and 0.47 eV, respectively. Deeper electron traps around the antisite defect and presumably higher concentration of these defects in LuAG with respect to YAG result in a more severe delay in energy delivery to the Ce^{3+} centers in LuAG host and can thus explain the more severe LY degradation in the Lu-based garnet structure.

An interesting and technologically feasible possibility to decrease substantially trapping phenomena related to antisite defects in bulk crystals was recently announced. In mixed $Lu_3Ga_2Al_3O_{12}$:Pr single crystal the TSL peaks in the 100–200 K temperature interval were completely suppressed, UV luminescence of the host disappeared at RT, and slower components and the

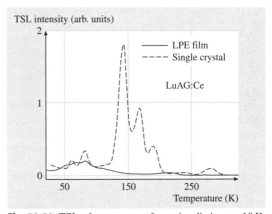

Fig. 50.20 TSL glow curves after x-irradiation at 10 K: comparison of LuAG:Ce LPE film and single crystal, see also [50.177]

α coefficient of scintillation decay were substantially lowered [50.179]. Even if the exact explanation of this effect has yet to be found, selective embedding of Ga^{3+} ions at octahedral lattice sites and related modification of the bottom of the conduction band due to lower-lying Ga-related energy levels [50.180] might be responsible for the observed effect.

Optimization of YAG- or LuAG-based fast scintillators by aliovalent ion codoping has not succeeded thus far. An attempt to codope LuAG:Ce with Zr^{4+} did not result in suppression of deep traps, at variance with what occurs in YAP:Ce [50.39, 144]. Deep traps related to 80 and 280 °C TSL peaks were markedly increased and only TSL peaks above 300 °C were suppressed [50.181]. It is interesting to note that recent atomistic simulations of defect creation in $RE_3Al_5O_{12}$ [50.182] point to a different response of the garnet lattice to tetravalent ion doping with respect to perovskite lattice [50.145, 146]. While in the latter case the production of interstitial oxygen was found as the least energy-requiring reaction, in the former case the production of RE cation vacancies seems to be the most energetically favorable process. Such a result may indicate that a different technological approach must be taken to improve aluminum garnet scintillators further.

50.2.4 Ce-Doped Silicate Single Crystals

The scintillation characteristics of Ce-doped rare-earth (RE) oxyorthosilicates were reported for the first time more than 20 years ago. After the first study, which revealed the promising properties of Gd_2SiO_5:Ce (GSO) [50.183], numerous investigations were devoted to several other compounds of this material series, such as Lu_2SiO_5:Ce (LSO), Y_2SiO_5:Ce (YSO), (Y_2)-$Lu_2Si_2O_7$:Ce (YPS and LPS), and several mixed compounds (most notably $(Gd)_{2-x}$- or $Lu_{2-x}Y_xSiO_5$:Ce; GYSO and LYSO, respectively). Single crystals were predominantly grown by Czochralski technique using an iridium crucible, which is necessary due to the very high melting point of these compounds (in the range ≈ 1950–2150 °C). Radiofrequency (RF) induction heating was used in order to decrease the formation of polycrystalline regions, voids, and cracks [50.184, 185]. The optimization of all the manufacturing process enabled the growth of large single crystals (approximately 70–80 mm diameter and 200 mm length) with very good and reproducible scintillation performance [50.186–189]. Furthermore, material preparation in bulk, powders, and thin-film forms was pursued by other routes such as the pulling-down technique [50.190], the sol–gel method [50.191, 192], and pulsed laser deposition [50.193]. In all cases promising scintillation characteristics were reported.

GSO possesses the monoclinic $P2_1/c$ crystal structure, while LSO and YSO possess the monoclinic $C2/c$ structure [50.194, 195]. The LSO structure is depicted in Fig. 50.21: it features SiO_4 tetrahedra and non-Si-bonded O atoms surrounded by four Lu atoms in a distorted tetrahedron. Band-structure and density-of-states calculations on YSO and YPS were recently performed [50.196].

In undoped crystals, in addition to intrinsic emission bands, further emission is clearly detected in the 250–400 nm region and interpreted as due to self-trapped excitons and holes [50.197]. Optical studies of Ce-doped samples showed two Ce^{3+} emission bands related to two different Ce^{3+} crystallographic sites in GSO, LSO, and YSO [50.198]. Due to its higher scintillation efficiency LSO was more thoroughly investigated; its PL spectra at 11 K are reported in Fig. 50.22 [50.199]. The doublet emission peaking at 393 and 427 nm was ascribed to Ce_1 whose luminescence is not quenched up to RT. On the other hand,

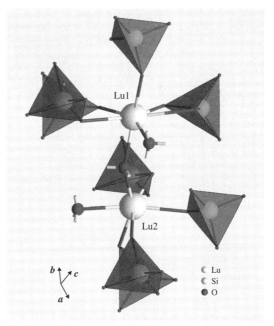

Fig. 50.21 The structure of LSO, with two Lu sites displayed and surrounding SiO_4 tetrahedra. Three oxygen ions with cut bonds form OLu_4 tetrahedra and do not participate in Si–O bonds

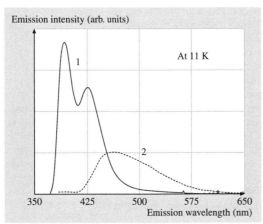

Fig. 50.22 Emission spectra of LSO at 11 K. The two emission spectra have excitation wavelength of 356 nm (*solid line*) and 376 nm (*dashed line*). Curves 1 and 2 refer to Ce_1 and Ce_2 centers, respectively (after [50.199])

the composite emission peaking at 460 nm, detected well only below 80 K, was ascribed to the Ce_2 center. The ionization processes of excited Ce^{3+} energy levels were studied by means of parallel photoconductivity and photoluminescence excitation–emission experiments [50.200], and by the microwave resonator technique [50.201], revealing three different delocalization processes of direct photoionization, tunneling, and thermal ionization of the lowest relaxed $5d_1$ level, found 0.45 eV below the conduction band.

The presence of two distinct Ce^{3+} emissions is in accordance with the existence of two Lu^{3+} crystallographic sites and with the substitutional incorporation of Ce for Lu.

Detailed information about incorporation of Ce^{3+} ions into Lu_2SiO_5 was obtained from ESR studies [50.202]. The g-value of a rare-earth ion is closely related to the wavefunction of its electronic ground state and therefore strongly depends on the point symmetry of the crystallographic site. Therefore ESR is indeed a good tool to investigate Ce^{3+} site symmetry and occupancy in crystal lattice.

Lu ions reside at two crystallographic sites with low (C_1) symmetry, Lu_1 and Lu_2, with seven and six oxygen neighbors, respectively. Ce^{3+} ions are expected to substitute only for Lu^{3+} ions due to their similar ionic radii (Ce^{3+}: 0.103 nm; Lu^{3+}: 0.085 nm; Si^{4+}: 0.026 nm). ESR measurements support this assumption since Ce^{3+} ions are found at both lutetium sites. As the ESR intensity is directly linked to the concentration of paramagnetic ions, the relative Ce^{3+} concentration in each site could be determined. The values are: $Ce_1 = 95\%$ and $Ce_2 = 5\%$. The most substituted site is attributed to the larger Lu_1 site with $6+1$ oxygen neighbors. ESR spectra of LSO showing two Ce^{3+} resonance lines are shown in Fig. 50.23b. Moreover, in Fig. 50.23a the ESR spectrum of lutetium pyrosilicate ($Lu_2Si_2O_7$, LPS, with space group $C2/m$; see [50.203]) is also shown. In this case only one resonance signal is detected due to Ce^{3+} substituting for Lu^{3+} ion at the lone C_2 symmetry site.

In the $(RE)_2SiO_5$ material group, LSO presents the best figure of merit due to its highest density (7.4 g/cm^3), short decay time (about 40 ns), high light output (about 3.5–4 times that of BGO), and satisfactory energy resolution (7.75%) [50.187]. Another important advantage is the absence of shallow traps, as indicated by the very low TSL intensity below RT [50.204], which is consistent with the absence of slower components in scintillation decay. On account of these characteristics, LSO found large-scale application, substituting for BGO in medical positron emission tomography (PET) systems [50.205]. Yttrium-admixed (several percent) LSO (LYSO) is an interesting alter-

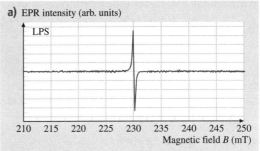

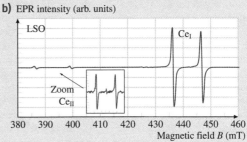

Fig. 50.23a,b ESR spectra at 12 K of Ce^{3+} in (**a**) LPS and (**b**) LSO crystals with the magnetic field B parallel to the a-axis (after [50.102]) ($\pm 5\%$). The spectra were recorded at 9.5 GHz with a microwave power of 20 mW; modulation amplitude was 1 mT (after [50.202])

native to LSO because of its more favorable growth properties due to a lower melting point, lower formation of inclusions, lower cost of starting materials, and comparable or even better scintillation performance [50.186]. GSO has clearly inferior light output (Table 50.1) and scintillation decay also shows slower components. Recently, a positive effect of Zr codoping was reported, which increases its LY by about 20% [50.206]. Due to the lower density (6.7 g/cm^3) its attenuation length is 40% longer than that of LSO. However, its application in PET cameras is being considered as well [50.207].

In the case of LSO, it was soon recognized that it exhibits a fairly strong afterglow [50.209]. It consists of a light-emission signal which decreases exponentially following the end of an excitation pulse, and that can persist for several minutes up to hours at RT. This is a detrimental property of the crystal, since it causes background instability during repeated measurements.

Fundamental studies aimed at the comprehension of the microscopic physical mechanism governing afterglow were thus carried out in order to find possible technological solutions. The activation energy of the process was found to be approximately 1 eV [50.209]; the is in accordance with the calculated trap depth of a TSL peak at 375 K (using a heating rate of 6 K/s) so that afterglow appears to be due to RT carrier detrapping from the trap responsible for this peak followed by radiative recombination at Ce^{3+} luminescent centers. Actually the 375 K peak is the first of a series of as many as six peaks observed in the glow curve above RT, whose spectral emission coincides with the Ce^{3+} 5d–4f transition [50.209]. Annealing experiments in reducing or oxidizing atmosphere led to the suggestion that traps could be related to oxygen vacancies [50.210]. The presence of oxygen-related defects following irradiation was suggested from optical absorption measurements [50.211]; oxygen vacancies, together with STEs and STHs, were proposed to be responsible for TSL peaks below RT [50.197, 212]. A very nice confirmation of the intrinsic origin of traps was the finding that all oxyorthosilicates possessing $C2/c$ structure (YSO, LSO, YbSO, ErSO) are characterized by similar TSL glow curves above RT, irrespective of their doping [50.208, 211], so that crystal structure seems to play an important role in trap properties. In Fig. 50.24 the glow curves of LSO, LSO:Ce, YSO:Ce, and YbSO:Ce are compared. The favorable absence of afterglow in LPS, accompanied by the lack of TSL glow peaks close to RT [50.213], can thus be explained in terms of its different $C2/m$ structure, leading to different Ce^{3+}–oxygen (and thus oxygen vacancy) configurations and distances.

It is interesting to note that Pr^{3+} doping was also attempted in order to obtain a faster scintillation response of LSO-based scintillator. Fast Pr^{3+} 5d–4f photoluminescence has a maximum at 273 nm and decay time of about 6–7 ns at RT [50.214]. However, significant ionization of the $5d_1$ relaxed excited state occurs at RT, because the $5d_1$ level was found to be closer to the conduction band (0.28 eV) compared with the Ce^{3+}

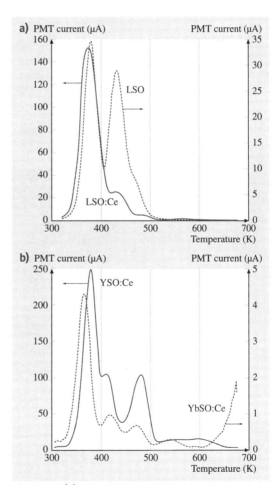

Fig. 50.24 (a) X-ray induced TSL glow curves of undoped and Ce-doped LSO taken by heating the samples at 5 °C/s. (b) X-ray-induced TSL glow curves of YSO : Ce and YbSO : Ce. The sharp upturn above 650 °C for the weak YbSO : Ce TSL is due to black-body radiation emanating from the silver sample holder (after [50.208])

center (0.45 eV). This ionization process significantly degrades LY values of LSO:Pr and prevents its practical application.

In conclusion, Ce-doped LSO, LYSO, and GSO proved to have excellent scintillating performances and are presently key materials for medical diagnostic PET applications; their further improvement should be based on better control of intrinsic point defects, especially those including oxygen vacancies.

50.2.5 Ce-Doped Rare-Earth Halide Single Crystals

During the last decade research on fast and efficient scintillators has found new very interesting candidates in the family of rare-earth (mostly La and Lu) halides. These are Ce-doped bromides, chlorides, and iodides such as LaX_3:Ce and LuX_3:Ce, where X = Cl, Br or I. Some of their characteristics are listed in Table 50.2 where scintillation time decay, LY, and energy resolution are given for a specific Ce concentration.

Single crystals of Ce-doped rare-earth (RE) halides are usually obtained by vertical Bridgman growth technique using binary RE halides as starting materials. These starting materials can be prepared by several different routes [50.217, 223, 224]; for example, the preparation of RE iodides is performed by direct sealing of the metal with slight excess of iodine under vacuum in a silica ampoule, and heating up to the iodine melting point. On the other hand, the so-called *ammonium halide route* can be employed for the preparation of rare-earth chlorides and bromides. This process consists of two main steps. First, the dissolution of RE oxides in HCl or HBr with the addition of ammonium halide is

Table 50.2 Structural and optical characteristics of Ce-doped RE halides. Numbers in parenthesis in the scintillation decay time column represent the percentage of total intensity governed by the indicated decay time, while slower components characterize the decay of the remaining intensity. Times in parenthesis in the LY column represent the time gates with which the LY was evaluated. Original references are shown in square brackets

Crystal	Space group	Density (g/cm³)	Band gap (eV)	Ce 5d–4f em. (nm)	Ce 4f–5d abs. (nm)	Ce conc. (mol%)	Scintillation decay time (ns)	LY (10^3 phot/MeV)	Energy res. (%)
$LaCl_3$:Ce	Hexagonal $P6_3/m$ (C_{6h}^2) [50.215]	3.86 [50.215]	7 [50.216]	337, 358 [50.215]	243, 250, 263, 274, 281 [50.215]	10	24 (60%) [50.217]	50 [50.217]	3.1 [50.217]
$LaBr_3$:Ce	Hexagonal $P6_3/m$ (C_{6h}^2) [50.215]	5.29 [50.218]	5.6 [50.219]	355, 390 [50.219]	260, 270, 284, 299, 308 [50.219]	5	16 (100%) [50.217]	70 [50.217]	2.6 [50.217]
LaI_3:Ce	$PuBr_3$ $Cmcm$ [50.215]	5.614 [50.220]	3.3 [50.220]	452, 502 [50.220]	420 [50.219]	0.5	19 ($T < 150$ K) [50.220]	≈ 0 (RT) [50.220] 16 ($T < 100$ K)	–
$LuCl_3$:Ce	Monoclinic $C2/m$ [50.215]	4.00 [50.215]		374, 400 [50.215]	215, 235 [50.215]	0.45	50 (25%) [50.215]	1.3 (0.5 μs) 5.7 (10 μs) [50.215]	16 [50.215]
$LuBr_3$:Ce	Rombohedral $R-3$ [50.215]	5.17 [50.215]		408, 448 [50.215]	310–400, 230 [50.215]	0.76	32 (10%) [50.215]	10 (0.5 μs) [50.215] 24 (10 μs)	5 [50.215]
LuI_3:Ce	Hexagonal $R-3$ [50.222]	5.6 [50.221]		475, 520 [50.221]	≈ 300, 390, 419 [50.221]	0.5 2	< 50 ns (50%) [50.222]	42 (0.5 μs) [50.222] 51 (10 μs) 58 (0.5 μs) [50.222] 71 (10 μs)	4.7 [50.222] –

achieved

$$M_2O_3 + 6HX + 6NH_4X \rightarrow 2(NH_4)_3MX_6 + 3H_2O, \tag{50.6}$$

where M stands for the RE element (for instance La or Lu) and X stands for Cl or Br.

Subsequently, the ternary salt is heated under vacuum (typically up to 400 °C) and decomposed to obtain the final RE halide

$$(NH_4)_3MX_6 \rightarrow MX_3 + 3NH_4X. \tag{50.7}$$

Purification of the binary halides by sublimation under high vacuum is finally performed, allowing oxyhalides to be separated as involatile residues. The obtained powders can be used as starting materials for Bridgman growth. The crystals grown are highly hygroscopic. They must be sealed in appropriate ampoules for handling and characterization. Recently, Bridgman growth from commercially available anhydrous binary halide powders was performed and crystals with very good optical and scintillation characteristics were obtained [50.224].

While Pr^{3+} and Nd^{3+} doping of RE halides were considered for laser applications [50.226, 227], only the Ce^{3+} ion appears to be suitable as a dopant of these matrices to obtain fast and efficient scintillators. The radioluminescence spectra of Ce-doped samples are dominated by an intense 5d–4f Ce^{3+} emission. Other emissions related to the host crystal were detected as well. The assignment of such bands to specific centers and the study of their role in the energy transfer processes towards Ce^{3+} were soon the subject of systematic investigations. In this respect the most detailed study was performed on $LaCl_3$:Ce [50.215, 216, 219, 225, 228, 229]. Figure 50.25 displays the x-ray excited emission of $LaCl_3$:Ce^{3+} at different Ce^{3+} concentrations [50.225]. Apart from the Ce^{3+} doublet peaking at 337 and 358 nm, a shoulder attributed to STE emission is observed at about 400 nm. The excitonic nature of the 400 nm emission was established by an optically detected magnetic resonance investigation which revealed the existence of two excited triplet states attributed to *out-of-plane* STEs [50.230], one peaking at about 310 nm (superimposed to the Ce^{3+} emission) and the second at about 420 nm, similar to the shoulder seen in Fig. 50.25. On the other hand, no stable self-trapped holes (V_k centers) were found due to their immediate recombination with an electron from the conduction band resulting in STE state creation.

The mutual intensities of the Ce^{3+} and STE emissions are found to change with temperature (Fig. 50.26).

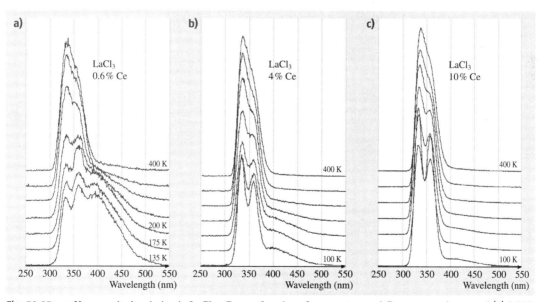

Fig. 50.25a–c X-ray excited emission in $LaCl_3$: Ce as a function of temperature and Ce concentration: panel (**a**) 0.06% Ce; panel (**b**) 4% Ce; panel (**c**) 10% Ce. Unless otherwise indicated spectra are separated by 50 K temperature intervals (after [50.225])

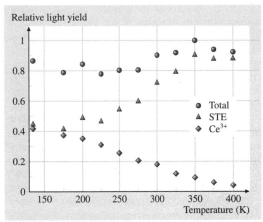

Fig. 50.26 Temperature dependence of the Ce^{3+}, STE, and total luminescence in $LaCl_3$:0.57% Ce, derived from x-ray-induced emission spectra. The error bars are comparable with the dimensions of the data points (after [50.215])

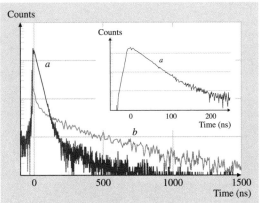

Fig. 50.27 Scintillation decay time spectra of (*a*) $LaBr_3$: 0.5%Ce and (*b*) pure $LaBr_3$ at RT under ^{137}Cs γ-ray excitation. The decay of the doped sample is approximately single-exponential with decay time of 30 ns. The faster component in the decay of pure $LaBr_3$ is probably due to Ce^{3+} trace impurities; the longer component is due to STE luminescence. The *inset* shows the spectrum of $LaBr_3$: 0.5%Ce on a shorter time scale (after [50.233])

Namely, STE emission decreases for $T > 200$ K, while an opposite dependence is detected for the Ce^{3+} 5d–4f transition. The anticorrelation between STE and Ce^{3+} luminescence was interpreted by assuming the occurrence of energy transfer towards Ce^{3+} ions through STE diffusion. Specifically, soon after irradiation, thermalized electrons and holes form STEs, which can migrate and be trapped at Ce sites before their radiative de-excitation. STE mobility increases with increasing temperature in accordance with the observed temperature dependences in Fig. 50.26 [50.229]. The nonexponential character of the Ce^{3+} scintillation time decay [50.215, 229] was interpreted as due to the existence of two energy localization mechanisms at Ce^{3+} sites, namely (i) prompt trapping of holes and electrons by cerium ions, leading to a fast decay component determined by the intrinsic cerium 5d–4f emission lifetime, and (ii) the above-mentioned thermally assisted energy transfer from STE to cerium ions, leading to a slow scintillation decay governed by STE lifetime in the microsecond timescale. Room-temperature STE lifetime (3.5 µs) was derived from the scintillation time decay of an undoped $LaCl_3$ sample [50.229]. Moreover, a possible role of defects acting as trap levels in the slow scintillation decay processes was taken into account. Such trap levels were investigated in a few TSL studies [50.231, 232]. These studies showed the presence of several TSL peaks in $LaCl_3$ and $LaBr_3$ both below and above RT. However, TSL properties have not yet been clearly correlated with slow decay components.

The best scintillation performances in terms of LY, decay time, and energy resolution are displayed by $LaBr_3$:Ce [50.218, 233, 234]. In Fig. 50.27 scintillation time decay spectra of $LaBr_3$:Ce are shown, which feature a fast and approximately single exponential component. In this crystal, efficient direct energy localization at Ce^{3+} ions occurs and an intermediary role of STE migration processes seems to be not relevant, probably due to the shorter STE decay time and higher STE migration rate evidenced in bromides with respect to chlorides [50.219]. Moreover, a higher number of electron–hole pairs are generated with respect to $LaCl_3$ due to a smaller bandgap (Table 50.2), resulting in a very high LY of 70 000 phot/MeV.

Even smaller bandgap, shorter STE decay time, and higher STE migration rate characterize the heaviest iodides [50.219]. However, other drawbacks arise in the case of LaI_3:Ce. For example, the small bandgap value and the small energy difference (0.1–0.2 eV) between the position of the Ce 5d level and the bottom of the conduction band seem to be responsible for its very poor scintillation properties at RT [50.220]. At $T > 120$ K 5d electrons can be thermally activated to the conduction band and luminescence becomes quenched. A more favorable situation is encountered in the case of LuI_3 [50.221, 222, 235] due to the slightly higher bandgap value and longer emission wavelength, so that the energy difference between the Ce^{3+} 5d level and the conduction band is greater. Thermally activated transition into the conduction band does not occur and a quite high LY (70 000 phot/MeV, Table 50.1) was reported. Optimization of LuI_3:Ce is still needed as it presently shows intense slow decay components; moreover crystal inhomogeneities were also reported [50.222].

In conclusion, Ce-doped RE halides constitute a very interesting material group in which scintillation properties vary strongly due to slightly different microscopic properties. At the present time $LaBr_3$:Ce possesses the best scintillator performances, due to a favorable combination of STE lifetime, STE migration rate, bandgap value, and position of Ce^{3+} energy levels in the forbidden bandgap.

50.2.6 Optical Ceramics and Microstructured Materials

Optical ceramics are transparent or translucent materials constituted by tight aggregating crystallite micrograins, each randomly oriented with respect to its neighbors. They have been under development as an alternative to single-crystal materials to provide bulk optical elements in the case where single crystals cannot form or when ceramic materials show superior properties, for example, in terms of achievable concentration or homogeneity of the dopant. The associated technology has developed greatly, mainly within the last two decades; in the case of cubic material, technological progress enabled bulk elements that were visually indistinguishable from single crystals and in terms of some parameters (doping profile) even clearly superior. Among the developed materials, mainly YAG:Nd and Y_2O_3:Nd reached the highest degree of perfection; their primary application is in the field of solid-state lasers [50.236, 237]. The size of the single-crystal grains is on the order of a few tens of microns, interface thickness is reported to be only about 1–2 nm, and the residual volume of pores is several ppm.

It is worth noting that the application demands are clearly higher in the case of scintillators compared with laser optical ceramics. In the latter case the scattering losses due to pores (voids) and grain interfaces limit the material performance. The already achieved reduction of the pore volume down to a few ppm and interface thickness to a few nm reported in [50.236, 237] is enough to make them fully comparable to their single-crystal analogs. However, in the case of scintillators, even point defects at the atomic scale can seriously limit material performance due to the introduction of trapping levels in the material forbidden gap. If we consider a grain size on the order of a few tens of μm and thickness of the interface layer of 1–2 nm, the expected concentration of the interface-related (trapping) states may easily reach several hundred ppm. Such a concentration can noticeably influence the characteristics of the transport stage due to retrapping of the migrating carriers or even the introduction of nonradiative traps, which can seriously degrade the scintillation response and/or overall efficiency parameters.

Development of optical ceramics for scintillator applications was triggered by the needs of computer tomography (CT) medical imaging. A review of the results achieved, mainly for the manufacturing and characterization of $(Y,Gd)_2O_3$:Eu, Gd_2O_2S:Pr,Ce,F, and $Gd_3Ga_5O_{12}$:Cr,Ce ceramics, was reported by *Greskovich* and *Duclos* [50.238]; see also [50.239–241] for more details about the particular materials. In the latter two materials, codoping with Ce^{3+} is used to reduce the afterglow through the capture of holes at Ce ions, $Ce^{3+} \rightarrow Ce^{4+}$ conversion, and subsequent nonradiative recombination with migrating electrons and/or temporary Eu^{2+} centers. Moreover, codoping of $(Y,Gd)_2O_3$:Eu [50.242] with Pr^{3+} appeared to be an efficient tool to reduce the afterglow (by a mechanism similar to that of the Ce ions above). Annealing in a controlled atmosphere can further diminish or passivate trapping states in the forbidden gap related to the grain interfaces. Eu^{3+}- or Tb^{3+}-doped Lu_2O_3 are worth mentioning among new optical ceramics suggested for scintillator applications and based on the slow 4f–4f transitions of rare-earth ions. True optical ceramic Lu_2O_3:Eu samples were synthesized, their performance was tested in comparison with CsI:Tl, and a large-area scintillation screen was prepared [50.243, 244]. While providing 60% of the emission intensity of CsI:Tl, a better match to the CCD photodetector sensitivity was concluded. Unfortunately, Lu_2O_3:Eu was found to suffer of noticeable afterglow for times longer than 6 ms [50.245].

Recently, fast and potentially less expensive optical ceramics based on Ce-doped YAG have been reported in the literature [50.246]. Nowadays, Ce-doped YAG optical ceramics are available from Baikowski, Japan Ltd., the production technology of which should be analogous to that of YAG:Nd described in [50.236]. Figure 50.28a,b shows a comparison of radioluminescence and scintillation decay characteristics in equally shaped samples ($\varnothing 15 \times 1$ mm) of an industrial standard-quality single-crystal YAG:Ce produced by CRYTUR, Ltd., Czech Republic and the aforementioned opti-

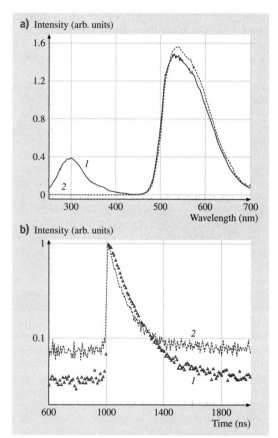

site defects are absent in the former system. Normalized scintillation decays are shown in Fig. 50.28b. The ceramic material shows faster decay in its initial part, the decay time of which is closer to the photoluminescence lifetime of the Ce^{3+} center (about 60–65 ns). This can be explained by a higher Ce concentration and the absence of Y_{Al} antisite defects. However, a higher content of very slow scintillation components is reflected in lower amplitude-to-background ratio. The presence of such very slow decay components is usually related to traps monitored by TSL measurements somewhat below RT. Indeed, the ceramic system shows higher TSL intensity above approximately 190 K [50.140], possibly due to the presence of traps at the grain interfaces. Recently, also LuAG:Ce optical ceramics were prepared by a coprecipitation route [50.247] (Fig. 50.29) and the scintillation performance was compared with that of LuAG:Ce single crystal [50.248]. Similar to

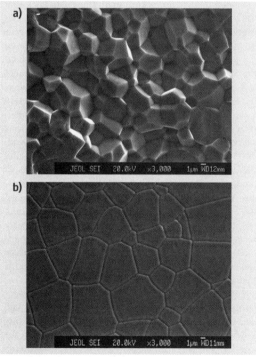

Fig. 50.28 (a) Radioluminescence spectra (Mo cathode, 35 kV) of YAG:Ce single-crystal (*1*) and ceramic (*2*) scintillators at room temperature. (b) Spectrally unresolved scintillation decay of YAG : Ce single-crystal (*1*) and ceramic (*2*) scintillators at RT, with excitation by 511 keV photons of ^{22}Na radioisotope. Approximation by a sum of exponentials yields decay time of 119 ns (*1*) and 85 ns (*2*) for the dominant component

cal ceramics [50.140]. While the radioluminescence intensity in the Ce^{3+} band at 550 nm is comparable in both samples (Fig. 50.28a), the host 300–350 nm band is absent in the ceramic sample. This UV emission is ascribed to a luminescence center based on an antisite Y_{Al} defect (Sect. 50.2.3) and its formation is crucially dependent on the preparation temperature. As the preparation temperatures of YAG-based optical ceramics are noticeably lower with respect to the single crystal [50.236], it can be concluded that the Y_{Al} anti-

Fig. 50.29a,b Scanning electron microscopy (SEM) photographs of LuAG:Ce ceramics sintered at 1800 °C for 10 h under vacuum: (a) the fracture surface; (b) the polished surface after thermal etching at 1400 °C for 1.5 h [50.247] (courtesy of Xue-Jian Liu, Shanghai Institute of Ceramics)

YAG:Ce optical ceramics, high radioluminescence efficiency was found. However, noticeably lower LY and the presence of more intense slow components in the scintillation decay point to retrapping phenomena at the grain interfaces, which is supported also by measured TSL characteristics.

Microstructured materials can be classified as single-crystal-based systems grown with tailored spatial morphology. These materials were developed to meet the demand of increased x-ray absorption, while keeping high spatial resolution in the x-ray imaging screens. Few scintillation materials, namely CsBr:Tl [50.249], CsI:Na [50.250], and CsI:Tl [50.251–253], can be prepared by vacuum evaporation in the form of long (up to 1–2 mm) and thin (several μm diameter) needles, which are densely packed and optically isolated (Fig. 50.30). A needle layer can be directly deposited on the photodetector and, due to the light-guiding effect in the needles, the high spatial resolution is preserved with increased layer thickness, ensuring higher x-ray absorption. While this concept has been used for a long time for CsI:Na in image intensifiers [50.254], where the CsI:Na is at the inner side of the photocathode, i.e., in vacuum, it could not be used in the open atmosphere due to its extreme hygroscopicity. Only the development of this growth morphology for much less hygroscopic CsI:Tl, and the availability of large-area position-sensitive semiconductor detectors (CCD, a-Si:H panels) enabled the construction of a new generation of flat-panel detectors, which provide a qualitative upgrade in many kinds of medical, industrial, and scientific imaging systems [50.255].

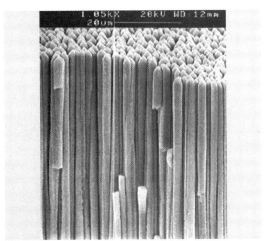

Fig. 50.30 Vapor-deposited column-shaped CsI : Tl scintillation crystals of very smooth structure. Diameter ≈ 3 μm, length > 0.5 mm; see also [50.20] (courtesy of Philips Research Laboratories, Aachen)

50.3 Future Prospects

Acceleration of research and development in the field of scintillation materials over the last 20 years has been motivated mainly by new demanding applications in the medical, industrial, scientific, and security imaging sectors. Further development and exploitation of powerful x-ray sources, e.g., synchrotron radiation, the availability of new radiopharmaceuticals significantly enhancing PET imaging capability in oncology [50.256], and the wider exploitation/monitoring of radioisotopes in different branches of human activity continue to contribute to increasing interest in fast and efficient scintillators.

Currently, there is a need for high-density, fast, and high-light-yield scintillators for PET. While LSO:Ce and GSO:Ce were successfully introduced and are replacing the previously used BGO [50.205–207] the search for a second material to be used in combined PET detectors to correct for the depth-of-interaction error continues [50.257]. Furthermore, in PET detectors sufficiently fast and bright Ce- or Pr-doped materials enable evaluation of time-of-flight information related to the coincidence detection of two 511 keV photons from the rising edge of the scintillation decay. Such information can limit the number of random coincidences and increase the signal-to-noise ratio [50.258]. The candidates are (Lu/Y)AP:Ce or LuAG-based scintillators, which are under intense investigation. Undoubtedly, all of these materials could be further improved since the nature of defects participating in the scintillation conversion and degrading their performance is not yet fully understood.

As the majority of applications use x-ray sources with energies below 150 keV, there is significant market potential for very high light yield, medium density, fast scintillators, where the latest developments led to the discovery of the group of Ce-doped rare-earth halides. Also in this case, understanding of material point defects is at the early stages and improvements in crystal quality are expected in the near future.

The rapidly developing field of optical ceramics may bring completely new material systems into consideration, which cannot be prepared in the form of single crystals. The recent success of slow scintillation ceramics based on Tb^{3+}-, Eu^{3+}- or Pr^{3+}-doped materials in CT medical imaging has clearly shown the potential of this manufacturing technology. Even in the case of YAG:Ce or LuAG:Ce materials, which can be relatively easily grown in single-crystal form, optical ceramics may offer an interesting alternative due to the absence of specific (e.g., antisite) defects, which degrade scintillation performance of the single crystal. On the other hand, one should keep in mind that grain boundaries and interfaces in the ceramics introduce other trapping states with undesirable consequences as described in Sect. 50.2.6.

It is interesting to mention recent literature reports dealing with phosphors prepared from materials of smaller bandgap (sulfides, selenides, tellurides) due to their expected higher conversion efficiency. In these materials the emission center can be based on an exciton localized close to a suitable doped impurity ion. Very high efficiency and decay time of about 1 μs at RT were reported for ZnTe:O [50.259]. Apart from the powder form, these materials can be prepared in the form of thin films as well using a novel sol–gel technology, which offers better controlled synthesis conditions and more perfect microcrystal grain surfaces. Thin-film scintillators were developed for special applications requiring high two-dimensional (2-D) resolution in monitoring tiny objects. High-quality single-crystalline films can be grown by liquid-phase epitaxy technique, which was demonstrated in the case of YAG and LuAG-based scintillators [50.260, 261] or ZnO [50.262].

Renewed interest in the scintillation characteristics of direct-bandgap semiconductors based on ZnO or other binary compounds has arisen recently [50.30, 31, 263] due to their superfast emission with decay times below 1 ns. Emission in these materials is due to Wannier excitons, free or localized at defects/doped ions. In the latter case it shows a low energy shift and still faster decay times of the order of tens of picoseconds.

Donor–acceptor recombination luminescence in suitably doped systems was also reported [50.264]. Such materials can show an unbeatable combination of superfast scintillation response and (intrinsically) high light yield. The problem of low Stokes shift and en-

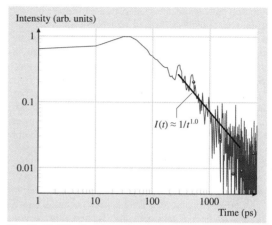

Fig. 50.31 RT photoluminescence decay of ZnO:In, Li liquid-phase epitaxy grown film. $\lambda_{exc} = 260$ nm, 150 fs laser pulse, $\lambda_{em} = 420–430$ nm. The *solid line* shows the $I(t)$ dependence proportional to inverse time typical of donor–acceptor pair recombination luminescence

hanced emission reabsorption in bulk elements might be solved by the aforementioned exciton localization or exploitation of fast donor–acceptor recombination luminescence (Fig. 50.31) [50.32]. A new class of so-called *quantum scintillators* based on quantum dots (nanostructures) of direct-gap semiconductors embedded in a suitable host was announced recently [50.265, 266]. These materials exploit a quantum size effect in free exciton emission, which can enhance the efficiency and luminescence speed in such nanostructured materials with respect to the bulk analogues.

A separate group of scintillation materials is under development for neutron detection. These materials must contain elements with high cross-sections for neutron capture (^{6}Li, ^{10}B, ^{155,157}Gd) and ideally should have a low effective Z-number, rendering them insensitivity to γ-rays. These constraints substantially limit the number of candidate material compositions [50.267]. Such scintillation detectors are needed in new safety precautions to be employed at airports, seaports, and country borders to defend against terrorist activities and to detect illicit trafficking or inadvertent circulation of nuclear materials or radioactive sources and chemical warfare agents. Development of such monitoring portals is conditional on the availability of sufficiently intense portable neutron sources, which are under development as well.

50.4 Conclusions

In this chapter the broad field of scintillation materials has been introduced, albeit limited to selected single-crystal groups, optical ceramics, and microstructured materials. The variety of materials considered or currently under development is due to the range of demands of the ever-growing number of practical applications and the need to tailor scintillation materials to each of them.

Attention was paid not only to listing the materials and their scintillation characteristics, but also to addressing the issue of specific material defects and their relationship with the manufacturing technology used. Material defects and irregularities often give rise to energy levels in the host forbidden gap. Such energy levels can constitute trapping states, which serve as centers of nonradiative recombination or simply as delaying elements in the process of energy transport to the emission centers. Consequently, both the scintillator efficiency and timing characteristics are degraded. Understanding the underlying physical mechanisms of energy transfer and storage and the role of particular material defects is of crucial importance for bringing material performance close to the intrinsic limits. The necessity to use correlated experiments based on time-resolved spectroscopy, thermoluminescence, and magnetic resonance to understand the nature of defects has been demonstrated in several examples.

Finally, it is worth noting that another area of progress in the realization of advanced detector design concerns the rapid development of the field of semiconductor photodetectors, which are an indispensable part of scintillation detectors.

References

50.1   W.C. Röntgen: Über eine neue Art von Strahlen, Sitz Ber. Phys. Med. Ges. Würzb **9**, 132–141 (1895), in German

50.2   W.C. Röntgen: On a new kind of rays, Science **3**, 227–231 (1896)

50.3   H. Becquerel: Sur les radiations invisibles emises par les corps phosphorescents, Com. Rend. **122**, 501–503 (1896), in French

50.4   M. Curie: Rayons emises par les composes de l'uranium et du thorium, Com. Rend. **126**, 1101–1103 (1898), in French

50.5   P. Curie, M. Curie: Sur une substance nouvelle radio-active, continue dans la pechblende, Com. Rend. **127**, 175–178 (1898), in French

50.6   L. Ozawa, M. Itoh: Cathode ray tube phosphors, Chem. Rev. **103**, 3835–3855 (2003)

50.7   R. Hofstätter: The detection of gamma-rays with thallium-activated sodium iodide crystals, Phys. Rev. **75**, 796–810 (1949)

50.8   W. Van Sciver, R. Hofstätter: Scintillations in thallium-activated CaI_2 and CsI, Phys. Rev. **84**, 1062–1063 (1951)

50.9   M.J. Weber: Inorganic scintillators: today and tomorrow, J. Lumin. **100**, 35–45 (2002)

50.10 M.J. Weber, R.R. Monchamp: Luminescence of $Bi_4Ge_3O_{12}$: spectral and decay properties, J. Appl. Phys. **44**, 5495–5499 (1973)

50.11 C.W.E. van Eijk, J. Andriessen, P. Dorenbos, R. Visser: Ce^{3+} doped inorganic scintillators, Nucl. Instrum. Methods Phys. Res. A **348**, 546–550 (1994)

50.12 C.W.E. van Eijk: Inorganic-scintillator development, Nucl. Instrum. Methods Phys. Res. A **460**, 1–14 (2001)

50.13 G. Blasse, B.C. Grabmaier: *Luminescent Materials* (Springer, Berlin 1994)

50.14 G.F. Knoll: *Radiation Detection and Measurement* (Wiley, New York 2000)

50.15 P.A. Rodnyi: *Physical Processes in Inorganic Scintillators* (CRC, Boca Raton 1997)

50.16 S. Shionoya, W.M. Yen (Eds.): *Phosphor Handbook* (CRC, Boca Raton 1998)

50.17 L.H. Brixner: New x-ray phosphors, Mater. Chem. Phys. **16**, 253–281 (1987)

50.18 M. Ishii, M. Kobayashi: Single crystals for radiation detectors, Prog. Cryst. Growth Charact. Mater. **23**, 245–311 (1992)

50.19 A.J. Wojtowicz: Rare-earth-activated wide bandgap materials for scintillators, Nucl. Instrum. Methods Phys. Res. A **486**, 201–207 (2002)

50.20 C.W.E. van Eijk: Inorganic scintillators in medical imaging, Phys. Med. Biol. **47**, R85–R106 (2002)

50.21 M. Nikl: Scintillation detectors for x-rays, Meas. Sci. Technol. **17**, R37–R54 (2006)

50.22 M. Nikl: Wide band gap scintillation materials: Progress in the technology and material understanding, Phys. Status Solidi (a) **178**, 595–620 (2000)

50.23 D.J. Robbins: On predicting the maximum efficiency of phosphor systems excited by ionizing-radiation, J. Electrochem. Soc. **127**, 2694–2702 (1980)

50.24 A. Lempicki, A.J. Wojtowicz, E. Berman: Fundamental limits of scintillator performance, Nucl. Instrum. Methods Phys. Res. A **333**, 304–311 (1993)

50.25 P.A. Rodnyi, P. Dorenbos, C.W.E. van Eijk: Energy loss in inorganic scintillators, Phys. Status Solidi (b) **187**, 15–29 (1995)

50.26 A.N. Vasil'ev: Polarization approximation for electron cascade in insulators after high-energy excitation, Nucl. Instrum. Methods Phys. Res. B **107**, 165–171 (1996)

50.27 J.A. Shepherd, S.M. Gruner, M.W. Tate, M. Tecotzky: Study of afterglow in x-ray phosphors for use on fast-framing charge-coupled device detectors, Opt. Eng. **36**, 3212–3222 (1997)

50.28 S. Kubota, J. Ruan, M. Itoh, S. Hashimoto, S. Sakuragi: A new type of luminescence mechanism in large band-gap insulators: Proposal for fast scintillation materials, Nucl. Instrum. Methods Phys. Res. A **289**, 253–260 (1990)

50.29 M. Nikl, N. Solovieva, K. Apperson, D.J.S. Birch, A. Voloshinovskii: Scintillators based on aromatic dye molecules doped in a sol-gel glass host, Appl. Phys. Lett. **86**, 101914 (2005)

50.30 S.E. Derenzo, M.J. Weber, M.K. Klintenberg: Temperature dependence of the fast, near-band-edge scintillation from CuI, HgI_2, PbI_2, ZnO:Ga and CdS:In, Nucl. Instrum. Methods Phys. Res. A **486**, 214–219 (2002)

50.31 J. Wilkinson, K.B. Ucer, R.T. Williams: Picosecond excitonic luminescence in ZnO and other wide-gap semiconductors, Radiat. Meas. **38**, 501–505 (2004)

50.32 D. Ehrentraut, H. Sato, Y. Kagamitani, A. Yoshikawa, T. Fukuda, J. Pejchal, K. Polak, M. Nikl, H. Odaka, K. Hatanaka, H. Fukumura: Fabrication and luminescence properties of single-crystalline, homoepitaxial zinc oxide films doped with tri- and tetravalent cations prepared by liquid phase epitaxy, J. Mater. Chem. **16**, 3369–3374 (2006)

50.33 M. Nikl: Energy transfer phenomena in the luminescence of wide band-gap scintillators, Phys. Status Solidi (a) **202**, 201–206 (2005)

50.34 M. Nikl, K. Nitsch, K. Polak, E. Mihóková, I. Dafinei, E. Auffray, P. Lecoq, P. Reiche, R. Uecker: Slow components in the photoluminescence and scintillation decays of $PbWO_4$ single crystals, Phys. Status Solidi (b) **195**, 311–323 (1996)

50.35 P. Dorenbos, J.P.M. de Haas, C.W.E. van Eijk: Non-proportionality in the scintillation response and the energy resolution obtainable with scintillation crystals, IEEE Trans. Nucl. Sci. **42**, 2190–2202 (1995)

50.36 C.D. Brandle: Czochralski growth of oxides, J. Cryst. Growth **264**, 593–604 (2004)

50.37 T. Fukuda, R. Rudolph, S. Uda: *Fiber Crystal Growth from the Melt* (Springer, Berlin 2004)

50.38 A. Yoshikawa, M. Nikl, G. Boulon, T. Fukuda: Challenge and study for developing of novel single-crystalline optical materials using micro-pulling down method, Opt. Mater. **30**, 6–10 (2007)

50.39 M. Nikl, J.A. Mareš, J. Chval, E. Mihóková, N. Solovieva, M. Martini, A. Vedda, K. Blažek, P. Maly, K. Nejezchleb, P. Fabeni, G.P. Pazzi, V. Babin, K. Kalder, A. Krasnikov, S. Zazubovich, C. D'Ambrosio: An effect of Zr^{4+} co-doping of YAP:Ce scintillator, Nucl. Instrum. Methods Phys. Res. A **486**, 250–253 (2002)

50.40 C. Brecher, A. Lempicki, S.R. Miller, J. Glodo, E.E. Ovechkina, V. Gaysinskiy, V.V. Nagarkar, R.H. Bartram: Suppression of afterglow in CsI:Tl by codoping with Eu^{2+} – I: Experimental, Nucl. Instrum. Methods Phys. Res. A **558**, 450–457 (2006)

50.41 J. Hlinka, E. Mihóková, M. Nikl, K. Polak, J. Rosa: Energy transfer between A_T and A_X minima in KBr:Tl, Phys. Status Solidi (b) **175**, 523–540 (1993)

50.42 S.W.S. McKeever: *Thermoluminescence of Solids* (Cambridge University Press, Cambridge 1985)

50.43 R. Chen, S.W.S. McKeever: *Theory of Thermoluminescence and Related Phenomena* (World Scientific, Singapore 1997)

50.44 J.A. Weil, J.R. Bolton, J.E. Wertz: *Electron Paramagnetic Resonance: Elementary Theory and Practical Applications* (Wiley, New York 1994)

50.45 J.-M. Spaeth, J.R. Niklas, R.H. Bartram: *Structural Analysis of Point Defects in Solids: An Introduction to Multiple Magnetic Resonance Spectroscopy* (Springer, Berlin 1992)

50.46 V.V. Laguta, M. Martini, A. Vedda, M. Nikl, E. Mihóková, P. Boháček, J. Rosa, A. Hofstaetter, B.K. Meyer, Y. Usuki: Photoinduced Pb^+ center in $PbWO_4$: Electron spin resonance and thermally stimulated luminescence study, Phys. Rev. B **64**, 165102 (2001)

50.47 W. van Loo: Luminescence of lead molybdate and lead tungstate. I. Experimental, Phys. Status Solidi (a) **27**, 565–574 (1979)

50.48 W. van Loo: Luminescence of lead molybdate and lead tungstate. II. Discussion, Phys. Status Solidi (a) **28**, 227–235 (1979)

50.49 J.A. Groening, G. Blasse: Some new observations on the luminescence of $PbMoO_4$ and $PbWO_4$, J. Solid State Chem. **32**, 9–20 (1980)

50.50 R. Öder, A. Scharmann, D. Schwabe, B. Vitt: Growth and properties of $PbWO_4$ and $Pb(WO_4)_{1-x}(MoO_4)_x$ mixed crystals, J. Cryst. Growth **43**, 537–540 (1978)

50.51 V.G. Baryshevsky, M.V. Korzhik, V.I. Moroz, V.B. Pavlenko, A.S. Lobko, A.A. Fyodorov, V.A. Kachanov, V.L. Solovjanov, B.I. Zadneprovsky, V.A. Nefyodov, P.V. Nefyodov, B.A. Dorogovin, L.L. Nagornaja: Single crystals of tungsten compounds as promising materials for the total absorption detectors of the e.m. calorimeters, Nucl. Instrum. Methods Phys. Res. A **322**, 231–234 (1992)

50.52 M. Kobayashi, M. Ishii, Y. Usuki, H. Yahagi: Scintillation characteristics of $PbWO_4$ single crystals at room temperature, Nucl. Instrum. Methods Phys. Res. A **333**, 429–433 (1993)

50.53 K. Nitsch, M. Nikl, S. Ganschow, P. Reiche, R. Uecker: Growth of lead tungstate single crystal scintillators, J. Cryst. Growth **165**, 163–165 (1996)

50.54 K. Tanji, M. Ishii, Y. Usuki, M. Kobayashi, K. Hara, H. Takano, N. Senguttuvan: Crystal growth of PbWO$_4$ by the vertical Bridgman method: Effect of crucible thickness and melt composition, J. Cryst. Growth **204**, 505–511 (1999)

50.55 A.A. Annenkov, M.V. Korzhik, P. Lecoq: Lead tungstate scintillation material, Nucl. Instrum. Methods Phys. Res. A **490**, 30–50 (2002)

50.56 R. Mao, J. Chen, D. Shen, Z. Yin: Growth and uniformity improvement of PbWO$_4$ crystal with yttrium doping, J. Cryst. Growth **265**, 518–524 (2004)

50.57 A.N. Annenkov, E. Auffray, R. Chipaux, G.Y. Drobychev, A.A. Fedorov, M. Géléoc, N.A. Golubev, M.V. Korzhik, P. Lecoq, A.A. Lednev, A.B. Ligun, O.V. Missevitch, V.B. Pavlenko, J.-P. Peigneux, A.V. Singovski: Systematic study of the short-term instability of PbWO$_4$ scintillator parameters under irradiation, Radiat. Meas. **29**, 27–38 (1998)

50.58 P. Boháček, M. Nikl, J. Novak, Z. Malkova, B. Trunda, J. Rysavy, S. Baccaro, A. Cecilia, I. Dafinei, M. Diemoz, K. Jurek: Congruent composition of PbWO$_4$ single crystals, J. Electr. Eng. **50**(2/s), 38–40 (1999)

50.59 N. Lei, B. Han, X.Q. Feng, G. Hu, Y. Zhang, Z. Yin: La^{3+} distribution and annealing effects in La:PbWO$_4$ crystal, Phys. Status Solidi (a) **170**, 37–45 (1998)

50.60 A. Krasnikov, M. Nikl, S. Zazubovich: Localized excitons and defects in PbWO$_4$ single crystals: A luminescence and photo-thermally stimulated disintegration study, Phys. Status Solidi (b) **243**, 1727–1743 (2006)

50.61 V. Murk, M. Nikl, E. Mihóková, K. Nitsch: A study of electron excitations in CaWO$_4$ and PbWO$_4$ single crystals, J. Phys. Cond. Matter **9**, 249–256 (1997)

50.62 M. Kobayashi, M. Ishii, K. Harada, Y. Usuki, H. Okuno, M. Shimizu, T. Yazawa: Scintillation and phosphorescence of PbWO$_4$ crystals, Nucl. Instrum. Methods Phys. Res. A **373**, 333–346 (1996)

50.63 M. Böhm, A.E. Borsevich, G.Y. Drobychev, G.Y. Drobychev, A. Hofstaetter, O.V. Kondratiev, M.V. Korzhik, M. Luh, B.K. Meyer, J.P. Peigneux, A. Scharmann: Influence of Mo impurity on the spectroscopic and scintillation properties of PbWO$_4$ crystals, Phys. Status Solidi (a) **167**, 243–252 (1998)

50.64 V. Babin, P. Boháček, A. Krasnikov, M. Nikl, A. Stolovits, S. Zazubovich: Origin of green luminescence in PbWO$_4$ crystals, J. Lumin. **124**, 113–119 (2007)

50.65 G.P. Pazzi, P. Fabeni, M. Nikl, P. Bohacek, E. Mihokova, A. Vedda, M. Martini, M. Kobayashi, Y. Usuki: Delayed recombination luminescence in lead tungstate (PWO) scintillating crystals, J. Luminesc. **102/103**, 791–796 (2003)

50.66 M. Martini, G. Spinolo, A. Vedda, M. Nikl, K. Nitsch, V. Hamplova, P. Fabeni, G.P. Pazzi, I. Dafinei, P. Lecoq: Trap levels in PbWO$_4$ crystals: Correlation with luminescence decay kinetics, Chem. Phys. Lett. **260**, 418–422 (1996)

50.67 A.N. Annenkov, E. Auffray, A.E. Borisevich, G.Y. Drobyshev, A.A. Fedorov, O.V. Kondratiev, M.V. Korzhik, P. Lecoq: Slow components and afterglow in PWO crystal scintillations, Nucl. Instrum. Methods Phys. Res. A **403**, 302–312 (1998)

50.68 E. Mihóková, M. Nikl, P. Boháček, V. Babin, A. Krasnikov, A. Stolovich, S. Zazubovich, A. Vedda, M. Martini, T. Grabowski: Decay kinetics of the green emission in PbWO$_4$:Mo, J. Lumin. **102–103**, 618–622 (2003)

50.69 A. Hofstaetter, A. Scharmann, D. Schwabe, B. Vitt: EPR of radiation induced MoO$_4^{3-}$-centres in lead tungstate, Z. Phys. B **30**, 305–311 (1978)

50.70 E. Auffray, I. Dafinei, P. Lecoq, M. Schneegans: Local trap centres in PbWO$_4$ crystals, Radiat. Eff. Defects Solids **135**, 841–845 (1995)

50.71 M. Springis, V. Tale, I. Tale: Nature of the blue luminescence bands in PbWO$_4$, J. Lumin. **72–74**, 784–785 (1997)

50.72 V.V. Laguta, J. Rosa, M.I. Zaritskii, M. Nikl, Y. Usuki: Polaronic (WO$_4$)$^{3-}$ centres in PbWO$_4$ single crystals, J. Phys. Cond. Matter **10**, 7293–7302 (1998)

50.73 M. Martini, G. Spinolo, A. Vedda, M. Nikl, Y. Usuki: Shallow traps in PbWO$_4$ studied by wavelength-resolved thermally stimulated luminescence, Phys. Rev. B **60**, 4653–4658 (1999)

50.74 V.V. Laguta, M. Martini, A. Vedda, E. Rosetta, M. Nikl, E. Mihokova, P. Boháček, J. Rosa, A. Hofstatter, B.K. Meyer, Y. Usuki: Photoinduced oxygen-vacancy related centers in PbWO$_4$: Electron spin resonance and thermally stimulated luminescence study, Radiat. Eff. Defects Solids **157**, 1025–1031 (2002)

50.75 A. Hofstaetter, M.V. Korzhik, V.V. Laguta, B.K. Meyer, V. Nagirnyi, R. Novotny: The role of defect states in the creation of intrinsic (WO$_4$)$^{3-}$ centers in PbWO$_4$ by sub-bandgap excitation, Radiat. Meas. **33**, 533–536 (2001)

50.76 V.V. Laguta, M. Martini, A. Vedda, E. Rosetta, M. Nikl, E. Mihóková, J. Rosa, Y. Usuki: Electron traps related to oxygen vacancies in PbWO$_4$, Phys. Rev. B **67**, 205102 (2003)

50.77 M. Nikl, P. Boháček, K. Nitsch, E. Mihokova, M. Martini, A. Vedda, S. Crocci, G.P. Pazzi, P. Fabeni, S. Baccaro, B. Borgia, I. Dafinei, M. Diemoz, G. Organtini, E. Auffray, P. Lecoq, M. Kobayashi, M. Ishii, Y. Usuki: Decay kinetics and thermoluminescence PbWO$_4$:La^{3+}, Appl. Phys. Lett. **71**, 3755–3757 (1997)

50.78 S. Baccaro, P. Boháček, B. Borgia, A. Cecilia, I. Dafinei, M. Diemoz, M. Ishii, O. Jarolimek, M. Kobayashi, M. Martini, M. Montecchi, M. Nikl, K. Nitsch, Y. Usuki, A. Vedda: Influence of La^{3+}-doping on radiation hardness and

50.78 ... thermolumines-cence characteristics of $PbWO_4$, Phys. Status Solidi (a) **160**, R5–R6 (1997)

50.79 S. Baccaro, P. Boháček, A. Cecilia, S. Croci, I. Dafinei, M. Diemoz, P. Fabeni, M. Ishii, O. Jarolimek, M. Kobayashi, M. Martini, M. Montecchi, M. Nikl, G. Organtini, G.P. Pazzi, J. Rosa, Y. Usuki, A. Vedda: The influence of defect states on scintillation characteristics of $PbWO_4$, Radiat. Eff. Defects Solids **150**, 15–19 (1999)

50.80 V.V. Laguta, M. Martini, F. Meinardi, A. Vedda, A. Hofstaetter, B.K. Mayer, M. Nikl, E. Mihóková, J. Rosa, Y. Usuki: Photoinduced $(WO_4)^{3-}$–La^{3+} center in $PbWO_4$: Electron spin resonance and thermally stimulated luminescence study, Phys. Rev. B **62**, 10109–10114 (2000)

50.81 V.V. Laguta, A. Vedda, D. Di Martino, M. Martini, M. Nikl, E. Mihóková, J. Rosa, Y. Usuki: Electron capture in $PbWO_4$:Mo and $PbWO_4$:Mo,La single crystals: ESR and TSL study, Phys. Rev. B **71**, 235108 (2005)

50.82 A. Hofstaetter, R. Öder, A. Scharman, D. Schwabe, B. Vitt: Paramagnetic resonance and thermoluminescence of the $PbWO_4/PbMoO_4$ mixed crystal system, Phys. Status Solidi (b) **89**, 375–380 (1978)

50.83 M. Nikl, P. Boháček, E. Mihóková, N. Solovieva, A. Vedda, M. Martini, G.P. Pazzi, P. Fabeni, M. Kobayashi: Complete characterization of doubly doped $PbWO_4$:Mo,Y scintillators, J. Appl. Phys. **91**, 2791–2797 (2002)

50.84 M. Kobayashi, Y. Usuki, M. Ishii, T. Yazawa, K. Hara, M. Tanaka, M. Nikl, K. Nitsch: Improvement in transmittance and decay time of $PbWO_4$ scintillating crystals by La-doping, Nucl. Instrum. Methods Phys. Res. A **399**, 261–268 (1997)

50.85 M. Kobayashi, Y. Usuki, M. Ishii, T. Yazawa, K. Hara, M. Tanaka, M. Nikl, S. Baccaro, A. Cecilia, M. Diemoz, I. Dafinei: Improvement of radiation hardness of $PbWO_4$ scintillating crystals by La-doping, Nucl. Instrum. Methods Phys. Res. A **404**, 149–156 (1998)

50.86 Y.B. Abraham, N.A.W. Holzwarth, R.T. Williams, G.E. Matthews, A.R. Tackett: Electronic structure of oxygen-related defects in $PbWO_4$ and $CaMoO_4$ crystals, Phys. Rev. B **64**, 245109 (2001)

50.87 M. Kobayashi, S. Sugimoto, Y. Yoshimura, Y. Usuki, M. Ishii, N. Senguttuvan, K. Tanji, M. Nikl: A new heavy and radiation-hard Cherenkov radiator based on $PbWO_4$, Nucl. Instrum. Methods Phys. Res. A **459**, 482–493 (2001)

50.88 Y.L. Huang, W.L. Zhu, X.Q. Feng, Z.Y. Man: The effects of La^{3+} doping on luminescence properties of $PbWO_4$ single crystal, J. Solid State Chem. **172**, 188–193 (2003)

50.89 W. Li, X.Q. Feng, Y. Huang: Characteristics of the optical absorption edge and defect structures of La^{3+}-doped $PbWO_4$ crystals, J. Phys. Cond. Matter **16**, 1325–1333 (2004)

50.90 T. Chen, T.Y. Liu, Q.R. Zhang, F.F. Li, D.S. Tian, X.Y. Zhang: First principles study of the La^{3+} doping $PbWO_4$ crystal for different doping concentrations, Phys. Lett. A **363**, 477–481 (2007)

50.91 M. Nikl, K. Nitsch, S. Baccaro, A. Cecilia, M. Montecchi, B. Borgia, I. Dafinei, M. Diemoz, M. Martini, E. Rosetta, G. Spinolo, A. Vedda, M. Kobayashi, M. Ishii, Y. Usuki, O. Jarolimek, R. Uecker: Radiation induced formation of color centres in $PbWO_4$ single crystals, J. Appl. Phys. **82**, 5758–5762 (1997)

50.92 T. Liu, Q. Zhang, X. Mi, X.Q. Feng: A new absorption band and the decomposition of the 350 nm absorption band of $PbWO_4$, Phys. Status Solidi (a) **184**, 341–348 (2001)

50.93 A. Annenkov, E. Auffray, M. Korzhik, P. Lecoq, J.P. Peigneux: On the origin of the transmission damage in lead tungstate crystals under irradiation, Phys. Status Solidi (a) **170**, 47–62 (1998)

50.94 Q. Lin, X.Q. Feng, Z. Man, Y. Zhang, Z. Yin, Q. Zhang: Origin of the radiation-induced 420 nm color center absorption band in $PbWO_4$ crystals, Solid State Commun. **118**, 221–223 (2001)

50.95 Q. Deng, Z. Yin, R.Y. Zhu: Radiation-induced color centers in La-doped $PbWO_4$ crystals, Nucl. Instrum. Methods Phys. Res. A **438**, 415–420 (1999)

50.96 S. Burachas, Y. Saveliev, M. Ippolitov, V. Manko, V. Lomonosov, A. Vasiliev, A. Apanasenko, A. Vasiliev, A. Uzunian, G. Tamulaitis: Physical origin of coloration and radiation hardness of lead tungstate scintillation crystals, J. Cryst. Growth **293**, 62–67 (2006)

50.97 M. Kobayashi, Y. Usuki, M. Ishii, M. Itoh, M. Nikl: Further study on different dopings into $PbWO_4$ single crystals to increase the scintillation light yield, Nucl. Instrum. Methods Phys. Res. A **540**, 381–394 (2005)

50.98 M. Nikl, P. Boháček, A. Vedda, M. Martini, G.P. Pazzi, P. Fabeni, M. Kobayashi: Efficient medium-speed $PbWO_4$:Mo,Y scintillator, Phys. Status Solidi (a) **182**, R3–R5 (2000)

50.99 A. Annenkov, A. Borisevitch, A. Hofstaetter, M. Korzhik, V. Ligun, P. Lecoq, O. Missevitch, R. Novotny, J.P. Peigneux: Improved light yield of lead tungstate scintillators, Nucl. Instrum. Methods Phys. Res. A **450**, 71–74 (2000)

50.100 M. Nikl, P. Boháček, E. Mihóková, N. Solovieva, A. Vedda, M. Martini, G.P. Pazzi, P. Fabeni, M. Kobayashi: Complete characterization of doubly doped $PbWO_4$:Mo,Y scintillators, J. Appl. Phys. **91**, 2791–2797 (2002)

50.101 J.A. Mareš, A. Beitlerova, P. Boháček, M. Nikl, N. Solovieva, C. D'Ambrosio: Influence of non-stoichiometry and doping on scintillating response of $PbWO_4$ crystals, Phys. Status Solidi (c) **2**, 73–76 (2005)

50.102 X. Liu, G. Hu, X. Feng, Y. Huang, Y. Zhang: Influence of PbF_2 doping on scintillation properties of $PbWO_4$

single crystals, Phys. Status Solidi (a) **190**, R1–R3 (2002)

50.103 M. Nikl, P. Boháček, E. Mihóková, N. Solovieva, A. Vedda, M. Martini, G.P. Pazzi, P. Fabeni, M. Ishii: Enhanced efficiency of PbWO$_4$:Mo,Nb scintillator, J. Appl. Phys. **91**, 5041–5044 (2002)

50.104 M.J. Weber: Optical spectra of Ce^{3+} and Ce^{3+}-sensitized fluorescence in YAlO$_3$, J. Appl. Phys. **44**, 3205–3208 (1973)

50.105 T. Takeda, T. Miyata, F. Muramatsu, T. Tomiki: Fast decay UV phosphor-YAlO$_3$:Ce, J. Electrochem. Soc. **127**, 438–444 (1980)

50.106 E. Autrata, P. Schauer, J. Kvapil, J. Kvapil: A single crystal of YAlO$_3$:Ce^{3+} as a fast scintillator in SEM, Scanning **5**, 91–96 (1983)

50.107 M. Nikl, A. Vedda, V.V. Laguta: Energy transfer and storage processes in scintillators: The role and nature of defects, Radiat. Meas. **42**, 509–514 (2007)

50.108 J. Kvapil, J. Kvapil, J. Kubelka, R. Autrata: The role of iron ions in YAG and YAP, Cryst. Res. Technol. **18**, 127–131 (1983)

50.109 G.-S. Li, X.-B. Guo, J. Lu, Z.-Z. Shi, J.-H. Wu, Y. Chen, J.-F. Chen: Application of several new procedures to improve the quality of Czochralski grown Nd^{3+}:YAlO$_3$ crystals, J. Cryst. Growth **118**, 371–376 (1992)

50.110 D.I. Savytskii, L.O. Vasylechko, A.O. Matkovskii, I.M. Solskii, A. Suchocki, D.Y. Sugak, F. Wallrafen: Growth and properties of YAlO$_3$:Nd single crystals, J. Cryst. Growth **209**, 874–882 (2000)

50.111 G.-S. Li, Z.-Z. Shi, J.-H. Wu, Y. Chen, J.-F. Chen, H. Yang: Growth of large size yttrium aluminum perovskite (YAP) laser crystals without light scattering centers, J. Cryst. Growth **119**, 363–367 (1992)

50.112 J. Kvapil, B. Manek, B. Perner, J. Kvapil, R. Becker, G. Ringel: The role of argon in yttrium aluminates, Cryst. Res. Technol. **23**, 549–554 (1988)

50.113 A.G. Petrosyan, G.O. Shirinyan, C. Pédrini, C. Dujardin, K.L. Ovanesyan, R.G. Manucharyan, T.I. Butaeva, M.V. Derzyan: Bridgman growth and characterization of LuAlO$_3$–Ce^{3+} scintillator crystals, Cryst. Res. Technol. **33**, 241–248 (1998)

50.114 J.A. Mareš, N. Cechova, M. Nikl, J. Kvapil, R. Kratky, J. Pospisil: Cerium-doped RE^{3+}AlO$_3$ perovskite scintillators: spectroscopy and radiation induced defects, J. Alloy. Comp. **275–277**, 200–204 (1998)

50.115 A.G. Petrosyan, G.O. Shyrinyan, K.L. Ovanesyan, C. Pédrini, C. Dujardin: Bridgman single crystal growth of Ce-doped (Lu$_{1-x}$Y$_x$)AlO$_3$, J. Cryst. Growth **198–199**, 492–496 (1999)

50.116 J. Trummer, E. Auffray, P. Lecoq, A.G. Petrosyan, P. Sempere-Roldan: Comparison of LuAP and LuYAP crystal properties from statistically significant batches produced with two different growth methods, Nucl. Instrum. Methods Phys. Res. A **551**, 339–351 (2005)

50.117 T. Tomiki, H. Ishikawa, T. Tashiro, M. Katsuren, A. Yonesu, T. Hotta, T. Yabiku, M. Akamine, T. Futemma, T. Nakaoka, I. Miyazato: Ce^{3+} centres in YAlO$_3$ (YAP) single crystals, J. Phys. Soc. Jpn. **64**, 4442–4449 (1995)

50.118 V.G. Baryshevsky, M.V. Korzhik, V.I. Moroz, V.B. Pavlenko, A.A. Fyodorov, S.A. Smirnova, O.A. Egorycheva, V.A. Kachanov: YAlO$_3$: Ce-fast-acting scintillators for detection of ionizing radiation, Nucl. Instrum. Methods Phys. Res. B **58**, 291–293 (1991)

50.119 S.I. Ziegler, J.G. Rogers, V. Selivanov, I. Sinitzin: Characteristics of the new YAlO$_3$:Ce compared with BGO and GSO, IEEE Trans. Nucl. Sci. **40**, 194–197 (1993)

50.120 J.A. Mareš, A. Beitlerova, M. Nikl, N. Solovieva, C. D'Ambrosio, K. Blažek, P. Maly, K. Nejezchleb, F. de Notaristefani: Scintillation response of Ce-doped or intrinsic scintillating crystals in the range up to 1 MeV, Radiat. Meas. **38**, 353–357 (2004)

50.121 C. Dujardin, C. Pédrini, J.C. Gacon, A.G. Petrosyan, A.N. Belsky, A.N. Vasil'ev: Luminescence properties and scintillation mechanisms of cerium- and praseodymium-doped lutetium orthoaluminate, J. Phys. Cond. Matter **9**, 5229–5243 (1997)

50.122 E.G. Gumanskaya, M.V. Korzhik, S.A. Smirnova, V.B. Pavlenko, A.A. Fedorov: Interconfiguration luminescence of Pr^{3+} ions in Y$_3$Al$_5$O$_{12}$ and YAlO$_3$ single crystals, Opt. Spectrosc. **72**, 155–159 (1992), in Russian

50.123 W. Drozdowski, A.J. Wojtowicz, D. Wisniewski, T. Lukasiewicz, J. Kisielewski: Scintillation properties of Pr-activated LuAlO$_3$, Opt. Mater. **28**, 102–105 (2006)

50.124 M. Zhuravleva, A. Novoselov, A. Yoshikawa, J. Pejchal, M. Nikl, T. Fukuda: Crystal growth and scintillation properties of Pr-doped YAlO$_3$, Opt. Mater. **30**, 171–173 (2007)

50.125 L. van Pieterson, M. Heeroma, E. de Heer, A. Meijerink: Charge transfer luminescence of Yb^{3+}, J. Lumin. **91**, 177–193 (2000)

50.126 M. Nikl, A. Yoshikawa, T. Fukuda: Charge transfer luminescence in Yb^{3+}-containing compounds, Opt. Mater. **26**, 545–549 (2004)

50.127 M. Nikl, N. Solovieva, J. Pejchal, J.B. Shim, A. Yoshikawa, T. Fukuda, A. Vedda, M. Martini, D.H. Yoon: Very fast Yb$_x$Y$_{1-x}$AlO$_3$ single crystal scintillators, Appl. Phys. Lett. **84**, 882–884 (2004)

50.128 J.B. Shim, A. Yoshikawa, T. Fukuda, J. Pejchal, M. Nikl, N. Sarukura, D.H. Yoon: Growth and charge transfer luminescence of Yb^{3+}-doped YAlO$_3$ single crystals, J. Appl. Phys. **95**, 3063–3068 (2004)

50.129 H.R. Asatryan, J. Rosa, J.A. Mareš: EPR studies of Er^{3+}, Nd^{3+} and Ce^{3+} in YAlO$_3$ single crystals, Solid State Commun. **104**, 5–9 (1997)

50.130 C. Dujardin, C. Pedrini, W. Blanc, J.C. Gâcon, J.C. van't Spijker, O.W.V. Frijns, C.W.E. van Eijk, P. Dorenbos, R. Chen, A. Fremout, F. Tallouf, S. Tavernier, P. Bruyndonckx, A.G. Petrosyan: Optical and scintillation properties of large $LuAlO_3:Ce^{3+}$ crystals, J. Phys. Condens. Matter **10**, 3061–3073 (1998)

50.131 A.J. Wojtowitz, J. Glodo, A. Lempicki, C. Brecher: Recombination and scintillation processes in $YAlO_3$:Ce, J. Phys. Cond. Matter **10**, 8401–8415 (1998)

50.132 A. Vedda, M. Martini, F. Meinardi, J.A. Mareš, E. Mihóková, J. Chval, M. Dusek, M. Nikl: Tunnelling process in thermally stimulated luminescence of mixed $Lu_xY_{1-x}AlO_3$:Ce crystals, Phys. Rev. B **61**, 8081–8086 (2000)

50.133 J. Kvapil, Jos. Kvapil, B. Perner, B. Manek, K. Blažek, Z. Hendrich: Nonstoichiometric defects in YAG and YAP, Cryst. Res. Technol. **20**, 473–478 (1985)

50.134 D. Sugak, A. Matkovskii, D. Savitskii, A. Durygin, A. Suchocki, Y. Zhydachevskii, I. Solskii, I. Stefaniuk, F. Wallrafen: Growth and induced color centers in $YAlO_3$–Nd single crystals, Phys. Status Solidi (a) **184**, 239–250 (2001)

50.135 J.A. Mareš, M. Nikl, E. Mihóková, N. Solovieva, K. Blažek, K. Nejezchleb, P. Maly, J. Pejchal, V. Mucka, M. Pospisil, A. Vedda, M. Martini, S. Baccaro: Radiation induced absorption color centers and damage in $YAlO_3$:Ce and $YAlO_3$:Ce,Zr scintillators, Radiat. Eff. Defects Solids **157**, 677–681 (2002)

50.136 O.F. Schirmer, K.W. Blazey, W. Berlinger, R. Diehl: ESR and optical absorption of bound-small polarons in $YAlO_3$, Phys. Rev. B **11**, 4201–4218 (1975)

50.137 V.Y. Zorenko, A.S. Voloshinovskii, I.V. Konstankevych, G.B. Striganyuk, V.I. Gorbenko: Exciton luminescence of $YAlO_3$ single crystals and single-crystal films, Opt. Spectrosc. **98**, 555–558 (2005)

50.138 V.Y. Zorenko, A.S. Voloshinovskii, I.V. Konstankevych: Luminescence of F^+ and F centers in $YAlO_3$, Opt. Spectrosc. **96**, 532–537 (2004)

50.139 V.V. Laguta, M. Nikl, A. Vedda, E. Mihokova, J. Rosa, K. Blazek: The hole and electron traps in the $YAlO_3$ single crystal scintillator, Phys. Rev. B **80**, 045114 (2009)

50.140 E. Mihóková, M. Nikl, J.A. Mareš, A. Beitlerová, A. Vedda, K. Nejezchleb, K. Blažek, C. D'Ambrosio: Luminescence and scintillation properties of YAG:Ce single crystal and optical ceramics, J. Lumin. **126**, 77–80 (2006)

50.141 V.V. Laguta, A.M. Slipenyuk, J. Rosa, M. Nikl, A. Vedda, K. Nejezchleb, K. Blažek: Electron spin resonance study of Mo^{3+} centers in $YAlO_3$, Radiat. Meas. **38**, 735–738 (2004)

50.142 S.A. Basun, T. Danger, A.A. Kaplyanskii, D.S. McClure, K. Petermann, W.C. Wong: Optical and photoelectrical studies of charge-transfer processes in $YAlO_3$:Ti crystals, Phys. Rev. B **54**, 6141–6149 (1996)

50.143 K. Blažek, A. Krasnikov, K. Nejezchleb, M. Nikl, T. Savikhina, S. Zazubovich: Luminescence and defect creation in Ce^{3+}-doped $YAlO_3$ and $Lu_{0.3}Y_{0.7}AlO_3$ crystals, Phys. Status Solidi (b) **242**, 1315–1323 (2005)

50.144 M. Nikl, E. Mihokova, V. Laguta, J. Pejchal, S. Baccaro, A. Vedda: Radiation damage processes in complex-oxide scintillators, Eur. Symp. Opt. Optoelectron. Damage VUV, EUV, X-ray Opt., ed. by L. Juha, R.H. Sobierajski, H. Wabnitz (2007)

50.145 C.R. Stanek, M.R. Levy, K.J. McClellan, B.P. Uberuaga, R.W. Grimes: Defect structure of ZrO_2-doped rare earth perovskite scintillators, Phys. Status Solidi (b) **242**, R113–R115 (2005)

50.146 C.R. Stanek, K.J. McClellan, M.R. Levy, R.W. Grimes: Defect behavior in rare earth $REAlO_3$ scintillators, J. Appl. Phys. **99**, 113518 (2006)

50.147 M. Kokta: Growth of oxide laser crystals, Opt. Mater. **30**, 1–5 (2007)

50.148 R. Autrata, P. Schauer, Jos. Kvapil, J. Kvapil: A single crystal of YAG – New fast scintillator in SEM, J. Phys. E **11**, 707–708 (1978)

50.149 M. Moszynski, T. Ludziewski, D. Wolski, W. Klamra, L.O. Norlin: Properties of the YAG:Ce scintillator, Nucl. Instrum. Methods Phys. Res. A **345**, 461–467 (1994)

50.150 A. Lempicki, M.H. Randles, D. Wisniewski, M. Balcerzyk, C. Brecher, A.J. Wojtowitz: $LuAlO_3$:Ce and other aluminate scintillators, IEEE Trans. Nucl. Sci. **42**, 280–284 (1995)

50.151 M. Nikl, E. Mihóková, J.A. Mareš, A. Vedda, M. Martini, K. Nejezchleb, K. Blažek: Traps and timing characteristics of $LuAG:Ce^{3+}$ scintillator, Phys. Status Solidi (b) **181**, R10–R12 (2000)

50.152 M.J. Weber: Nonradiative decay from 5d states of rare earths in crystals, Solid State Commun. **12**, 741–744 (1973)

50.153 Y. Kuwano, K. Suda, N. Ishizawa, T. Yamada: Crystal growth and properties of $(Lu,Y)_3Al_5O_{12}$, J. Cryst. Growth **260**, 159–165 (2004)

50.154 J. Kvapil, Jos. Kvapil, B. Manek, B. Perner: Czochralski growth of YAG:Ce in a reducing protective atmosphere, J. Cryst. Growth **52**, 542–545 (1981)

50.155 D. Mateika, E. Volkel, J. Haisma: Lattice-constant-adaptable crystallographics. II. Czochralski growth from multicomponent melts of homogeneous mixed-garnet crystals, J. Cryst. Growth **102**, 994–1013 (1990)

50.156 D.S. Hamilton, S.K. Gayen, G.J. Pogatshnik, R.D. Ghen, W.J. Miniscalco: Optical absorption and photoionization measurements from the excited states of $Ce^{3+}:Y_3Al_5O_{12}$, Phys. Rev. B **39**, 8807–8815 (1989)

50.157 M. Nikl, V.V. Laguta, A. Vedda: Energy transfer and charge carrier capture processes in wide-

50.157 band-gap scintillators, Phys. Status Solidi (a) **204**, 683–689 (2007)

50.158 M. Nikl, J.A. Mareš, N. Solovieva, J. Hybler, A. Voloshinovskii, K. Nejezchleb, K. Blažek: Energy transfer to the Ce^{3+} centers in $Lu_3Al_5O_{12}$:Ce scintillator, Phys. Status Solidi (a) **201**, R41–R44 (2004)

50.159 W. Drozdowski, T. Lukasiewicz, A.J. Wojtowicz, D. Wisniewski, J. Kisielewski: Thermoluminescence and scintillation of praseodymium-activated $Y_3Al_5O_{12}$ and $LuAlO_3$ crystals, J. Cryst. Growth **275**, e709–e714 (2005)

50.160 M. Nikl, H. Ogino, A. Krasnikov, A. Beitlerova, A. Yoshikawa, T. Fukuda: Photo- and radioluminescence of Pr-doped $Lu_3Al_5O_{12}$ single crystal, Phys. Status Solidi (a) **202**, R4–R6 (2005)

50.161 H. Ogino, A. Yoshikawa, M. Nikl, K. Kamada, T. Fukuda: Scintillation characteristics of Pr-doped $Lu_3Al_5O_{12}$ single crystals, J. Cryst. Growth **292**, 239–242 (2006)

50.162 S.R. Rotman, C. Warde, H.L. Tuller, J. Haggerty: Defect property correlations in garnet crystals. V. Energy transfer in luminescent yttrium aluminum-yttrium iron garnet solid solutions, J. Appl. Phys. **66**, 3207–3210 (1989)

50.163 C.Y. Chen, G.J. Pogatshnik, Y. Chen, M.R. Kokta: Optical and electron paramagnetic resonance studies of Fe impurities in yttrium aluminum garnet crystals, Phys. Rev. B **38**, 8555–8561 (1988)

50.164 I. Kamenskikh, C. Dujardin, N. Garnier, N. Guerassimova, G. Ledoux, V. Mikhailin, C. Pedrini, A. Petrosyan, A. Vasil'ev: Temperature dependence of the charge transfer and f-f luminescence of Yb^{3+} in garnets and YAP, J. Phys. Condens. Matter **17**, 5587–5594 (2005)

50.165 V.V. Laguta, A.M. Slipenyuk, M.D. Glinchuk, M. Nikl, J. Rosa, A. Vedda, K. Nejezchleb: Paramagnetic impurity defects in LuAG and LuAG:Sc single crystals, Opt. Mater. **30**, 79–81 (2007)

50.166 H.R. Lewis: Paramagnetic resonance of Ce^{3+} in YAG, J. Appl. Phys. **37**, 739–741 (1966)

50.167 V.V. Laguta, A.M. Slipenyuk, M.D. Glinchuk, I.P. Bykov, Y. Zorenko, M. Nikl, J. Rosa: Paramagnetic impurity defects in LuAG:Ce thick film scintillators, Radiat. Meas. **42**, 835–838 (2007)

50.168 C.R. Stanek, K.J. McClellan, M.R. Levy, C. Milanese, R.W. Grimes: The effect of intrinsic defects on $RE_3Al_5O_{12}$ garnet scintillator performance, Nucl. Instrum. Methods Phys. Res. A **579**, 27–30 (2007)

50.169 M. Springis, A. Pujats, J. Valbis: Polarization of luminescence of colour centres in YAG crystals, J. Phys. Cond. Matter **3**, 5457–5461 (1991)

50.170 A. Pujats, M. Springis: The F-type centres in YAG crystals, Radiat. Eff. Defects Solids **155**, 65–69 (2001)

50.171 M.K. Ashurov, Y.K. Voronko, V.V. Osiko, A.A. Sobol, M.I. Timoshechkin: Spectroscopic study of stoichiometric deviation in crystals with garnet structure, Phys. Status Solidi (a) **42**, 101–110 (1977)

50.172 V. Lupei, A. Lupei, C. Tiseanu, S. Georgescu, C. Stoicescu, P.M. Nanau: High resolution optical spectroscopy of YAG:Nd: A test for structural and distribution models, Phys. Rev. B **51**, 8–17 (1995)

50.173 Y. Zorenko, V. Gorbenko, I. Konstankevych, A. Voloshinovskii, G. Stryganyuk, V. Mikhailin, V. Kolobanov, D. Spassky: Single-crystalline films of Ce-doped YAG and LuAG phosphors: advantages over bulk crystals analogues, J. Lumin. **114**, 85–94 (2005)

50.174 Y. Zorenko, V. Gorbenko, A. Voloshinovskii, G. Stryganyuk, V. Mikhailin, V. Kolobanov, D. Spassky, M. Nikl, K. Blažek: Exciton-related luminescence in LuAG:Ce single crystals and single crystalline films, Phys. Status Solidi (a) **202**, 1113–1119 (2005)

50.175 V. Murk, N. Yaroshevich: Exciton and recombination processes in YAG crystals, J. Phys. Cond. Matter **7**, 5857–5864 (1995)

50.176 N.N. Ryskin, P. Dorenbos, C.W.E. van Eijk, S.K. Batygov: Scintillation properties of $Lu_3Al_{5-x}Sc_xO_{12}$ crystals, J. Phys. Cond. Matter **6**, 10423–10434 (1994)

50.177 M. Nikl, E. Mihóková, J. Pejchal, A. Vedda, Y. Zorenko, K. Nejezchleb: The antisite Lu_{Al} defect-related trap in $Lu_3Al_5O_{12}$:Ce single crystal, Phys. Status Solidi (b) **242**, R119–R121 (2005)

50.178 M. Nikl, E. Mihokova, J. Pejchal, A. Vedda, M. Fasoli, I. Fontana, V.V. Laguta, V. Babin, K. Nejezchleb, A. Yoshikawa, H. Ogino, G. Ren: Scintillator materials achievements, opportunities, and puzzles, IEEE Trans. Nucl. Sci. **55**, 1035–1041 (2008)

50.179 M. Nikl, J. Pejchal, E. Mihóková, J.A. Mareš, H. Ogino, A. Yoshikawa, T. Fukuda, A. Vedda, C. D'Ambrosio: Antisite defect-free $Lu_3(Ga_xAl_{1-x})_5O_{12}$:Pr scintillator, Appl. Phys. Lett. **88**, 141916 (2006)

50.180 Y.-N. Xu, W.Y. Ching, B.K. Brickeen: Electronic structure and bonding in garnet crystals $Gd_3Sc_2Ga_3O_{12}$, $Gd_3Sc_2Al_3O_{12}$, and $Gd_3Ga_3O_{12}$ compared to $Y_3Al_3O_{12}$, Phys. Rev. B **61**, 1817–1824 (2000)

50.181 A. Vedda, D. Di Martino, M. Martini, V.V. Laguta, M. Nikl, E. Mihóková, J. Rosa, K. Nejezchleb, K. Blažek: Thermoluminescence of Zr-codoped $Lu_3Al_5O_{12}$:Ce crystals, Phys. Status Solidi (a) **195**, R1–R3 (2003)

50.182 C.R. Stanek, K.J. McClellan, M.R. Levy, R.W. Grimes: Extrinsic defect structure of $RE_3Al_5O_{12}$ garnets, Phys. Status Solidi (b) **243**, R75–R77 (2006)

50.183 K. Takagi, T. Fukazawa: Cerium-activated Gd_2SiO_5 single crystal scintillator, Appl. Phys. Lett. **42**, 43–45 (1983)

50.184 T. Utsu, S. Akiyama: Growth and applications of Gd_2SiO_5:Ce scintillators, J. Cryst. Growth **109**, 385–391 (1991)

50.185 C.L. Melcher, R.A. Manente, C.A. Peterson, J.S. Schweizer: Czochralski growth of rare earth oxyorthosilicate single crystals, J. Cryst. Growth **128**, 1001–1005 (1993)

50.186 D.W. Cooke, K.J. McClellan, B.L. Bennett, J.M. Roper, M.T. Whittaker, R.E. Münchausen, R.C. Sze: Crystal growth and optical characterization of cerium-doped $Lu_{1.8}Y_{0.2}SiO_5$, J. Appl. Phys. **88**, 7360–7362 (2000)

50.187 C.L. Melcher, M.A. Spurrier, L. Eriksson, M. Eriksson, M. Schmand, G. Givens, R. Terry, T. Homant, R. Nutt: Advances in the scintillation performances of LSO:Ce single crystals, IEEE Trans. Nucl. Sci. **50**, 762–766 (2003)

50.188 M. Jie, G. Zhao, X. Zeng, L. Su, H. Pang, X. He, J. Xu: Crystal growth and optical properties of $Gd_{1.99-x}Y_xCe_{0.01}SiO_5$ single crystals, J. Cryst. Growth **277**, 175–180 (2005)

50.189 J.D. Zavartsev, S.A. Koutovoi, A.I. Zagumennyi: Czochralski growth and characterization of large $Ce^{3+}:Lu_2SiO_5$ single crystals co-doped with Mg^{2+} or Ca^{2+} or Tb^{3+} for scintillators, J. Cryst. Growth **275**, e2167–e2171 (2005)

50.190 B. Hautefeuille, K. Lebbou, C. Dujardin, J.M. Fourmigue, L. Grosvalet, O. Tillement, C. Pédrini: Shaped crystal growth of Ce^{3+}-doped $Lu_{2(1-x)}Y_{2x}SiO_5$ oxyorthosilicate for scintillator applications by pulling-down technique, J. Cryst. Growth **289**, 172–177 (2006)

50.191 C. Mansuy, R. Mahiou, J.-M. Nedelec: A new sol-gel route to Lu_2SiO_5 (LSO) scintillator: Powders and thin films, Chem. Mater. **15**, 3242–3244 (2003)

50.192 C. Mansuy, J.-M. Nedelec, R. Mahiou: Molecular design of inorganic scintillators: From alcoxides to scintillating materials, J. Mater. Chem. **14**, 3274–3280 (2004)

50.193 J.-K. Lee, R.E. Münchausen, J.-S. Lee, Q.X. Jia, M. Nastasi, J.A. Valdez, B.L. Bennett, D.W. Cooke, S.Y. Lee: Structure and optical properties of Lu_2SiO_5:Ce phosphor thin films, Appl. Phys. Lett. **89**, 101905 (2006)

50.194 J. Felsche: Rare earth silicates of the type $RE_2[SiO_4]O$, Naturwissenschaften **11**, 565–566 (1971)

50.195 T. Gustafsson, M. Klinterberg, S.E. Derenzo, M.J. Weber, J.O. Thomas: Lu_2SiO_5 by single-crystal x-ray and neutron diffraction, Acta Crystallogr. C **57**, 668–669 (2001)

50.196 W.Y. Ching, L. Ouyang, Y.-N. Xu: Electronic and optical properties of Y_2SiO_5 and $Y_2Si_2O_7$ with comparison to α-SiO_2 and Y_2O_3, Phys. Rev. B **67**, 245108 (2003)

50.197 D.W. Cooke, B.L. Bennett, R.E. Münchausen, J.-K. Lee, M.A. Nastasi: Intrinsic ultraviolet luminescence from Lu_2O_3, Lu_2SiO_5 and Lu_2SiO_5:Ce^{3+}, J. Lumin. **106**, 125–132 (2004)

50.198 H. Suzuki, T.A. Tombrello, C.L. Melcher, J.S. Schweizer: UV and gamma-ray excited luminescence of cerium-doped rare-earth oxyorthosilicates, Nucl. Instrum. Methods Phys. Res. A **320**, 263–272 (1992)

50.199 H. Suzuki, T.A. Tombrello, C.L. Melcher, J.S. Schweizer: Light emission mechanism of $Lu_2(SiO_4)O$:Ce, IEEE Trans. Nucl. Sci. **40**, 380–383 (1993)

50.200 E. van der Kolk, S.A. Basun, G.F. Imbush, W.M. Yen: Temperature dependent spectroscopic studies of the electron delocalization dynamics of excited Ce ions in the wide band gap insulator, Lu_2SiO_5, Appl. Phys. Lett. **83**, 1740–1742 (2003)

50.201 M.-F. Joubert, S.A. Kazanskii, Y. Guyot, J.-C. Gâcon, C. Pédrini: Microwave study of photoconductivity induced by laser pulses in rare-earth-doped dielectric crystals, Phys. Rev. B **69**, 165217 (2004)

50.202 L. Pidol, O. Guillot-Noël, A. Kahn-Harari, B. Viana, D. Pelenc, D. Gourier: EPR study of Ce^{3+} ions in lutetium silicate scintillators $Lu_2Si_2O_7$ and Lu_2SiO_5, J. Phys. Chem. Solids **67**, 643–650 (2006)

50.203 F. Brethean-Raynal, M. Lance, P. Charpin: Crystal data for $Lu_2Si_2O_7$, J. Appl. Cryst. **14**, 349–350 (1981)

50.204 P. Szupryczynski, C.L. Melcher, M.A. Spurrier, M.P. Maskarinec, A.A. Carey, A.J. Wojtowicz, W. Drozdowski, D. Wisniewski, R. Nutt: Thermoluminescence and scintillation properties of rare earth oxyorthosilicate scintillators, IEEE Trans. Nucl. Sci. **51**, 1103–1110 (2004)

50.205 W.W. Moses: Trends in PET imaging, Nucl. Instrum. Methods Phys. Res. A **471**, 209–214 (2001)

50.206 N. Shimura, M. Kamada, A. Gunji, S. Yamana, T. Usui, K. Kurashige, H. Ishibashi, N. Senguttuvan, S. Shimizu, K. Sumiya, H. Murayama: Zr-doped GSO:Ce single crystals and their scintillation performance, IEEE Trans. Nucl. Sci. **53**, 2519–2522 (2006)

50.207 S. Yamamoto, K. Matsumoto, M. Senda: Development of a GSO positron/single-photon imaging detector, Phys. Med. Biol. **51**, 457–469 (2006)

50.208 D.W. Cooke, B.L. Bennett, K.J. McClellan, J.M. Roper, M.T. Whittaker: Similarities in glow peak positions and kinetics parameters of oxyorthosilicates: evidence for unique intrinsic trapping sites, J. Lumin. **92**, 83–89 (2001)

50.209 P. Dorenbos, C.W.W. van Eijk, A.J.J. Bos, C.L. Melcher: Afterglow and thermoluminescence properties of Lu_2SiO_5:Ce scintillation crystals, J. Phys. Cond. Matter **6**, 4167–4180 (1994)

50.210 R. Visser, C.L. Melcher, J.S. Schweizer, H. Suzuki, T.A. Tombrello: Photostimulated luminescence and thermoluminescence of LSO scintillators, IEEE Trans. Nucl. Sci. **41**, 689–693 (1994)

50.211 D.W. Cooke, B.L. Bennett, R.E. Münchausen, K.J. McClellan, J.M. Roper, M.T. Whittaker: Intrinsic trapping sites in rare-earth and yttrium oxyorthosilicates, J. Appl. Phys. **86**, 5308–5310 (1999)

50.212 D.W. Cooke, B.L. Bennett, K.J. McClellan, R.E. Münchausen, J.R. Tesmer, C.J. Wetteland: Luminescence, emission spectra and hydrogen content of crystalline Lu_2SiO_5:Ce^{3+}, Philos. Mag. B **82**, 1659–1670 (2002)

50.213 L. Pidol, A. Kahn-Harari, B. Viana, B. Ferrand, P. Dorenbos, J.T.M. de Haas, C.W.E. van Eijk, E. Virey: Scintillation properties of $Lu_2Si_2O_7:Ce^{3+}$, a fast and efficient scintillator crystal, J. Phys. Cond. Matter **15**, 2091–2102 (2003)

50.214 M. Nikl, H. Ogino, A. Yoshikawa, E. Mihóková, J. Pejchal, A. Beitlerova, A. Novoselov, T. Fukuda: Fast 5d–4f luminescence of Pr^{3+} in Lu_2SiO_5 single crystal host, Chem. Phys. Lett. **410**, 218–221 (2005)

50.215 O. Guillot-Noël, J.T.M. De Haas, P. Dorenbos, C.W.E. Van Eijk, K. Krämer, H.U. Güdel: Optical and scintillation properties of cerium-doped $LaCl_3$, $LuBr_3$ and $LuCl_3$, J. Lumin. **85**, 21–35 (1999)

50.216 J. Andriessen, O.T. Antonyak, P. Dorenbos, P.A. Rodnyi, G.B. Stryganyuk, C.W.E. van Eijk, A.S. Voloshinovskii: Experimental and theoretical study of the spectroscopic properties of Ce^{3+} doped $LaCl_3$ single crystals, Opt. Comm. **178**, 355–363 (2000)

50.217 K.W. Krämer, P. Dorenbos, H.U. Güdel, C.W.E. van Eijk: Development and characterization of highly efficient new cerium doped rare earth halide scintillator materials, J. Mater. Chem. **16**, 2773–2780 (2006)

50.218 E.V.D. Van Loef, P. Dorenbos, C.W.E. van Eijk: High energy resolution scintillator: Ce^{3+} activated $LaBr_3$, Appl. Phys. Lett. **79**, 1573–1575 (2001)

50.219 E.V.D. Van Loef, P. Dorenbos, C.W.E. van Eijk, K. Krämer, H.U. Güdel: Influence of the anion on the spectroscopy and scintillation mechanism in pure and Ce^{3+}-doped K_2LaX_5 and LaX_3 (X = Cl, Br, I), Phys. Rev. B **68**, 045108 (2003)

50.220 A. Bessiere, P. Dorenbos, C.W.E. van Eijk, K.W. Krämer, H.U. Güdel, C. de Mello Donega, A. Meijerink: Luminescence and scintillation properties of the small band gap compound $LaI_3:Ce^{3+}$, Nucl. Instrum. Methods Phys. Res. A **537**, 22–26 (2005)

50.221 J. Glodo, K.S. Shah, M. Klugerman, P. Wong, B. Higgins, P. Dorenbos: Scintillation properties of $LuI_3:Ce$, Nucl. Instrum. Methods Phys. Res. A **537**, 279–281 (2005)

50.222 M.D. Birowosuto, P. Dorenbos, C.W.E. van Eijk: Scintillation properties of $LuI_3:Ce^{3+}$-high light yield scintillators, IEEE Trans. Nucl. Sci. **52**, 1114–1118 (2005)

50.223 G. Meyer: The ammonium chloride route to anhydrous rare earth chlorides – The example of YCl_3, Inorg. Synth. **25**, 146–150 (1989)

50.224 W.M. Higgins, J. Glodo, E. Van Loef, M. Klugerman, T. Gupta, L. Cirignano, P. Wong, K.S. Shah: Bridgman growth of $LaBr_3:Ce$ and $LaCl_3:Ce$ crystals for high-resolution gamma ray spectrometers, J. Cryst. Growth **287**, 239–242 (2006)

50.225 P. Dorenbos: Scintillation mechanism in Ce^{3+} doped halide scintillators, Phys. Status Solidi (a) **202**, 195–200 (2005)

50.226 J.S. Chivian, W.E. Case, D.D. Eden: The photon avalanche: A new phenomenon in Pr^{3+}-based quantum counters, Appl. Phys. Lett. **35**, 124 (1979)

50.227 N. Pelletier-Allard, R. Pelletier: Multiphoton excitations in neodymium chlorides, Phys. Rev. B **36**, 4425 (1987)

50.228 E.V.D. Van Loef, P. Dorenbos, C.W.E. van Eijk, K. Krämer, H.U. Güdel: Scintillation properties of $LaCl_3:Ce^{3+}$ crystals: Fast, efficient, and high energy resolution scintillators, IEEE Trans. Nucl. Sci. **48**, 341–345 (2001)

50.229 E.V.D. Van Loef, P. Dorenbos, C.W.E. van Eijk: The scintillation mechanism in $LaCl_3:Ce^{3+}$, J. Phys. Cond. Matter **15**, 1367–1375 (2003)

50.230 U. Rogulis, S. Schweizer, J.-M. Spaeth, E.V.D. Van Loef, P. Dorenbos, C.W.E. van Eijk, K.W. Krämer, H.U. Güdel: Magnetic resonance investigations of $LaCl_3:Ce^{3+}$ scintillators, Radiat. Eff. Defects Solids **157**, 951–955 (2002)

50.231 S.M. Kuzakov: Electron-hole traps and thermoluminescence of the $LaCl_3:TR$ single crystals, Rad. Prot. Dosim. **33**, 115–117 (1990)

50.232 J. Glodo, K.S. Shah, M. Klugerman, P. Wong, B. Higgins: Thermoluminescence od $LaBr_3:Ce$ crystals, Nucl. Instrum. Methods Phys. Res. A **537**, 93–96 (2005)

50.233 E.V.D. Van Loef, P. Dorenbos, C.W.E. van Eijk, K. Krämer, H.U. Güdel: Scintillation properties of $LaBr_3:Ce^{3+}$ crystals: fast, efficient and high-energy-resolution scintillators, Nucl. Instrum. Methods Phys. Res. A **486**, 254–258 (2002)

50.234 P. Dorenbos, E.V.D. Van Loef, A.P. Vink, E. van der Kolk, C.W.E. van Eijk, K.W. Krämer, H.U. Güdel, W.M. Higgins, K.S. Shah: Level location and spectroscopy of Ce^{3+}, Pr^{3+}, Er^{3+}, and Eu^{2+} in $LaBr_3$, J. Lumin. **117**, 147–155 (2006)

50.235 K.S. Shah, J. Glodo, M. Klugerman, W. Higgins, T. Gupta, P. Wong, W.W. Moses, S.E. Derenzo, M.J. Weber, P. Dorenbos: $LuI_3:Ce$ – a new scintillator for gamma ray spectroscopy, IEEE Trans. Nucl. Sci. **51**, 2302–3205 (2004)

50.236 J. Lu, K. Ueda, H. Yagi, T. Yanagitani, Y. Akiyama, A.A. Kaminskii: Neodymium doped yttrium aluminum garnet $Y_3Al_5O_{12}$ nanocrystalline ceramics – a new generation of solid state laser and optical materials, J. Alloy. Compd. **341**, 220–225 (2002)

50.237 V. Lupei, A. Lupei, A. Ikesue: Single crystal and transparent ceramic Nd-doped oxide laser materials: a comparative spectroscopic investigation, J. Alloy. Compd. **380**, 61–70 (2004)

50.238 C. Greskovich, S. Duclos: Ceramic scintillators, Ann. Rev. Mater. Sci. **27**, 69–88 (1997)

50.239 B.C. Grabmaier, W. Rossner: New scintillators for x-ray computed tomography, Nucl. Tracks Rad. Meas. **21**, 43–45 (1993)

50.240 W. Rossner, M. Ostertag, F. Jermann: Properties and applications of gadolinium oxysulfide based ceramic scintillators. In: *Physics and Chemistry of Luminescent Materials: Proc. 7th Int. Symp*, Vol. 98-24, ed. by C.W. Struck, K.C. Mishra, B. DiBartolo (Electrochemical Society, Pennington 1998), 187–194

50.241 R. Hupke, C. Doubrava: The new UFC-detector for CT-imaging, Phys. Medica **XV**, 315–318 (1999)

50.242 S.J. Duclos, C.D. Greskovich, R.J. Lyons, J.S. Vartuli, D.M. Hoffman, R.J. Riedner, M.J. Lynch: Development of the HiLight scintillator for computed tomography medical imaging, Nucl. Instrum. Methods Phys. Res. A **505**, 68–71 (2003)

50.243 A. Lempicki, C. Brecher, P. Szupryczynski, H. Lingertat, V.V. Nagarkar, S.V. Tipnis, S.R. Miller: A new lutetia-based ceramic scintillator for x-ray imaging, Nucl. Instrum. Methods Phys. Res. A **488**, 579–590 (2002)

50.244 V.V. Nagarkar, S.R. Miller, S.V. Tipnis, A. Lempicki, A. Brecher, H. Lingertat: A new large area scintillator screen for x-ray imaging, Nucl. Instrum. Methods Phys. Res. B **213**, 250–254 (2004)

50.245 C. Brecher, R.H. Bartram, A. Lempicki: Hole traps in Lu_2O_3:Eu ceramic scintillators. I. Persistent afterglow, J. Lumin. **106**, 159–168 (2004)

50.246 E. Zych, C. Brecher, A.J. Wojtowitz, H. Lingertat: Luminescence properties of Ce-activated YAG optical ceramic scintillator materials, J. Lumin. **75**, 193 (1997)

50.247 H.-L. Li, X.-J. Liu, R.-J. Xie, Y. Zeng, L.-P. Huang: Fabrication of transparent cerium-doped lutetium aluminum garnet ceramics by Co-precipitation routes, J. Am. Ceram. Soc. **89**, 2356 (2006)

50.248 M. Nikl, J.A. Mareš, N. Solovieva, H. Li, X. Liu, L. Huang, I. Fontana, M. Fasoli, A. Vedda, C. D'Ambrosio: Scintillation characteristics of $Lu_3Al_5O_{12}$:Ce optical ceramics, J. Appl. Phys. **101**, 033515 (2007)

50.249 B. Schmitt, M. Fuchs, E. Hell, W. Knupfer, P. Hackenschmied, A. Winnacker: Structured alkali halides for medical applications, Nucl. Instrum. Methods Phys. Res. B **191**, 800–804 (2002)

50.250 K. Oba, M. Ito, M. Yamaguchi, M. Tanaka: A CsI(Na) scintillation plate with high spatial-resolution, Adv. Electron. Electron. Phys. **74**, 247–255 (1988)

50.251 H. Washida, T. Sonoda: High resolution phosphor screen for x-ray image intensifier, Adv. Electron. Electron. Phys. **52**, 201–207 (1979)

50.252 C.M. Castelli, N.M. Allinson, K.J. Moon, D.L. Watson: High spatial resolution scintillator screens coupled to CCD detectors for x-ray imaging applications, Nucl. Instrum. Methods Phys. Res. A **348**, 649–653 (1994)

50.253 V.V. Nagarkar, T.K. Gupta, S.R. Miller, Y. Klugerman, M.R. Squillante, G. Entine: Structured CsI(Tl) scintillators for x-ray imaging applications, IEEE Trans. Nucl. Sci. **45**, 492–496 (1988)

50.254 C.B. Johnson, L.D. Owen: Image tube intensified electronic imaging. In: *Handbook of Optics*, Vol. 1 (McGraw-Hill, New York 1995) pp. 21.1–21.32

50.255 J.-P. Moy: Recent developments in x-ray imaging detectors, Nucl. Instrum. Methods Phys. Res. A **442**, 26–37 (2000)

50.256 P. Olivier: Nuclear oncology, a fast growing field of nuclear medicine, Nucl. Instrum. Methods Phys. Res. A **527**, 4–8 (2004)

50.257 T.F. Budinger, K.M. Brennun, W.W. Moses, S.E. Derenzo: Advances in positron tomography for oncology, Nucl. Med. Biol. **23**, 659–667 (1996)

50.258 W.W. Moses, S.E. Derenzo: Prospects for time-of-flight PET using LSO scintillator, IEEE Trans. Nucl. Sci. **46**, 474–478 (1999)

50.259 Z.T. Kang, C.J. Summers, H. Menkar, B.K. Wagner, R. Durst, Y. Diawara, G. Mednikova, T. Thorson: ZnTe:0 phosphor development for x-ray imaging applications, Appl. Phys. Lett. **88**, 111904 (2006)

50.260 A. Koch, C. Raven, P. Spanne, A. Snigirev: X-ray imaging with submicrometer resolution employing transparent luminescent screens, J. Opt. Soc. Am. A **15**, 1940–1951 (1998)

50.261 Y. Zorenko, I. Konstankevych, M. Globus, B. Grinyov, V. Lyubinskiy: New scintillation detectors based on oxide single crystal films for biological microtomography, Nucl. Instrum. Methods Phys. Res. A **505**, 93–96 (2003)

50.262 D. Ehrentraut, H. Sato, M. Miyamoto, T. Fukuda, M. Nikl, K. Maeda, I. Niikura: Fabrication of homoepitaxial ZnO films by low-temperature liquid-phase epitaxy, J. Cryst. Growth **287**, 367–371 (2006)

50.263 J.S. Neal, L.A. Boatner, N.C. Giles, L.E. Halliburton, S.E. Derenzo, E.D. Bourret-Courchesne: Comparative investigation of the performance of ZnO-based scintillators for use as a-particle detectors, Nucl. Instrum. Methods Phys. Res. A **568**, 803–809 (2006)

50.264 B.K. Meyer, J. Sann, A. Zeuner: Lithium and sodium acceptors in ZnO, Superlattices Microstr. **38**, 344–348 (2005)

50.265 A. Kaplan, A. Sajwani, Z.Y. Li, R.E. Palmer, J.P. Wilcoxon: Efficient vacuum ultraviolet light frequency downconversion by thin films of CdSe quantum dots, Appl. Phys. Lett. **88**, 171105 (2006)

50.266 K. Shibuya, M. Koshimizu, H. Murakami, Y. Muroya, Y. Katsumura, K. Asai: Development of ultra-fast semiconducting scintillators using quantum confinement effect, Jpn. J. Appl. Phys. **43**, L1333–L1336 (2004)

50.267 C.W.E. van Eijk: Inorganic scintillators for thermal neutron detection, Radiat. Meas. **38**, 337–342 (2004)

51. Silicon Solar Cells: Materials, Devices, and Manufacturing

Mohan Narayanan, Ted Ciszek

The phenomenal growth of the silicon photovoltaic industry over the past decade is based on many years of technological development in silicon materials, crystal growth, solar cell device structures, and the accompanying characterization techniques that support the materials and device advances. This chapter chronicles those developments and serves as an up-to-date guide to silicon photovoltaic technology. Following an introduction to the technology in Sect. 51.1, an in-depth discussion of the current approaches to silicon material crystal growth methods for generating solar cell substrates is presented in Sect. 51.2. Section 51.3 reviews the current manufacturing techniques for solar cell devices and also presents the latest advances in device structures that achieve higher efficiency. Finally, a perspective on the technology and what might be expected in the future is summarized in Sect. 51.4.

51.1 Silicon Photovoltaics 1701
    51.1.1 Physics of a Solar Cell 1701
    51.1.2 The Photovoltaic Value Chain 1703
    51.1.3 The Photovoltaic Module 1703
    51.1.4 Commercial PV Technologies 1704

51.2 Crystal Growth Technologies
for Silicon Photovoltaics 1704
    51.2.1 Silicon Photovoltaics...................... 1704
    51.2.2 Single-Crystal Ingot Growth
        (CZ and FZ) 1705
    51.2.3 Multicrystalline Ingot Growth 1707
    51.2.4 Silicon Ribbon or Sheet Growth...... 1709
    51.2.5 PV Silicon Crystal Growth
        Approaches 1711

51.3 Cell Fabrication Technologies................. 1711
    51.3.1 Homojunction Devices 1711
    51.3.2 Enhancing Solar Cell Performance .. 1714
    51.3.3 Advanced Commercial Solar Cell
        Concepts 1714

51.4 Summary and Discussion 1715

References 1716

51.1 Silicon Photovoltaics

Solar cells convert sunlight into electricity via the photovoltaic effect. The photovoltaic (PV) effect was first reported in 1839 by Becquerel when he observed a light-dependent voltage between electrodes immersed in an electrolyte. However, nearly a century later in 1941, the effect was reported in silicon. In 1954, the first working solar cell module was announced. The photovoltaic industry has grown from producing a few kW in the 1960s to a multi-GW production in this decade. The success of the industry is mainly due to its ability to supply reliable and modular power, cost effectively, from a few W to multi-MW. With the market growing by nearly 20% per year for the past 10 years, the amount of silicon used in the PV industry is poised to be significantly more than that used in the semiconductor industry in this decade.

51.1.1 Physics of a Solar Cell

When incident sunlight (photon) is absorbed by a semiconductor, photon energy is transferred to the material. If the absorbed photon has sufficient energy, interband transition occurs and an electron is excited from the valance band into the conduction band. The valance state vacated by the electron is called a hole. The free electrons generated flow freely inside the material and can be drawn to the external world to be used as electricity.

The solar radiation spectrum can be broadly divided into three portions: (1) infrared, (2) visible, and (3) ultraviolet. The long-wavelength, infrared portion of the sun spectrum does not have the threshold energy needed to free electrons from silicon atoms and passes through the cell without interacting. The material is transparent to these long wavelengths. The ultraviolet (short-wavelength) portion of the sun spectrum has more than enough energy to create the electron–hole pairs. In the case of silicon, with a bandgap of 1.1 eV at room temperature, only photons with energy greater than 1.1 eV will exhibit the PV effect. The excess energy transferred to the charge carriers is dissipated as heat. Hence, only about 40% of the incident light energy is effectively used. The irradiance spectrum is an important consideration in designing solar cells, as solar cells respond differently to photons of different energy. The spectrum of sunlight changes continually as a function of time of the day, weather, and location. Solar cells are exposed to different spectra and hence need to operate under a variety of irradiances.

The solar cell is a p–n junction, like a large-area diode with metal contacts on either side, as illustrated by Fig. 51.1. The n-type portion (typically near the front of the device) has a high density of electrons and few

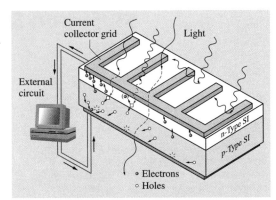

Fig. 51.1 Operation of a silicon solar cell (after [51.1])

holes, so generated electrons can travel easily in this region. The opposite is true in the p-type region, where holes travel easily. The built-in electric field near this junction of n- and p-type silicon causes photogenerated excess electrons to wander toward the grid on the front surface, while the holes wander toward the back contact. The electrons generated far away from the junction in the bulk tend to diffuse towards the junction if they are within the influence of this built-in electric field.

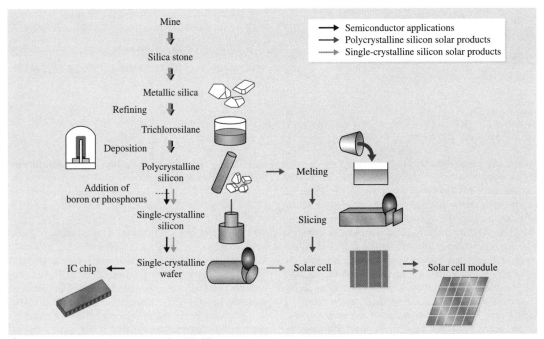

Fig. 51.2 PV silicon value stream (after [51.2])

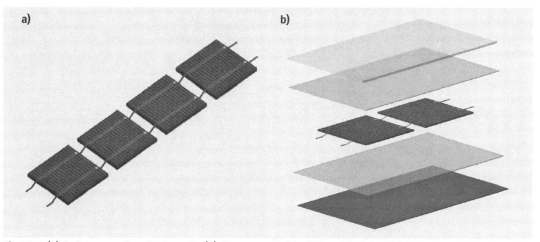

Fig. 51.3 (a) Series connection of solar cells. (b) Components of the PV module (after [51.7])

If the electrons survive their trip across the cell thickness, without recombining at defects or impurities, they are collected at the grid. Then they flow through an external circuit as current that can operate an electronic instrument or appliance. After that, they reenter the solar cell at the back contact to recombine with holes and the process repeats [51.3–6].

51.1.2 The Photovoltaic Value Chain

More than 80% of manufactured solar cells are based on a crystalline silicon (single-crystalline or multicrystalline) substrate. The value stream of the photovoltaic industry is shown in Fig. 51.2 [51.2].

High-purity polysilicon (produced by the conventional Siemens polysilicon process or by a fluidized bed process) along with recycled silicon is the dominant feedstock for the PV industry. Due to increased demand for PV systems, several programs to upgrade metallurgical-grade silicon are also being actively pursued. The feedstock is then converted into silicon wafers by casting or crystal growth followed by a wire-sawing process. Details of this process step are described in Sect. 51.2. The silicon substrate is converted into solar cells using technologies based on semiconductor device processing and surface-mount technology (SMT). The cell process technology (Sect. 51.4) mainly consists of wafer surface etching, junction formation, antireflection coating deposition, and metal contact formation. The individual solar cells are connected and assembled into the finished product: PV modules, which are integrated with system components, inverters, charge conditioners, batteries etc. and then installed at the site.

The crystalline silicon wafer accounts for about 40% of the cost of a PV module. There have been ongoing efforts to reduce the cost of PV modules: the use of thinner substrates to save the cost of silicon used, device research to increase the conversion efficiency of the module, high-volume manufacturing with inline process control to reduce the cost of manufacturing, etc.

51.1.3 The Photovoltaic Module

The PV module is designed to generate electricity for many years (usually more than 25 years). It has to operate over a wide range of daily and seasonal variations of sunlight and weather. It must be mechanically rigid to provide a means for mounting the module into its service structure. It must be dielectrically rigid in order to protect living beings from coming into contact with dangerous electrical voltages, as well as to prevent system failure through short circuit.

The photovoltaic module (as shown in Fig. 51.3a,b) consists of a number of solar cells electrically connected to each other with metal interconnects and supported by a rigid superstrate. The interconnected cells are encapsulated by one or more layers of polymeric materials like ethyl vinyl acetate (EVA) and Tedlar, which provide electrical insulation. The encapsulation provides mechanical rigidity to the brittle cells and the flexible interconnection. In addition, it offers chemical protection as a moisture barrier. The encapsulated cell assembly, called a laminate, is then attached to a tempered, low-iron glass superstrate. An electrical junction box is provided at the rear of the module to harness the electricity generated by the module.

51.1.4 Commercial PV Technologies

The commercial success of PV is largely due to the proven reliability and long lifetime (> 25 years) of crystalline silicon modules. Accelerated environmental testing, based on correlation of the specific failure mechanism with outdoor field test results, is used to certify the module performance in the field. If the module passes the environmental tests it will likely perform reliably for more than 20 years.

More than 85% of all modules sold today are based on crystalline-silicon solar cells. Several factors have contributed to the choice of crystalline silicon: high cell conversion efficiencies of 15–20%; availability of commercial equipment from the semiconductor and SMT industries; extensive volume of knowledge on silicon device physics, established feedstock technologies, abundant supply of the source material (sand), etc. Other PV technologies include devices based on amorphous silicon (a-Si), cadmium telluride (CdTe), copper indium diselenide (CIS), and gallium arsenide (GaAs). While a-Si-based devices suffer from lower efficiencies, devices based on GaAs and CdTe have shown high conversion efficiencies. However, the high cost of the material, scarce raw material availability, and toxic nature of the process and material are major challenges facing these technologies for multi-GW manufacturing. Here, only silicon-based PV issues will be discussed in detail.

51.2 Crystal Growth Technologies for Silicon Photovoltaics

51.2.1 Silicon Photovoltaics

The history of crystalline and multicrystalline silicon growth for PV applications starts with, and is closely aligned with, the methods utilized in the semiconductor industry. From the first solar cell produced at Bell Labs in 1954 [51.8] on Czochralski (CZ)-grown silicon through to the development of modern high-efficiency cells, the prominent integrated circuit (IC) industry crystal growth methods have served the PV industry well. The highest-efficiency PV cells and modules commercially available today continue to use the CZ method and, to a lesser extent, the float-zone (FZ) method.

Early in IC development, it was realized that elimination of ingot wafering held many advantages, including simplified surface preparation, thinner wafers, and improved feedstock utilization. The dendritic web Si ribbon growth method, originated in 1962 [51.9], was briefly explored commercially by Dow Corning Corporation in the mid 1960s with an eye toward IC use as well as applications in the fledgling PV industry. Despite more than 30 years of development for PV use, largely by Westinghouse, the web method ultimately proved nonviable for PV production throughput needs in the rapidly growing PV industry. When Tyco Laboratories announced the growth of shaped sapphire crystals by the edge-defined film-fed (EFG) method in 1971 [51.10], Dow Corning researchers quickly realized that this method might also be useful for PV and they published the first growth of silicon ribbons by EFG, using graphite capillary dies, in 1972 [51.11]. A few years later, in 1975, researchers at IBM published the first growth of silicon tubes by EFG [51.12] for potential application as the active, internally cooled, PV receiver in linear parabolic concentrators. The closed tubular shape eliminates edge instability problems and is easier to grow than ribbons. The later, larger, hybrid consisting of a closed tubular shape, but with multiple flat walls, is the current EFG production method for PV wafers. The wafer blanks are laser-cut from the flat tube walls. While dendritic web growth produces single-crystal ribbon surfaces, EFG growth yields multicrystalline wafers.

The dendritic web method, in a sense, spawned another multicrystalline ribbon growth method in which foreign filaments replace the continuously propagating edge dendrites. This simplifies thermal control considerably. The method was introduced as edge-supported pulling (ESP) [51.13] in 1980, using carbon-based edge filaments incorporated into the growing ribbon. After many years of dormancy, it was resurrected by Evergreen Solar as string ribbon. The string ribbon and flat-sided tubular EFG growth methods are in use today and have made strides in improving throughput by effectively increasing the width of material grown in a single apparatus through multiple wide faces (EFG) and multiple ribbons (string ribbon). This was not possible with the more complex dendritic web growth method. Another class of silicon sheet growth methods that we will include in this discussion, although there is currently no large commercial application, is that in which the sheet moves rapidly in a direction orthogonal to the direction of heat removal. This class of growth methods has shown remarkable throughput potential (more than 40 MW of PV wafer material per machine per year), but so far has suffered from relatively large sheet thickness, small grain size, high impurity content, and low cell efficiencies.

Casting of multicrystalline silicon shapes has been conducted for over 80 years. In the mid 1970s, attention was turned to this method as a way of producing PV wafer material. *Fischer* and *Pschunder* reported casting of silicon into graphite molds for PV applications [51.15] in 1976. The technique of melting and then resolidifying silicon in the same graphite container (thermal expansion-matched to silicon), with bottom seeding, was introduced in 1979 [51.16]. Multicrystalline ingot growth has become the dominant method for PV wafer production and is most often conducted by melting and then directionally solidifying (DS) the Si material in the same silicon-nitride-coated, silica-based, square crucible with small height-to-width aspect ratio. This allows more columnar growth than taller crucibles. One semicontinuous multicrystalline ingot casting method introduced in 1985 [51.17, 18] uses electromagnetic repulsion to keep silicon from adhering to water-cooled metal finger *walls*. Electromagnetic casting (EMC) currently produces > 2000 kg silicon ingots at a much higher throughput rate than conventional directional solidification in silica crucibles.

We will consider the above silicon growth methods that are presently in use for PV wafer manufacturing in the following sections. The categories include single-crystal ingots, multicrystalline ingots, and multicrystalline ribbons or sheets.

51.2.2 Single-Crystal Ingot Growth (CZ and FZ)

Since CZ growth is the main silicon growth method of the IC industry, it is quite well documented in other literature, and we need not go into it in detail here. A basic schematic of the process is shown in Fig. 51.4, along with one for FZ growth. Here we will focus primarily on the aspects of growth pertinent to PV applications. CZ and FZ ingots are round, so the active area of a module relative to the total area is lower than for cast or ribbon wafers, unless the ingots are trimmed to a square or rounded-square cross-section. For optimum perfor-

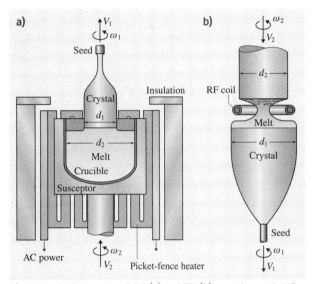

Fig. 51.4a,b Schematic of CZ (**a**) and FZ (**b**) crystal growth (after (**a**) [51.14], (**b**) [51.1])

mance, the ingots must be both single crystalline and dislocation free. Multicrystalline CZ or FZ ingots are more highly stressed than *cast* multicrystalline ingots, and the grain boundaries are more electrically active, resulting in poorer cell efficiency. If ingot growth is initiated single crystalline but not dislocation free, the ingots soon become multicrystalline (an exception is the special case of using a tricrystalline seed [51.19]). The requirement for dislocation-free growth means that care must be taken to avoid particulate matter, high impurity levels, sudden thermal changes, and other perturbations that could disrupt dislocation-free growth. Impurities in CZ growth (and in directional solidification of multicrystalline ingots) obey the normal freezing equation, $C/C_0 = k(1-f)^{k_0-1}$, if there is thorough mixing in the liquid. C is the impurity concentration in the ingot where a fraction f of the melt has been solidified, and k is the effective segregation coefficient of the impurity (always less than the ideal segregation coefficient k_0); C_0 is the initial average impurity concentration in the melt before solidification commences. Impurity segregation is quite effective in the early-grown portion of the ingot where $C \approx kC_0$, but impurity concentrations rise monotonically to levels > C_0 near the termination of growth, as shown in Fig. 51.5. Boron hardly segregates at all, since $k \approx 0.9$. This is advantageous for giving a flat doping profile along the ingot, but it also dictates that no excess boron can be present in the feed-

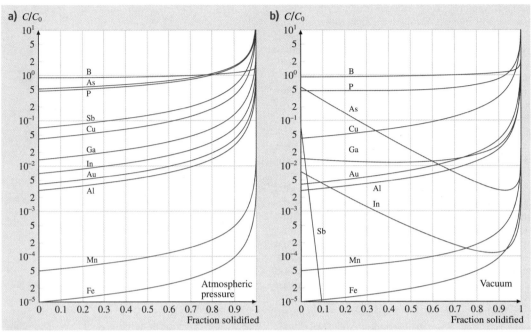

Fig. 51.5a,b Distribution of selected impurities as a function of ingot fraction solidified for CZ or directional solidification carried out in near-atmospheric pressure (**a**) and in vacuum (**b**) (after [51.14])

stock, since it cannot be removed by segregation during growth.

The vapor pressure and hence the effective evaporation coefficient of a particular impurity also can affect the observed impurity distribution in the ingot. The evaporation of an impurity depends on the growth rate and exposed melt surface area [51.20]. The approximate effect for selected impurities is shown graphically on the right-hand side of Fig. 51.5 for growth in a vacuum. By selecting an appropriate partial vacuum level, the distribution of dopants such as P or Ga can be made nearly uniform along most of the ingot length. The importance of a vacuum for reduction of P content in the material is also evident, although the magnitude of such reduction is small.

In FZ growth, or other semicontinuous growth processes, the impurity distribution, given by *Peizulaev's* equation [51.21], is more complex, especially if both effective segregation k and effective evaporation g coefficients are incorporated and if more than one growth pass is made on the silicon rod

$$\frac{C_n(x)}{C_0} = \left[\frac{k}{(k+g)}\right]^n \left[1 - (1-k-g)Z_n \, e^{-(k+g)x}\right],$$

where

$$Z_n = n - \sum_{s=1}^{n-1}(n-s)(k+g)^{s-1}$$
$$\times e^{-s(k+g)} \left[\frac{(s+x)^{s-2}}{s!}\right]$$
$$\times \{(s-1)x + (s+x)[1-(k+g)x]\},$$

where n is the number of times the ingot is zone-melted, and $C_n(x)$ is the impurity concentration at position x along the ingot (x is in units of melt zone length) after n solidification passes. Here, C_0 is the initial concentration of the impurity in the feed rod (assumed uniform).

Figure 51.6, analogous to Fig. 51.5 for CZ growth, gives the distributions of selected impurities as a function of position along a FZ ingot or other semicontinuously grown silicon form where continuous melt replenishment is used. A total length equivalent to 20 melt zones is depicted.

Note that the characteristic shape of the impurity profiles differs from the CZ case. However, judicious use of a partial vacuum can again allow flatter profiles for impurities that evaporate, and can allow phosphorous removal to some extent.

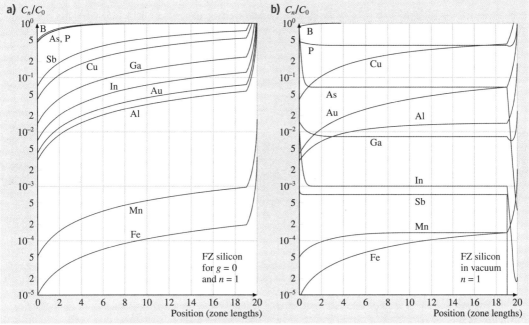

Fig. 51.6a,b Distribution of selected impurities as a function of position along ingot for FZ or other semicontinuous solidification processes with continuous melt replenishment carried out in near-atmospheric pressure (**a**) and in vacuum (**b**) (after [51.20])

For CZ growth near atmospheric pressure, the flattest doping profiles that can be attained are those shown in Fig. 51.5, where all dopant is mixed into the initial melt volume. On the other hand, for FZ growth and other semicontinuous processes, if the dopant has a small segregation coefficient (Ga for example), there is an advantage to inserting the dopant in the first melt zone rather than having it uniformly distributed along the feed rod. This is called pill doping and results in a flat doping profile. Impurities with large segregations coefficients (B, P) must be added continuously with consumption of the feed rod.

Typically, CZ growth is conducted with an argon or argon/partial vacuum ambient and the gas flows down over the seed and melt interface, then over the rim of the crucible and down toward the bottom of the growth furnace. In this way, the copious amounts of SiO_x that form from dissolution of the quartz crucible are swept away from the growth region. FZ growth is typically conducted in a slight overpressure of argon. While growth can be done in a vacuum, some evaporation of silicon will result and cause growth disruption problems if solid Si deposits flake off the chamber or radiofrequency (RF) coil surfaces and fall into the melt zone.

51.2.3 Multicrystalline Ingot Growth

Using a container to shape the growing ingot instead of manipulation of the solid–liquid–vapor boundary greatly simplifies control of the growth process, allows much larger ingots, and allows ingots to have a square cross-section. All of these factors improve productivity. A schematic diagram of the directional solidification (DS) or *casting* process is shown in Fig. 51.7. Silicon can be melted in the same container in which it is solidified, eliminating the top vessel in Fig. 51.7, which is designated directional solidification. Alternatively, it can be melted in one container and poured into a second container for solidification, as implied in the figure. This might properly be called casting with directional solidification. However, the terms DS and casting tend to be used interchangeably for either process.

Either induction (shown in the figure) or resistance heating can be used. The heat source, insulation, and heat sink are designed to provide a nearly unidirectional

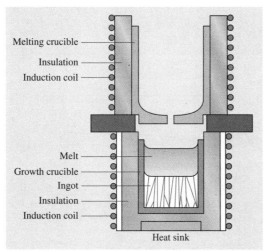

Fig. 51.7 Schematic diagram of the directional solidification process for multicrystalline silicon (after [51.22])

temperature gradient to drive columnar growth, and some relative motion may be imparted to components as growth progresses. Square-cross-section, flat-bottomed silica crucibles contain the liquid silicon. They are coated with silicon nitride to minimize sticking of the ingot to the crucible, and are a one-use item. Because of the crucible–ingot contact, single-crystal growth is not generally achieved, although some attempts at seeding have been reported, as mentioned earlier.

The same impurity distribution equation applies here as for CZ growth, and the general impurity profiles follow those in Fig. 51.5 (melt residence time and melt surface area exposed to the ambient modify the details of the profiles in the case of a vacuum ambient). There are additional impurity phenomena to contend with in this process that are not present for CZ growth. Because the melt and the grown ingot are in contact with the container for an extended period of time, impurities can both dissolve and diffuse into the melt and later diffuse into the sides and bottom of the ingot from the crucible. Metallic impurities in the melt are strongly segregated to the last layer to freeze at the top of the ingot. During cool down of the massive ingot, these impurities can diffuse back into the top region of the ingot. These phenomena, along with small grain sizes at the ingot bottom and sides, require that a portion of the bottom, sides, and top of the ingot be cropped away because the minority charge carrier there is too low for adequate PV performance.

In the electromagnetic casting process (EMC) method, container walls comprised of water-cooled, electrically conductive *fingers* provide the sidewall melt confinement during a semicontinuous casting process. Silicon is prevented from making substantial contact with these walls by Biot–Savart law repulsion between the fingers and the melt. This is visualized in Fig. 51.8a, where a helical induction coil, a cylindrical array of fingers, and the Si melt are depicted in a top sectional view. The current directions at one instant in the RF cycle are shown. The induced currents in the fingers and in the melt are in opposite directions, giving rise to the repulsive force. In PV ingot production, a square array of fingers is used and new material is fed into the top of

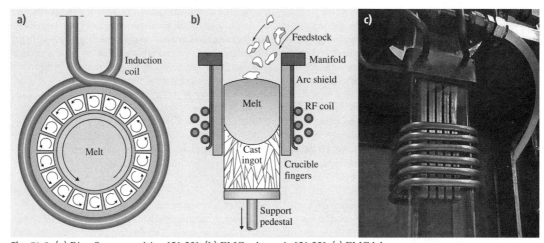

Fig. 51.8 (a) Biot–Savart repulsion [51.23], (b) EMC schematic [51.22], (c) EMC laboratory apparatus

the container while the ingot is continuously withdrawn from the bottom. A schematic of the process is shown in Fig. 51.8b, and a small, laboratory-scale apparatus is shown in Fig. 51.8c.

The impurity distribution profiles in EMC ingots follow the Peizulaev equation, since the configuration is essentially like float zoning, and the curves of Fig. 51.6 apply. Again, the details of the profiles under vacuum will depend on melt residence time and melt surface area exposed to the ambient.

In current production, ingots weighing 1 metric ton are grown with a $35 \times 35 \text{ cm}^2$ cross-section and length of ≈ 4 m, at a throughput (with sawing) of $500 \text{ m}^2/\text{day}$ – approximately seven times faster than any other ingot growth method. The ingots are generally cleaner than DS ingots, but have smaller grain size, with the PV conversion efficiency being about the same.

51.2.4 Silicon Ribbon or Sheet Growth

Many approaches to ribbon or sheet growth of silicon have been made over the years. Only two are in large-scale use at the time of this writing. They are the EFG and string ribbon methods mentioned in the introduction. Other methods that had been pursued until recently are dendritic web growth and a form of horizontal ribbon growth on substrate (RGS) where melt is solidified on a moving substrate that is later detached for reuse. There are indications that RGS growth may reemerge on the PV scene, so this section will summarize the EFG, string ribbon, and RGS techniques.

The EFG technique utilizes a die that is wetted by liquid silicon. The liquid rises up capillary channels in the die to reach the die top, and connects to a seed. As the seed is pulled upward, the top portion of the liquid meniscus continually solidifies, adding growth to the seed. The bottom of the meniscus spreads to the outer edges of the die top and is pinned there. The rising of liquid up the capillary channel is governed by Laplace's equation, which relates the pressure difference Δp across a liquid surface to the two principal radii of curvature of the surface R_1 and R_2, and the surface tension γ as $\Delta p = \gamma(1/R_1 + 1/R_2) = \rho g h$. The liquid will rise vertically to a height h; ρ is the liquid density, and g is the acceleration due to gravity. For capillary channels comprised of plates or annular tubes, one radius of curvature R_2 is approximately infinite and the other is $R_1 = s/2\cos\theta$, where s is the plate or tube separation and θ is the wetting angle, so $h = (2\gamma/\rho g s)\cos\theta$. For silicon in graphite dies, $\theta \sim 0$ and $h \sim 2\gamma/\rho g s$.

Heights of several centimeters to 10 cm could be used, but for a number of practical reasons the height is kept quite small (impurity segregation is improved because back-mixing of the rejected impurities is better, the meniscus is higher and there is less susceptible to freeze-out to the die, the meniscus is more stable, and pick up of SiC particles on the ribbon surface is reduced).

A schematic of EFG growth using an annular die is shown in Fig. 51.9a. EFG wafer growth for PV use is similar, with the tube having eight flat faces. Sections of octagonal tubes with 10 and 12.5 cm faces are shown in Fig. 51.9b. After growth, the tubes are transported to a laser-cutting station, which removes rectangular cell blanks from the tube faces.

Typically, induction heating is used and the tubes grow in an argon ambient at about 2 cm/min linear pulling speed. FZ-type impurity profile equations apply, since the process is continuous with melt replenishment.

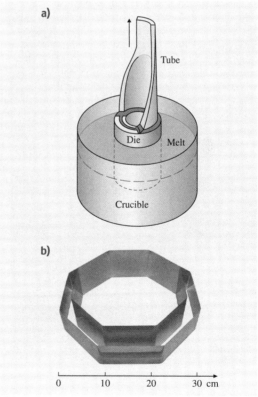

Fig. 51.9a,b Schematic of EFG tube growth (**a**); 10 and 12.5 cm face octagonal tube sections (**b**) [51.12]

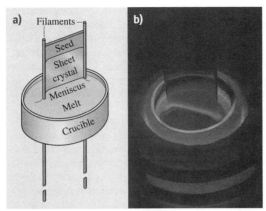

Fig. 51.10a,b Schematic of string ribbon growth (**a**) and (**b**) a growing ribbon [51.24]

However, the effective *zone length* is extremely large. Also, the fast pulling speed, lack of rotation, and capillary die restriction of back-mixing of impurities rejected at the solid–liquid interface back into the melt all reduce the effective segregation.

The string ribbon growth process is shown schematically in Fig. 51.10, along with its original ESP implementation using a quartz crucible and graphite filament *strings*. Two thin filaments enter the bottom of the crucible through small openings, pass upward through the melt, and are bridged above the melt by a seed plate. A liquid meniscus rises up to the seed plate and continually solidifies onto it as the string–seed plate assembly is moved upward at about 2 cm/min.

Carbon-based strings are typically used. They are easily wetted by the liquid silicon, and become an integral, imbedded part of the ribbon, but have the disadvantage of nucleating many small grains near the ribbon edges. The central portion of the ribbon, however, is reasonably large-grained. Other string materials (based, e.g., on alumina or quartz) can be used and nucleate much fewer grains at the edge [51.24], but the contact to the ribbon is less definite and the thermal-expansion mismatch causes them to partially or totally break away from the ribbon edge during cooling. Figure 51.11 shows the development of grain structure for graphite filaments and a graphite *seed* plate, as well as the lack of grain nucleation at the edge when quartz filaments are used. Graphite can be used for the crucible, as well as quartz.

A drawback of both EFG and string ribbon growth is the slow areal throughput. EFG addresses this by effectively growing eight ribbons simultaneously via the eight-sided tube structure. In modern string ribbon production [51.26], a long and narrow graphite mesa crucible is used to retain the continuously replenished melt by capillarity on its top surface. The long edges of the mesa are close and parallel to the ribbons, and the edges define the base of the meniscus from which the ribbons grow. Multiple pairs of strings (four in the present *quad* growth system) are used to delineate the edges of the ribbons. The mesa crucible thus plays a similar role to the die in EFG growth, the main difference being that the melt is replenished directly onto a region of the mesa top and flows laterally along it rather than arriving via distributed capillary channels as in EFG. The small volume of the melt zone reduces heater power requirements and hot-zone material requirements for growth, and the close proximity of the mesa edges to the ribbon helps to stabilize flat ribbon growth.

Impurity and dopant distribution in this system is basically governed by the same equation as for FZ growth, with some ribbon-to-ribbon variations in concentrations due to the path length differences from the replenishment point to each ribbon.

Large-area solid–liquid interface growth with heat removal perpendicular to the pulling direction has the potential to be the highest throughput silicon sheet growth method. A number of variants of the process have been tried in the past, but none are currently in large-scale commercial production. A representative schematic, in this case of the ribbon growth on sub-

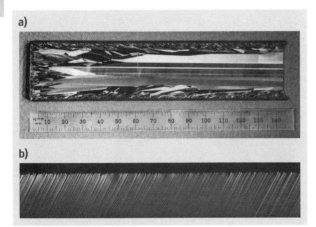

Fig. 51.11 (**a**) Development of grain structure in a 32 mm wide ribbon grown with graphite filaments and *seed* plate [51.25]. (**b**) A 10 mm long portion of the edge of a 20 mm wide ribbon grown with quartz filaments, showing no grain nucleation [51.24]

strate (RGS) process, is shown in Fig. 51.12. Some approaches utilize a moving substrate that is reusable. Others attempt to grow the ribbon off of the free melt surface in a crucible. The important aspect is that the length of the melt zone b be much greater than the sheet thickness t. Then fast pulling speeds can be achieved according to the equation

$$V_s = \frac{4\alpha K_m b}{(2K_m - \alpha t)tL\rho_m}\Delta T ,$$

where α is the effective coefficient of heat transfer, ΔT is the temperature gradient between melt and substrate (or free surface, if no substrate is used), L is latent heat of fusion, ρ_m is the density at the melting temperature, and K_m is the thermal conductivity at the melting temperature [51.27]. The equation predicts a 6 m/min growth rate at $\Delta T = 160\,°C$, and experimental pulling speeds near that value have been realized (Fig. 51.13). Horizontal growth of this type can be hundreds of times faster than vertical ribbon pulling, particularly if b and ΔT are maximized.

51.2.5 PV Silicon Crystal Growth Approaches

Of the many approaches that have been tried for PV silicon growth, only six are currently in commercial use. The traditional CZ method (and to a lesser extent, the FZ method) produces single-crystal silicon ingots that yield the highest-efficiency silicon solar cells. The DS and EMC multicrystalline ingot methods offer simpler operation and higher throughput (especially EMC) but a somewhat lower cell efficiencies. Ribbon growth eliminates the need for wafering and hence provides better feedstock utilization. The two methods in current use are EFG and string ribbon growth. They provide cell efficiencies similar to that of multicrystalline ingot. The family of horizontal, large solid–liquid interface

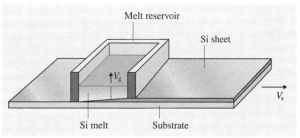

Fig. 51.12 Schematic of the RGS sheet growth process (after [51.22])

Fig. 51.13 Wide silicon sheet growing at 3 m/min (AstroPower, 2002)

techniques provides much faster sheet growth, but so far these have not been viable commercially, with cell performance limited by small grain size and high impurity content. Passivation of multiple, small grains has also proved difficult in thin-film silicon growth on substrates, and no commercially viable silicon thin-film growth approaches have been found yet.

51.3 Cell Fabrication Technologies

The main challenge of PV fabrication technologies is developing ways to cost-effectively mass-produce high-performing devices with the highest yield, reliability, and consistency. Capital efficiency, equipment efficiency, cost of production, and device performance have to be optimized to achieve these goals. In this section, in addition to the commercial cell fabrication technologies, a brief review of the advances in the silicon solar cell technologies currently being pursued by various researchers will be discussed. The basic device structure will be discussed and the commercial production tools and process will be highlighted.

51.3.1 Homojunction Devices

Solar cells manufactured by nine out of the top ten PV cell companies in 2005 were based on homojunction devices. In this structure, only one type of semiconductor

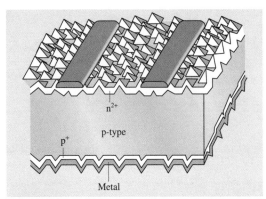

Fig. 51.14 Cross-section of a commercial silicon solar cell (after [51.28])

material, crystalline silicon, is used on both sides of the junction. The device structure is shown in Fig. 51.14.

A p-type crystalline silicon wafer (typically 225 μm thick and 156 mm × 156 mm in size) is used as the base substrate. After cleaning and/or texturing the front surface, n-type emitter is formed by a diffusion process (junction depth of about 0.3 μm). The front and back side are isolated by plasma or chemical etching or laser scribing. An antireflection coating of silicon nitride (about 75 nm thick and refractive index of about 2.10) is deposited using plasma-enhanced chemical vapor deposition, low-pressure chemical vapor deposition, or sputtering. The front (silver) and rear contacts (aluminum and silver) are screen-printed with metallic inks and subsequently fired to make good mechanical and electrical contact. The front metal coverage is typically 8%. The completed cells are then tested and classified according to the cell parameters. The commercial efficiencies of solar cells based on multi- and monocrystalline silicon are in the range 14.5–15.5 and 16.0–17.0%, respectively. The efficiency ranges are due to the material quality, cell design, and process tools. The efficiency of monocrystalline solar cells is higher as they can be more effectively surface-textured and the electronic quality of the material is better than that of multicrystalline silicon. In the following section, each of the process steps used in commercial cell sequence and the improvement techniques being evaluated will be described [51.29, 30].

Surface Damage Removal and Texturing

In this step, the surface damage resulting from sawing is etched off using hot, concentrated (10–30%) sodium hydroxide (NaOH) solution. Approximately 10 μm of silicon is removed during the process. About 200 wafers are etched in a batch process. Monocrystalline silicon substrates are subsequently textured using a low-concentration etch containing 2% sodium hydroxide and isopropyl alcohol (IPA). The (100) crystal planes are etched relatively faster than other planes. This results in the intersection of (111) planes and the exposed surface forms with tiny pyramids, about 3–5 μm in size. This process removes an additional 10 μm of silicon.

For the past 5 years, a new technique, isochemical texturing, has been commercialized for saw-damage removal and texturing for multicrystalline wafers. The process involves use of an acid-based etch solution [hydrofluoric (HF) and nitric acid (HNO_3)] which is a standard etch in semiconductor wafer preparation. The main difference is that the etch solution is kept at a constant temperature, below 10 °C. Only about 10 μm of silicon is removed, resulting in a substantial saving in raw material. Texturing of multicrystalline silicon wafers increases the conversion efficiency of the cells by reducing reflection and enabling advanced cell design parameters. A special feature of the process is that it can be performed as an inline process, hence enabling the use of thinner silicon substrates.

Emitter Formation

The semiconductor junction is formed by phosphorus diffusion across the entire front surface. The process is carried out in a tube furnace using $POCl_3$ as the dopant source. Inline belt furnace diffusion (using phosphoric acid solution as the dopant source) is increasingly being used to enable thin wafer processing. The semiconductor junction of the majority of commercial solar cells is about 0.3–0.5 μm and the surface concentration is about 5×10^{20} cm^{-3}. The sheet resistivity (a measure of lateral resistance in the n-type doped layer) of the commercial cells is about 50 Ω/square. The diffusion of dopant from a nearly infinite source into silicon at high temperature > 860 °C results in a complementary error function (erfc) or similar distribution of phosphorus (as shown in Fig. 51.15).

Excess phosphorus beyond the solid solubility limit is precipitated as inactive phosphorus in a silicon region called the *dead layer*. In this region the minority-carrier lifetime is significantly reduced or the generated carriers are recombined instantly. The high surface concentration of phosphorus and low sheet resistivity are not optimum conditions for maximizing the generation of carriers. However, widespread use of this approach is due to its compatibility with high-volume manufacturing tools such as screen-printing metallization practiced

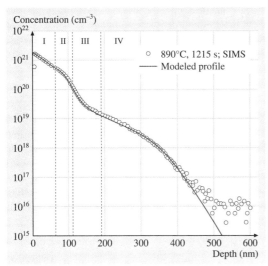

Fig. 51.15 Diffusion profile of the silicon surface (after [51.31]) (SIMS – secondary ion mass spectroscopy)

in the surface-mount assembly industry. There are several approaches being actively investigated to enhance the performance of solar cells by reducing the dead layer and developing screen-printing paste compatible with higher sheet resistivity and lower surface concentration of phosphorus [51.31].

Junction Isolation

During the diffusion step, the edge and the rear of the wafers also gets diffused. Hence, to prevent leakage paths, the front and the rear need to be isolated. There are several techniques used to achieve this in commercial solar cell manufacturing. One of the widely used techniques is a plasma etch by which the edges of coin-stacked wafers are etched. However, due to the textured surface of the wafer, some active area of the cell in the front surface is also etched. Another technique involves the use of a laser system, where the edges of the wafers (front or rear) are trenched or cleaved. During the past 3 years, a new technique has been commercialized, enabling thin wafer processing. This technique is based on chemically etching off the rear phosphorus junction without etching the front [51.33].

Phosphorus Glass Removal

In this step, the phosphorus glass formed during the diffusion step is removed using a dilute hydrofluoric acid etch. The glass is very thick (20–40 nm) and would affect the effectiveness of the antireflection layer that will be deposited later in the process. This step can be done as a batch or inline process.

Antireflection Coating Deposition

The goal of this step is to minimize the reflection of incident light, thereby increasing PV efficiency. The weighted average of reflected sunlight from a bare silicon surface is about 30%. Under glass in the encapsulated stage, this value can be reduced to about 15%. By texturing the surface the reflection can be reduced to 10%. In order to further decrease the reflectance, an antireflection (AR) coating is applied to solar cells. By selecting an appropriate film thickness and refractive index (RI) of the AR coating, the reflection can be reduced to below 4%. Apart from ability to reduce refection, the AR coating must be transparent, i.e., it should not absorb incident sunlight.

The most widely used antireflection coating is a silicon nitride film (about 75 nm thick with RI of about 2.10), deposited by plasma-enhanced chemical vapor deposition (PECVD). In addition to providing the antireflection properties, the silicon nitride deposition process introduces hydrogen which diffuses into bulk silicon. This hydrogenation improves the electronic quality of surface and bulk silicon by means of passivation [51.34].

Metallization and Contact Sintering

In this step, both top and rear contacts are printed and sintered. A typical front pattern is shown in Fig. 51.16.

The most widely used technique to deposit metal paste is screen printing. The top contact is a paste containing silver with organic and glass binders. On the

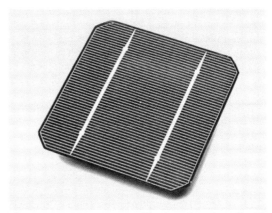

Fig. 51.16 Front contact metallization of a solar cell with antireflection coating [51.32]

rear, an aluminum paste is printed in all areas except where silver paste is applied to make contacts for the external circuit. The aluminum paste after sintering provides a p^+ surface (back surface field) and additional gettering of impurities in the bulk silicon. The pastes are applied sequentially with a drying step between each printing step. The printed wafers are then fired to make contact with the silicon. The front paste fires through the silicon nitride layer and makes contact to the n-type layer. If the sintering step is very aggressive (higher temperature, longer duration), the metal will make contact to the p-type bulk silicon, resulting in electrical shorting. On the other hand, if the paste is not sintered adequately the strength of the bond to the interconnection between two solar cells will be weak and will not have the required chemical and mechanical contact properties. Hence the selection of the paste, printing height, and the firing conditions are very important in determining the performance of the cell. This step determines many of the cell parameters and long-term performance of the solar cell. The strength of the interconnection between the contacts and the tabbing ribbon is affected by the paste/sintering characteristics and the soldering process [51.35].

Testing and Binning

In this last step, the solar cell is measured using a sun simulator at standard testing conditions (STC), i.e., irradiance of $1000\,\text{W/m}^2$ with an air mass 1.5 spectrum (AM 1.5) at $25\,°\text{C}$. STC is selected as it is easy to reproduce in the laboratory. The cell is tested against a calibrated cell. Cell performance in air will be different from its performance under encapsulated conditions due to the difference in optical coupling and the increased series resistance. The cells are sorted and binned according to their electrical performance and various mechanical and visual defects such as cosmetics, color nonuniformity, edge chips etc. The finished cells are packed for further processing into modules. The modules will operate at conditions other than STC, hence the module manufacturer provides temperature coefficients of voltage and current as well as predicted performance as a function of irradiance.

51.3.2 Enhancing Solar Cell Performance

The gap between the efficiency of laboratory cells and commercial cells is decreasing with the commercialization of several high-efficiency techniques. Efficient solar cell design involves maximization of carrier generation and carrier collection. The generation of carriers in a silicon solar cell depends on the electronic quality of substrates (minority-carrier lifetime), the active area (the area not covered by metal contact lines), spectral response, absence of dead layer, etc. The collection of carriers depends on bulk and surface passivation and the lateral and contact resistance.

The performance of conventional commercial cells is limited by the requirement of the metal pastes, the capability of the printing equipment, and surface recombination of the generated carriers. The front metal paste requires heavier phosphorus diffusion and increased surface concentration of phosphorus. This requirement affects current generation and current collection. The use of fine line-printing techniques, better screens, and better metal pastes reduces the shading loses due to metal coverage. The development of new pastes compatible with lower surface concentration of phosphorus and higher sheet resistivity will increase the current generation capability. The incorporation of a surface passivation layer will enhance the beneficial effects of cells with higher emitter sheet resistivity and thin cells. In addition, thin cells (less than $150\,\mu\text{m}$) require a back surface field which does not bow the cells. All of these topics are being actively pursued by several paste manufacturers in collaboration with cell technologists.

51.3.3 Advanced Commercial Solar Cell Concepts

In this section, a few of the commercial sequences which are not based on conventional screen-printing technology are presented. These sequences are applicable to limited types of substrates. However, several research institutions and PV companies are trying to incorporate some of the concepts of these cells into conventional screen-printing sequences.

Buried-Contact Cells

The buried-contact solar cell sequence is one approach which has been commercialized to overcome the limitation of screen printing. Cell efficiencies up to 18% (laboratory efficiency 22.7% record cell) have been demonstrated on monocrystalline (CZ) wafers [51.36]. In this sequence the grooves are formed on the top surface into which metal contacts are plated (Fig. 51.17).

Additional high-efficiency concepts incorporated in this sequence are surface passivation and a shallow emitter with deep diffusion under the contacts. In addition to an increased number of steps, the sequence involves several high-temperature steps. Hence it re-

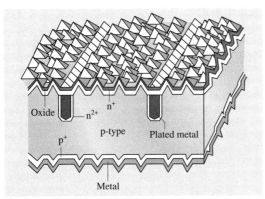

Fig. 51.17 Cross-section of a buried-contact solar cell (after [51.28])

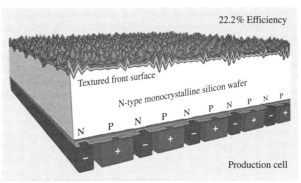

Fig. 51.18 Back-contact solar cell [51.37]

quires monocrystalline silicon wafers with low oxygen content. This limits the widespread commercialization of buried-contact solar cells.

Back-Contact Cells

A back-contact (interdigitated contact) cell sequence has been commercialized for high-lifetime n-type wafers. Cell efficiencies up to 22.2% have been demonstrated in large-scale production [51.38]. The sequence includes a textured front surface with antireflection coating, well-passivated surfaces, and screen-printed metal contacts on the rear. The special feature of the cell (shown in Fig. 51.18) is that all the contacts are on the rear, thereby reducing the shading losses and improving the cosmetic appearance of the cells.

HIT Cells

The heterojunction with intrinsic thin layer (HIT) cell sequence has been commercialized. Efficiencies of HIT cells up to 21.8% have been reported on n-type substrates (as shown in Fig. 51.19). The unique feature of

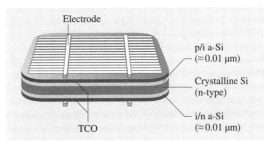

Fig. 51.19 HIT cell structure [51.37]

this sequence is the incorporation of a thin amorphous silicon layer on both surfaces of the solar cell [51.39]. These layers provide high-quality surface passivation to achieve a very low surface recombination velocity. An additional advantage of this design is that the cell can be bifacial, thereby converting backscattered light into useful power output. TCO in Fig. 51.19 refers to a thin-film conducting oxide.

It is exciting that many of these advanced cell concepts, once only achievable in laboratory-scale devices, are now being used successfully in manufacturing.

51.4 Summary and Discussion

The silicon photovoltaic industry has been on a rapid growth path over the past decade – on the order of 30–40% per year. As of 2007, the consumption of high-purity silicon for solar cells has exceeded the amount used for all other electronic applications. The rapid growth has presented challenges in all segments of the PV value chain (Fig. 51.2). However, as these challenges are met and the industry benefits from economies of scale, the technology becomes ever stronger and closer to meeting the cost and performance milestones that will establish it as a viable source of a significant portion of world energy needs.

The well-established technology base and ready availability, proven performance, and salubrity of silicon, coupled with economies of scale in larger factories, will likely allow Si to remain the dominant PV material for the foreseeable future. The rapidly increasing demand for polycrystalline silicon feedstock for PV use has caused a disruption in the demand/supply ratio, but this is not a fundamental problem, nor does it represent a fundamental shortage. It is, rather, a planning/forecasting issue which was exacerbated by: (1) investor caution resulting from the economic and high-technology downturn around the year 2000, and (2) investor skepticism of the viability and continued growth projections of silicon photovoltaic technology. Part of this investor reluctance results from uncertainties about whether there will be a future role of compound semiconductor thin films in photovoltaics. The performance and costs of the thin-film approaches have not so far been competitive with silicon. Furthermore, the Earth's crust contains 27.7% Si, in contrast to 0.00002% Cd, 0.00001% In, 0.000009% Se, and 0.0000002% Te (commonly used thin-film elements). So it is difficult to see how improved costs for thin-film PV can come to fruition despite the fact that they use a thinner layer of semiconductor material. The temporary shortage of polycrystalline feedstock has instigated numerous metallurgical approaches to the purification of silicon, in the hope that adequate purity for solar cells can be achieved at a lower cost than that of the Siemens process. However, this is a difficult challenge because 99.9999% or six-9s pure silicon is required for PV.

In single-crystal CZ ingot growth, we are likely to see increased effort to make hot zones more energy efficient, to grow larger diameters, and to achieve continuously melt-replenished long growth runs. There will be continuing effort to achieve more wafers per length of ingot, and to take advantage of potentially higher cell efficiencies afforded by thinner wafers when back surface fields are used in the cell design. Despite the potential advantages of FZ material, it is unlikely that its role in PV will increase significantly because of higher costs for the crack-free, long cylindrical feedstock it requires and the difficulty in producing the larger FZ diameters. Multicrystalline casting, directional solid-ification, and electromagnetic casting command the largest share of the Si PV market. This trend is likely to continue because the processes and equipment are simpler and the throughputs are higher (especially for electromagnetic casting) by a factor of 5–20.

Even though ribbon and sheet growth technologies have the advantage of minimal silicon consumption and elimination of wafering, it is difficult for them to compete effectively with the higher throughput of ingot technologies. Their challenge will be to increase areal throughput via wider ribbons, multiple ribbons or other approaches. The effective areal throughputs of ingot growth range between $30\,m^2$/day and $600\,m^2$/day per machine. By contrast, a capillary die octagon tube growth machine produces about $20\,m^2$/day and string-supported ribbon machines produce about $8-9\,m^2$/day. We will probably see continued progress in horizontally pulled, large-area solid–liquid interface sheets by some variant of the horizontal growth method because the throughput potential is enormous and one growth furnace could easily generate material for $35\,MW/a$ or more of solar cell production.

The future is expected to bring continued exploration of thin-layer Si growth approaches, in search of ones that have significant economic advantages over the best ingot and sheet techniques. Successful ones will have fast deposition rates, large grain sizes, high efficiencies (at least 14% production efficiency), compatibility with low-cost substrates, and amenability to low-cost cell fabrication schemes. It is not likely that production of thin-layer Si PV modules will account for a significant fraction of the mainstream PV market for at least 10 years, although they, like the ingot and sheet approaches, would have substantial advantages over many other thin-film PV approaches. These include the simple chemistry, the relative abundance of the Si starting material, compatibility with SiO_2 surface passivation, and relative lack of toxicity.

References

51.1 T.F. Ciszek: Silicon for solar cells. In: *Crystal Growth of Electronic Materials*, ed. by E. Kaldis (Elsevier Science, Amsterdam 1985) pp. 185–210

51.2 J.W. Pichel, M. Yang: 2005 Solar Year-end Review and 2006 Solar Industry Forecast Renewable Energy World (2006) http://www.renewableenergyworld.com/rea/news/infocus/story?id=41508 (last accessed October 8, 2007)

51.3 H.J. Möller: *Semiconductor for Solar Cells* (Artech House, London 1993)

51.4 M.A. Green: *Solar Cells* (Prentice Hall, Englewood Cliffs 1982)

51.5 A. Rohatgi: Road to coast-effective crystalline silicon photovoltaics, Proc. 3rd World Conf. Photovolt. Energy Convers., Osaka (2003)

51.6 S. Narayanan, J. Wohlgemuth: Cost-benefit analysis of high-efficiency cast polycrystalline silicon solar cell sequences, Prog. Photovolt. **2**(2), 121–128 (1994)

51.7 P.C. de Jong: PV Module Technology, Energy Research Centre of the Netherlands,

51.8 G.L. Pearson, C.S. Fuller, D.M. Chapin: A new silicon p–n junction photocell for converting solar radiation into electrical power, J. Appl. Phys. **25**, 676 (1954)

51.9 S.N. Dermatis, J.W. Faust: Process for producing an elongated unitary body of semiconductor material crystallizing in the diamond cubic lattice structure and the product so produced, US Patent 3129061 (1964)

51.10 H.E. LaBelle Jr.: Growth of controlled profile crystals from the melt, part II: edge-defined, film-fed growth, Mater. Res. Bull. **6**, 581 (1971)

51.11 T.F. Ciszek: Edge-defined, film-fed growth of silicon ribbons, Mater. Res. Bull. **7**, 731–737 (1972)

51.12 T.F. Ciszek: Melt growth of crystalline silicon tubes by a capillary action shaping technique, Phys. Status Solidi (a) **32**, 521–527 (1975)

51.13 T.F. Ciszek, J.L. Hurd: Melt growth of silicon sheets by edge-supported pulling. In: *Proc. Symposia on Electronic and Optical Properties of Polycrystalline or Impure Semiconductors and Novel Silicon Growth Methods*, ed. by K.V. Ravi, B. O'Mara. St. Louis, 1980, Electrochem. Soc. Proc. **80**(5), 213–222 (1980)

51.14 T.F. Ciszek: Electromagnetic and float-zone methods for high-purity silicon solidification. In: *Containerless Processing Techniques and Applications*, ed. by W.F. Hofmeister, R. Schiffman (The Minerals, Metals & Materials Society, Warrendale 1993) pp. 139–146

51.15 H. Fischer, W. Pschunder: Low cost solar cells based on large area unconventional silicon, IEEE 12th Photovolt. Spec. Conf. Rec. (IEEE, New York 1976) pp. 86–92

51.16 T.F. Ciszek, G.H. Schwuttke, K.H. Yang: Directionally solidified solar-grade silicon using carbon crucibles, J. Cryst. Growth **46**, 527–533 (1979)

51.17 T.F. Ciszek: Some applications of cold crucible technology for silicon photovoltaic material preparation, J. Electrochem. Soc. **132**, 963–968 (1985)

51.18 T.F. Ciszek: Method and Apparatus for Casting Conductive and Semiconductive Materials, US Patent 4572812 (1986)

51.19 A. Endros, G. Martinelli: Silicon Semiconductor Wafer Solar Cell and Process for Producing Said Wafer, US Patent 5702538 (1997)

51.20 T.F. Ciszek: A graphical treatment of combined evaporation and segregation contributions to impurity profiles for zone-refining in vacuum, J. Cryst. Growth **75**, 61–66 (1986)

51.21 S. Peizulaev: Segregation and evaporation of impurities during zone refining, Inorgan. Mater. **3**, 1329 (1967)

51.22 T.F. Ciszek: Silicon crystal growth for photovoltaics. In: *Crystal Growth Technology*, ed. by H.J. Scheel, T. Fukuda (Wiley, Sussex 2003) pp. 267–289

51.23 T.F. Ciszek: Some applications of cold crucible technology for silicon photovoltaic material preparation, J. Electrochem. Soc. **132**, 963 (1985)

51.24 T.F. Ciszek, J.L. Hurd, M. Schietzelt: Filament materials for edge-supported pulling of silicon sheet crystals, J. Electrochem. Soc. **129**, 2838–2843 (1982)

51.25 J.L. Hurd, T.F. Ciszek: Semicontinuous edge-supported pulling of silicon sheets, J. Cryst. Growth **59**, 499 (1982)

51.26 E.M. Sachs: Method and Apparatus for Crystal Growth, US Patent 20060249071 (2006)

51.27 H. Lange, I.A. Schwirtlich: Ribbon growth on substrate (RGS) – A new approach to high speed growth of silicon ribbons for photovoltaics, J. Cryst. Growth **104**, 108–112 (1990)

51.28 M.A. Green: *Crystalline Silicon Solar Cells* (World Scientific, New York 2001), http://www.worldscibooks.com/phy_etextbook/p139/p139_chap4.pdf

51.29 P. Manshanden, A.R. Burgers, A.W. Weeber: Wafer thickness, texture and performance of multicrystalline silicon solar cells, Solar Energy Mater. Solar Cell. **90**, 3165–3173 (2006)

51.30 B. González-Diaz, R. Guerrero-Lemus, D. Borchert, C. Hernández-Rodriguez, J.M. Martinez-Duart: Low-porosity porous silicon nanostructures on monocrystalline silicon solar cells, Physica E **38**, 215–218 (2007)

51.31 A. Bentzen, J.S. Christensen, B.G. Svensson, A. Holt: Understanding phosphorus emitter diffusion in silicon solar cell processing, Proc. 21st Eur. Photovolt. Sol. Energy Conf., Dresden (2006) pp. 1388–1391

51.32 US Department of Energy, http://www1.eere.energy.gov/solar/pv_systems.html (last accessed August 17, 2005)

51.33 D. Kray, S. Hopman, A. Spiegel, B. Richerzhagen, G.P. Willeke: Study on the edge isolation of industrial silicon solar cells with waterjet-guided laser, Sol. Energy Mater. Sol. Cell. **91**, 1638–1644 (2007)

51.34 A.G. Aberle: Surface passivation of crystalline silicon solar cells: a review, Prog. Photovolt. Res. Appl. **8**(5), 473–487 (2000)

51.35 M. Edwards, J. Bocking, J.E. Cotter, N. Bennett: Screen-print selective diffusions for high-efficiency industrial silicon solar cells, Prog. Photovolt. Res. Appl. **16**(1), 31–45 (2008)

51.36 J. Zhao: Recent advances of high-efficiency single-crystalline silicon solar cells in processing technologies and substrate materials, Sol. Energy Mater. Sol. Cell. **82**, 53–64 (2004)

51.37 R.M. Swanson: Photovoltaics: The path from niche to mainstream supplier of clean energy, Sol. Power 2006 Conf. (2006), http://www.tvworldwide.com/events/eqtv/061016/ppt/Swanson.pdf

51.38 K.R. McIntosh, M.J. Cudzinovic, D.D. Smith, W.P. Mulligan, R.M. Swanson: The choice of silicon wafer for the production of low-cost rear-contact solar cells, 3rd World Conf. Photovolt. Energy Convers., Osaka (2003) pp. 11–18

51.39 M. Tanaka, S. Okamoto, S. Tsuge, S. Kiyama: Development of HIT solar cells with more than 21% conversion efficiency and commercialization of highest performance HIT modules, Proc. 3rd World Conf. Photovolt. Energy Convers., Osaka (2003) pp. 955–958

52. Wafer Manufacturing and Slicing Using Wiresaw

Imin Kao, Chunhui Chung, Roosevelt Moreno Rodriguez

Wafer manufacturing (or wafer production) refers to a series of modern manufacturing processes of producing single-crystalline or poly-crystalline wafers from crystal ingot (or boule) of different sizes and materials. The majority of wafers are single-crystalline silicon wafers used in microelectronics fabrication although there is increasing importance in slicing poly-crystalline photovoltaic (PV) silicon wafers as well as wafers of different materials such as aluminum oxide, lithium niobate, quartz, sapphire, III–V and II–VI compounds, and others. Slicing is the first major post crystal growth manufacturing process toward wafer production. The modern wiresaw has emerged as the technology for slicing various types of wafers, especially for large silicon wafers, gradually replacing the ID saw which has been the technology for wafer slicing in the last 30 years of the 20th century. Modern slurry wiresaw has been deployed to slice wafers from small to large diameters with varying wafer thickness characterized by minimum kerf loss and high surface quality. The needs for slicing large crystal ingots (300 mm in diameter or larger) effectively with minimum kerf losses and high surface quality have made it indispensable to employ the modern slurry wiresaw as the preferred tool for slicing. In this chapter, advances in technology and research on the modern slurry wiresaw manufacturing machines and technology are reviewed. Fundamental research in modeling and control of modern wiresaw manufacturing process are required in order to understand the cutting mechanism and to make it relevant for improving industrial processes. To this end, investigation and research have been conducted for the modeling, characterization, metrology, and control of the modern wiresaw manufacturing processes to meet the stringent precision requirements of the semiconductor industry. Research results in mathematical modeling, numerical simulation, experiments, and composition of slurry versus wafer quality are presented. Summary and further reading are also provided.

52.1 From Crystal Ingots to Prime Wafers 1721
    52.1.1 Semiconductor Single-Crystalline Wafers 1721
    52.1.2 Alternative Wafer Production Processes 1722
    52.1.3 Substrate Manufacturing with a System-Oriented Approach .. 1723

52.2 Slicing: The First Postgrowth Process in Wafer Manufacturing 1726
    52.2.1 ID Saws 1726
    52.2.2 The Modern Wiresaw 1727
    52.2.3 Saws with Diamond-Impregnated Wires . 1727
    52.2.4 Others.. 1728
    52.2.5 Comparison of Slicing Technology ... 1728
    52.2.6 Wafer Manufacturing for Large Wafers 1729

52.3 Modern Wiresaw in Wafer Slicing 1730
    52.3.1 Definition of Modern Wiresaw 1730
    52.3.2 Modern Wiresaw Technology 1731
    52.3.3 Modeling and Control of the Modern Wiresawing Process . 1731

52.4 Conclusions and Further Reading 1733

References 1733

Wafer manufacturing (or *wafer production*) refers to a modern process of producing single- or polycrystalline wafers from crystal ingots (or *boules*) of different sizes and materials. The majority of wafers are single-crystalline silicon wafers used in microelectronics fabrication. Due to the increasing importance of the pho-

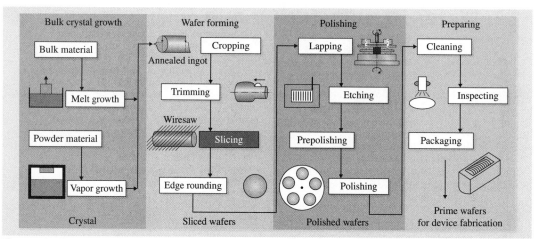

Fig. 52.1 Wafer manufacturing: from crystal ingots to prime wafers. Four categories of manufacturing operations are illustrated: bulk crystal growth, wafer forming from crystalline ingot, wafer polishing to produce prime wafers, and wafer preparing. Each category includes several operations

tovoltaic (PV) industry, polycrystalline silicon wafers (although some PV wafers are single-crystalline silicon) are being produced in increased volumes. Wafers of different materials such as aluminum oxide, lithium niobate, quartz, and others have also been produced.

The health of the wafer manufacturing industry has become a good economic indicator in this microelectronics and information age; for example, when the economy suffered a downturn in the USA around 2001, the wafer production industry also suffered a drop in production due to lower demands. On the other hand, when the economy was picking up a few years afterwards, demand for wafer production heralded this recovery before it happened. With greater reliance on microelectronics, computers, and information worldwide in the 21st century, this trend is expected to become stronger than ever.

Wafer manufacturing includes a series of processes, beginning with crystal growth and ending with prime wafers, as illustrated in Fig. 52.1, in which a process flow of wafer manufacturing with various categories of operations is shown [52.1]. Slicing is the first major postgrowth wafer-forming process, and is primarily accomplished using various technologies, discussed in Sect. 52.2. The first modern wiresaws were designed and built by HCT. The first generation of industrial wiresaws were employed to slice polysilicon photovoltaic (PV) wafers. Since the introduction of the technology of the slurry wiresaw in the early 1990s, this approach has seen a dramatic increase in usage

in the USA and worldwide, replacing thousands of inner-diameter (ID) saws, primarily for the following reasons:

- Acceptance of slurry wiresaws for the slicing of silicon wafers, especially for single-crystalline silicon wafers in the microelectronics industry at the turn of the century.
- In contrast to ID sawing, wiresaw slicing results in higher throughput and can slice large wafers of different crystal materials.
- Recent research advances in slurry wiresaw and slicing manufacturing processes [52.1–78] have made significant contributions to the understanding of fundamental manufacturing process modeling, promoted research and development (R&D), and provided insights into parameter optimization and process control.
- The versatility of wiresaws to slice wafers of various sizes made of different materials.

When the single-crystalline silicon wafer manufacturing industry finally adopted the wiresaw as the main slicing tool towards the end of the 1990s, it prompted a wholesale replacement of thousands of inner diameter (ID) saws which had been the only slicing tool for decades until then. Many wiresaw equipment producers also appeared on the horizon, as well as ancillary industries for the supply of consumables such as saw wire, abrasive grits, carrier fluid, etc. In addition to the silicon industry, wiresaws have also been employed to slice crystal wafers such as LiNbO$_3$, SiC, InP, quartz, sapphire, aluminum oxide, and others.

As slurry wiresaw technology continues to become more mature and widely employed, applications in other areas also arise. A case in point is the development in the application of wiresaw reported in July 2005 [52.59]: a special-purpose wiresaw designed by HCT for Corning Inc., which weighs 50 t with a height of 8 m (much bigger than regular industrial wiresaws), used for slicing fused silica for the production of liquid-crystal display (LCD) image mask blanks, used in the process by which makers of liquid-crystal displays imprint a pattern of electrical circuits onto the *mother glass* for computer and television displays. This example illustrates the potential of modern wiresaws in a variety of applications in the future due to the unique capability, configuration, and nature of the manufacturing process.

52.1 From Crystal Ingots to Prime Wafers

Modern *wafer manufacturing* (also called *wafer production*) includes a plethora of manufacturing operations by which prime wafers are produced from crystal ingots [52.1, 79]. For example, semiconductor devices are built on high-quality substrates and economics demands thin wafers (500–900 μm thick) of large area. Quality requirements of prime wafers include a high degree of crystalline perfection, low defect, low microstructure and subsurface damage, global surface planarization, uniform thickness, and very low residual stresses. Only when these demands are met can the full potential of the various semiconductors be realized. Figure 52.1 illustrates a flow of various processes and categories associated with the production of prime wafers from crystalline ingots.

The technology to produce semiconductor wafers is driven by many factors. New materials result in new applications that, in turn, inspire the design and engineering of new and higher-quality materials, new processing technologies, and special equipment. Broader applications of these wafers and economics demand continuous reduction in costs, improvements in quality, and innovations in materials and manufacturing technology. For example, both the size and quality of silicon substrates have been constantly increasing, and 400 mm-diameter wafer has already been used, although 300 mm wafers are the most well established today. The 400 mm silicon wafer has gradually become the standard in today's microfabrication industry, and will eventually replace 300 mm fab lines. In a similar way, the reduction in silicon wafer costs can spur exponential growth in the consumer market of solar cells.

In the following subsections, the process and operations of wafer production will be discussed.

52.1.1 Semiconductor Single-Crystalline Wafers

Typical semiconductor materials, such as silicon, lithium niobate, III–V compounds, II–VI compounds, and others, are hard and brittle. Silicon constitutes more than 90% of the total consumption of semiconductor materials, and has been extensively used in electronic and photovoltaic (PV) industries. The wafer manufacturing processes from ingot to prime wafers are illustrated in Fig. 52.1. As shown in the figure, slicing is the first major postgrowth process in wafer manufacturing. Recently, wiresaw slicing has emerged as a leading technology of wafer preparation in the semiconductor industry, especially for PV and large silicon wafers due to its capability of cutting ingots of large diameter (e.g., 300 mm and up) with small kerf loss and high yield. Nevertheless, the trend of using large wafers (with diameter of 300 mm or higher) as the future standard in microelectronic fabrication, projected by National Technology Roadmap for Semiconductors (NTRS), Semiconductor Industry Association (SIA), and Semiconductor Manufacturing Technology consortia (SEMATECH) [52.80–83], has virtually eliminated the conventional ID saw as a slicing tool, making the wiresaw an indispensable tool for wafer forming.

The importance of manufacture of silicon in the USA is illustrated by the statistics showing that in 1993 the US\$ 3 billion silicon resulted in an average of US\$ 6.67 billion device market per month and a total of US\$ 700 billion electronic equipment [52.80]. This figure has increased to sales of US\$ 10.97 billion device in 1998 [52.82] – a 65% increase over 6 years. Apart from the increase in quantity, quality requirements have also increased. Growing demand in chip surface area produced annually is expected with the expansion of microelectronics, telecommunication, microelectromechanical systems (MEMS), and medical applications. Accompanied by the increase in wafer diameter, there is an increasing degree of integration of components and a decreasing structural width.

The objective of wafer manufacturing (or *wafer production*) for semiconductor industry is to produce prime wafers ready for microelectronics or PV solar cell fabrication, from single- and polycrystalline ingots. Four categories are outlined in Fig. 52.1 and described in the following. For the fabrication of PV solar cells, wafers are often processed as-sliced, without going through the rigorous process in Fig. 52.1, which is typically used for the production of microelectronic-scale silicon prime wafers.

Crystal Growth

Silicon ingots are grown using various techniques of crystal growth as presented in the earlier chapters of this Handbook, including bulk melt growth, powder vapor growth, and others, as illustrated in Fig. 52.1. A casting process is often employed for polysilicon ingots, normally in square shape, for PV wafers. Typical single-crystalline silicon ingots assume a cylindrical shape when grown, up to many feet tall, depending on the process.

Once the crystal ingot is grown it is processed to form wafers and substrates. The processes can be broken into three main categories, as described in the following.

Wafer Forming

The second group of manufacturing operations, also called *wafer forming* as shown in Fig. 52.1, forms the shape of thin wafers. It includes the following operations:

- Cropping both ends of the ingot (especially the tail end, which tends to have higher impurity)
- Trimming it to have a cylindrical shape with consistent cross-sectional area, followed by grinding of the orientation flat or notch to identify the crystalline orientation
- Slicing the ingot into slices of wafers
- Edge-rounding to smooth and remove tiny fractures which may have been formed along the edges of the wafer during the forming operations

Wafer Polishing

The objectives of this group of manufacturing operations are (1) to remove surface waviness and subsurface damage from the forming operations, and (2) to polish wafers to high precision (submicron surface roughness) with a mirror surface finish. It includes a series of operations as follows:

- *Lapping:* This is typically the first post-slicing process, based on the mechanical free abrasive machining (FAM) process to remove the surface waviness after slicing. This involves abrasives suspended in slurry to perform the mechanical process of removing the surface roughness and subsurface damage. An initial global planarization is achieved after this process.
- *Grinding:* As the size of silicon wafers becomes larger (300 mm or larger), and slicing is performed by slurry wiresaws, there are increasing demands on the grinding of the surface after slicing operations. One of the main reasons for the employment of the grinding operation is the removal of the surface waviness produced by the slicing process. Grinding using diamond tools is a faster operation than lapping. Often the demands of double-sided wafers also require grinding to be performed with consideration of elimination of the elastic spring-back effect that has commonly been found in lapping of large wafers.

The ground wafers can be subject to subsequent lapping and polishing processes.

- *Polishing:* This is the final process in this category, to render well-polished wafers ready for photolithography and microelectronics fabrication to produce microelectronic chips. The standard process at this stage is chemical–mechanical polishing (CMP). Wafers at this stage are called prime wafers.

Wafer Preparing

The category of *wafer preparing* refers to steps through which the prime wafers are cleaned, inspected for defects to assure quality, and packaged in a boat ready to be shipped.

52.1.2 Alternative Wafer Production Processes

While the wafer manufacturing process outlined in Fig. 52.1 represents the general operations in wafer production, different operations in this production process may be skipped, simplified or augmented, depending on the substrate type and the requirements on surface finish. For example, the generic process can be slightly altered with epitaxial growth of a thin

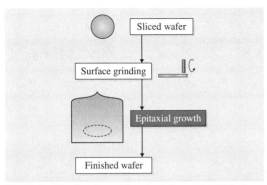

Fig. 52.2 Sliced wafers with epitaxial growth of thin layer of silicon for device fabrication. The sliced wafer is subject to surface grinding using a diamond grinding wheel, followed by the epitaxial growth process to render a finished wafer for device fabrication

layer (5–60 µm) of silicon. By so doing, the etching and CMP process can be bypassed to yield a finished wafer directly. An illustration of such a process is shown in Fig. 52.2. Certain logic gates and devices have been fabricated using these types of epitaxial wafers.

Another example is the processing of photovoltaic (PV) wafer. PV wafers are often employed as-sliced without further polishing after slicing.

52.1.3 Substrate Manufacturing with a System-Oriented Approach

The key processes and challenges in wafer manufacturing are summarized in Fig. 52.3, which shows the flow of operations discussed above. The values after each box represent typical surface roughness. Various issues related to each category of manufacturing operations are summarized in the figure for the manufacturing operations ranging from slicing, lapping, and polishing to inspecting, cleaning, and packaging.

Wafer Forming and Polishing Challenges
Tables 52.1 to 52.5 outline various important issues for consideration at every stage of the wafer production. In Table 52.1, the applications and corresponding issues in wafer forming and slicing are identified, along with research methodology. Table 52.2 outlines different applications in grinding, lapping, and polishing and relevant issues with some possible research approaches, including also considerations of materials and process control.

Wafer Preparing Challenges
Current cleaning, inspection, and metrology techniques used by industry were developed 10–20 years ago. Many recent innovations in metrology and laser-based methodology can satisfy modern demands and be

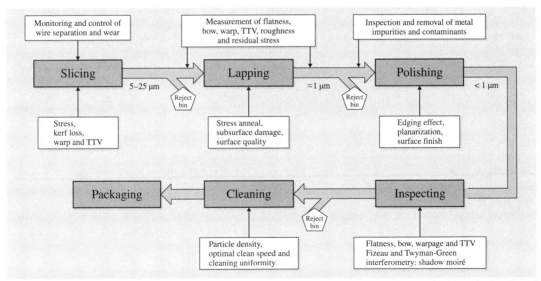

Fig. 52.3 Challenge and important process parameters in the wafer manufacturing, illustrated with typical scales of surface finish after each operation, as well as issues and challenges associated in each operation and process flow

Table 52.1 Wafer forming: potential applications and issues of consideration, as well as research approaches and methodology. Some items represent future considerations

Potential applications	Issues	Research methodology
Large-diameter substrate	Residual stress, warpage, TTV, kerf loss, dynamics, vibration	Solid and fracture mechanics modeling, stiffness control, stress relaxation, finite element method (FEM), subsurface damage
Apply cutting Si to other crystals	Different properties, process	Model-driven design, flexible manufacturing, optimized automation
Moderate number of slices (10–50): no current technology, cost-effective	Flexibility and versatility on number of slices	Model-based versatile design and fabrication in research and development
Cost reduction	Consumable, kerf loss, materials loss during forming process	Interconnected growth process control for more uniform crystal to reduce crop/trim loss; reduce kerf loss
Brittle, ductile, and very hard materials	Abrasive, wire speed, surface quality	Material- and size-independent substrate manufacturing strategies and equipment

Table 52.2 Applications and issues of consideration for lapping, grinding, and polishing in the wafer production process

Potential applications	Issues	Research methodology
Large-diameter substrate	Residual stress, warpage, TTV, kerf loss, dynamics, vibration	Modeling of stress influence, chemical versus mechanical reactions, requirements (0.35–0.13 μm), high level
Prime wafers of large diameter/size	Global planarization, parallelism of surfaces of wafers, quick removal of initial surface waviness and roughness	Geometry analysis, elastic analysis in grinding, double-sided grinding
Different materials, e.g., ultrahard materials	Abrasive, lapping speed, microcracks, defects, contaminants	Control strategies, holder design, expert system, experimental validation
Alternative processes for polishing, e.g., epitaxial growth	Process control, surface finish, orientation tolerance, flatness tolerance	Model-driven design, Taguchi methods, statistical process control (SPC), total quality management (TQM)

Table 52.3 Wafer preparation: the identification of potential applications as well as issues at hand with research approaches

Potential applications	Issues	Research methodology
Nonintrusive, online, multistaged, full-field, whole-wafer inspection	Warp, bow, TTV, size, methods, resolution of measurements	Capacitive probes, moiré, interferometry, and optical metrology
Next-generation cleaning tools	Clean, contaminants removal	Laser-assisted method, gas cluster ion-beam processor, superheated gas chamber

applied in the preparation of substrates. Table 52.3 outlines some issues identified and related research approaches.

Metrology, Inspection, and Qualify Control

Technology's simultaneous move towards smaller devices and larger wafer size has created new challenges

Table 52.4 Wafer manufacturing with a system-oriented approach: integrated issues with suggested research approaches and methodologies

System-oriented manufacturing	Important issues in the future	Research approaches and methodology
Real-time sensing and control of wiresaw manufacturing	Kerf loss, wire thinning and breakage, control, quality assurance	Modeling, dynamics, vibration, thermal and residual stresses, and high-level control strategies. Sensor fusion and metrology systems. Short-pulsed high-intensity laser-assisted cleaning
Integrated computer-based control systems and manufacturing automation	Interconnected processes with surface resolution from 0.5 to 0.18 μm slicing, lapping, CMP, cleaning, metrology, quality, SPC, TQM, reduce cost	Integrated engineering system. Size and material index. Substrate manufacturing systems. Implementation of optimized automation strategy manufacturing management, and expert system. Advanced control strategies, artificial neural network (ANN)
Experimental validation and benchmarking	Lab testing and new industrial equipment development. Yields and cost reduction	Taguchi experimental design. Interconnected methodology from modeling, simulation, to experiments
Industrial technology development in partnership with academics	Leadership and core competence	Industrial manufacturing leadership and knowledge base. Dissemination of research results and information database

for metrology, as desired levels of accuracy cannot be achieved by simply extending current technologies. A suite of new metrological techniques is thus required. Currently metrology is employed mostly in an offline fashion. To reach the technological goals set by NTRS, the role of metrology has to be redirected from offline sampling to inline and in situ control [52.4, 9, 12, 17, 53, 84]. The future research effort in metrology should emphasize inline or in situ measurement for quick process feedback and quality control. Table 52.4 presents system-oriented wafer manufacturing and identifies important issues of research for the future, along with possible methodology for solutions. Table 52.5 shows the identified metrology issues in wafer manufacturing and suggests several plausible research methodologies. The specific topics can include, but are not limited to, nonintrusive, inline, full-field wafer inspection; optical methods such as Moiré, Fizeau, and Twyman–Green interferometry can provide nonintrusive, inline, full-field measurement of flatness, bow, warpage, and total thickness variation (TTV). Using these techniques, measurement throughput will be significantly increased over current technologies, which translates into cost savings. Also, measurement accu-

Table 52.5 Identified metrology issues in wafer manufacturing and plausible research methodology

Metrology issues	Research methodology
Nonintrusive, inline, full-field flatness, bow, warpage, and TTV measurement	Shadow Moiré, reflective Moiré, Fizeau and Twyman–Green interferometry
Inline residual stress measurement	Moiré interferometry and others
Real-time wire separation inspection	Moiré interferometry and others
Real-time wire thinning inspection	Laser diffraction
Inline metal impurities and contamination inspection	Fiber-optic sensors based on photoluminescence, Raman scattering, photon correlation spectroscopy, and evanescent-wavefield technique

racy, repeatability, and reproducibility can be improved by eliminating scanning by performing measurements on an entire piece simultaneously. In addition, laser speckle technology and fiber-optic sensors can also provide inline and rapid measurement of surface roughness after slicing and lapping, which is important for process feedback and control.

Real-Time Inspection of Wiresaw Operations

Practitioners in the industry of wafer production using wiresawing to slice ingots are often faced with challenges of understanding and optimizing process parameters to improve accuracy and efficiency. From the process monitoring and control point of view, two issues have been identified as having a critical impact on the quality of sliced wafers: uneven separation of wires, and wire thinning during operation. To this end, moiré interferometry and laser diffraction methods, to name but two, can be used to measure these two parameters, which can be fed back to correct the wiresaw operation in real time [52.27, 39, 85].

Inline Residual Stress Measurement

During both crystal growth and wafer production, residual stresses are created that result in warpage (i.e., the potato chip effect) of the wafer slice. These stresses can be quantified by using the moiré method to measure the wafer warpage before and after the wafer annealing step that relieves the residual stress. This information can then be used to adjust, optimize, and improve the previous steps.

Inline Metal Impurities and Contamination Inspection

Integrated fiber-optic sensors can be designed to remotely sense metal impurities and contamination on the wafers. Material composition can be determined from the spectral response based on photoluminescence. The concentration of various species in a mix can be measured through Raman scattering, while the size of submicron particle contaminants (3–3000 nm) can be measured using photon correlation spectroscopy. A concentration of heavy metal ions as low as 1 ppb in processing baths can be estimated through absorption measurements using evanescent-wavefield sensors. Fiber-optic-based sensors can provide both characterization and feedback during various stages of high-quality wafer producing. Early detection and removal of contaminants will increase yield and reduce downtime.

52.2 Slicing: The First Postgrowth Process in Wafer Manufacturing

Slicing is the first postgrowth process in wafer manufacturing for wafer forming from a crystalline ingot. Different slicing technologies and equipment are discussed in the following.

52.2.1 ID Saws

In the early days of wafer production, inner diameter (ID) saws were employed. An ID saw utilizes the edge of its inner hole to slice through an ingot, as shown schematically in Fig. 52.4. As shown in Fig. 52.4, a thin annular steel blade with a round hole in the middle is coated along the ID edge with industrial diamond abrasive grits, and is stretched with very high tension to keep the blade surface taut. The ingot to be sliced is fed through the inner-diameter hole, held by the jig shown in the figure. As the ID blade traverses down, a slice of wafer is obtained when the blade slices through the

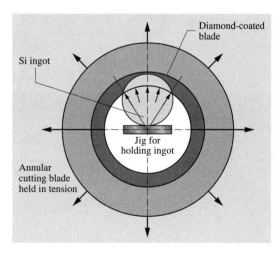

Fig. 52.4 Schematic of inner diameter (ID) saw for wafer slicing. The ingot, held by the jig, is fed through the inner opening of the ID saw with the inner edge coated by diamond abrasive grits. The annular steel blade is stretched with very high tension and rotates at high speed to slice through the ingot, as shown, by way of a fixed abrasive ploughing machining process ▶

ingot. Industrial motion control is employed for the operation and control to generate each slice of wafer.

As the diameter of the wafers in microelectronics application becomes larger, the inner diameter of the hole needs to be enlarged in order to accommodate the ingot and the jig. ID saws are practical for wafers up to 8 inches in diameter. For ingots of larger diameters, vibration becomes excessive, along with limitations of other practical issues, making ID saws unsuitable for wafer slicing.

52.2.2 The Modern Wiresaw

The modern slurry-based wiresaw has emerged as the slicing technology for semiconductor wafer production industry, as well as the preferred tool for slicing of ingots made of various materials. Since a century ago, wire has been utilized as a cutting tool to slice through stones and other materials. However, the literature on slicing using wiresaw in patent disclosures (for example [52.55, 57, 58, 62, 64, 71, 72, 86–89]) and research has only appeared in the last decade. Employed to slice single- and polycrystalline ingots to produce very thin (as thin as $200\,\mu m$) and thick wafers, the modern wiresaw is subjected to more stringent requirements to produce wafers with low total thickness variation (TTV), low warp, and low residual stresses.

A schematic of the modern wiresaw is illustrated in Fig. 52.5. A single steel wire is drawn from a supply spool and wound over three or four grooved cylindrical wire guides (four wire guides are illustrated in Fig. 52.5), to form the *wire web* surface on which the crystal ingot mounted on a holding jig is fed, with abrasive slurry manifolds feeding slurry with abrasive grits. As the wire moves with high speed, the ingot on the web is fed downward for slicing. Several hundred slices of wafers are produced simultaneously when the ingot traverses through the wire web to finish the slicing operation. More details of this modern wiresaw will be introduced in Sect. 52.3.

The modern wiresaw has sophisticated control of various parameters for successful precision operation, including control of wire tension [typical industrial proportional–integral–differential (PID) control], slurry temperature, wire speed, duty cycle of wire feed, ingot feed rate onto the wire web, and others. This is done through a computer-based monitoring and control unit to give operators greater flexibility and the ability to control and optimize the process parameters.

In addition, the wiresaw is capable of cutting much harder materials such as some III–V compounds and

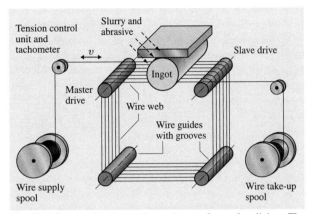

Fig. 52.5 Schematic of a modern wiresaw for wafer slicing. The schematic shows a wire supply spool with a single wire, winding over four grooved wire guides, collected by the take-up spool. The winding procedure provides wire webs as a surface of equally spaced wire segments onto which the ingot is fed to be sliced into wafers. Manifolds on both sides of the ingot pour abrasive slurry continuously onto the wire web surface

ceramics with proper abrasive grits and carrier. It has also been shown to work well for highly anisotropic materials such as lithium niobate.

52.2.3 Saws with Diamond-Impregnated Wires

Saws with diamond-impregnated wires are designed for slicing operation with different configuration and mechanisms [52.62, 90–93]. Such saws rely on the abrasives (typically industrial diamond) impregnated along the circumferential surface of a wire for slicing. The saws can be classified into rotary or reciprocating (linear) motions. Conceptually, this type of saws is akin to the ID saw because the diamond-impregnated wiresaw is essentially an ID saw with a linear instead of an annular cutting edge.

Fundamentally, however, the manufacturing process model of such saws is entirely different from that of the slurry wiresaw. This type of saw with diamond-impregnated wires removes materials by means of a ploughing process – a more brutal machining process than the rolling–indenting free abrasive machining process of the slurry wiresaw. In addition, the abrasive grits on the surface of a wire can be easily stripped off, leading to the loss of the ability for the wire to perform further slicing operations.

52.2.4 Others

As mentioned earlier in Sect. 52.1.2, as-sliced wafers can be ground first, followed by epitaxial growth process to produce wafers ready for device fabrication. In other cases, circular saws with the circumferential outer edge coated with diamond abrasive grits are also employed in slicing. This type of slicing tool is normally used for small samples, and typically is subject to a large degree of vibration and inaccuracy of surface finish.

52.2.5 Comparison of Slicing Technology

Modern wiresaw manufacturing is based on the so-called *free abrasive machining* (FAM) process with rolling–indenting, which removes materials via third-party free abrasives in slurry [52.1]. In contrast, the conventional inner diameter (ID) saw is based on the *ploughing machining* process which is characterized by the removal of materials with forceful media attached to the tool. The two schematics in Figs. 52.4 and 52.5 contrast these two manufacturing processes and arrangements.

The conventional ID saw in Fig. 52.4 cuts through the ingot to obtain one slice of wafer each time with an annular cutting blade stretched and held in high tension. The blade rotates at very high speed while feeding onto the ingot to cut through ingot with its diamond-coated inner edge. Analysis of stress and vibration in the ID saw with annular blades has been presented [52.75, 76, 94].

As shown in Fig. 52.5, and later in Fig. 52.9, the wiresaw consists of one wire moving either unidirectionally or bidirectionally on the surface of crystal ingot. The single wire is wound carefully on the wire guides with grooves of constant pitch to form a *wire web*, as shown in Fig. 52.5. The wire guides are rotated by a pair of master–slave drives, causing the entire wire web to move at high speed while carrying the abrasive slurry to remove material from the surface of ingot. The ingot is fed in the downward direction perpendicular to the wire web as shown in Fig. 52.5. The wire is maintained at constant tension during the cutting process. A spool of cold-drawn steel wire is used to continuously supply the wire necessary for cutting.

Depending upon the process control, parallel wire marks may be visible on the wafer surface after they are sliced. Typical surface of a wafer sliced by wiresaw resembles that of a lapped wafer. The wiresaw is capable of slicing wafers of large diameter as long as the distance between the wire guides is larger than the diameter of the ingot. With a wire web consisting of 200–400 strands of wound wire, a total of 200–400 wafers are produced simultaneously once the wire has finished traversing through the ingot. Typical kerf loss is the sum of the diameter of the wire, size of abrasives, and vibration amplitude of wire. The thickness of the wafer is controlled by the pitch of the grooves on the wire guides.

Comparison Between Modern Wiresaw and ID Saw

Comparisons of various properties between the modern wiresaw and ID saw are offered in Table 52.6. From the table, it is clear that the wiresaw has much higher throughput and yield with less kerf loss and surface damage. The wiresaw has found gradually increasing usage in slicing 200 mm and 300 mm single-crystalline silicon wafers and various shapes of polycrystalline

Table 52.6 Comparisons of various properties between the modern slurry wiresaw and ID saw. The comparison suggests that the wiresaw is a more favorable tool for slicing of large quantities of wafers with better surface quality than the ID saw

Property	Wiresaw	ID saw
Manufacturing process model	FAM/lapping	Ploughing/grinding
Typical cut surface features	Parallel wire marks	Chipping and fracture
Depth of damage	Uniform, 5–15 µm	Variable, 20–30 µm
Productivity (typical)	110–220 cm^2/h	10–30 cm^2/h
Total time per run	5–8 h (depending on ingot size)	About 15 min for each slice
Wafers per run	200–400 wafers	One wafer
Kerf loss	180–210 µm	300–500 µm
Minimum thickness of wafer	200 µm (typical)	350 µm
Maximum ingot diameter	300 mm or higher	Up to 200 mm

Table 52.7 Comparisons between the modern slurry wiresaw and the saw with diamond-impregnated wires. The comparison suggests that wiresaw is a more favorable tool for most slicing operations, while the saw with diamond-impregnated wires can be quite effective for slicing very hard wafers with small size

Property	Slurry wiresaw	Saw with diamond-impregnated wires
Manufacturing process model	FAM with rolling–indenting	Ploughing
Abrasive grits	Free abrasive; rejoin cutting proc.	Lost forever once stripped off surface
Depth of damage	Uniform, 5–15 μm	Variable, 20–30 μm
Cost of consumables	Lower (per wafer)	Higher (per wafer)
Slicing very hard wafers	Not very effective	Very effective
Kerf loss	180–210 μm	250–400 μm

wafers with thickness as small as 200 μm for photovoltaic applications. The wiresaw can also be employed to slice highly anisotropic crystals and other materials such as alumina (Al_2O_3) and quartz.

Comparison Between Modern Wiresaw and Saw with Diamond-Impregnated Wires

It is important to note that the modern slurry wiresaw is entirely different from the saw with diamond-impregnated wires in which the diamond grits coated on the surface of wire are used to remove the materials by the the shear-dominated process similar to the orthogonal machining process [52.90, 91]. Table 52.7 contrasts the wiresaw and saws with diamond-impregnated wires, showing that the wiresaw is suitable for general slicing operations, while saws with diamond-impregnated wires can be effective in slicing very hard or specialized wafers with small sizes.

52.2.6 Wafer Manufacturing for Large Wafers

After slicing wafers by wiresaw, the subsequent process is flattening to achieve a higher degree of parallelism and flatness of the wafer [52.2, 10, 19, 57, 74]. Both grinding and lapping are conventionally used for this process. Polishing is the next process to obtain a smoother wafer surface.

For the production of large, 300 mm silicon prime wafers using modern wiresaw for electronics fabrication, however, a special challenge with surface waviness of different wavelengths, known as the *nanotopography*, warrants special consideration. Such waviness across the surface of the large wafer is not easily dealt with using the conventional lapping operation because of the elasticity of the large wafer, causing the waviness to be flattened as the lapping tool is applied. Surface grinding, therefore, has become an attractive alternative for the initial removal of such surface waviness and roughness after slicing.

The nanotopography is illustrated in Fig. 52.6, as defined by SEMATECH. In the figure, different ranges of surface features are identified and categorized for the sake of distinguishing one from the other. Nevertheless, nanotopography and waviness, with amplitudes of surface roughness and wavelengths of waviness shown in Fig. 52.6, are typically found in large wafers.

Grinding in Wafer Manufacturing

After wafers are sliced by wiresaw, grinding is often employed for large wafers for flattening [52.66, 95, 96]. The advantages of grinding over lapping are (1) fully automatic cassette-to-cassette operation, (2) use of a fixed abrasive grinding wheel rather than loose abrasive slurry so the cost of consumables per wafer may be lower, and (3) higher throughput. However, the grinding process cannot entirely remove the waviness induced by wiresaw slicing, which is why grinding cannot totally replace the lapping process.

The grinding operation cannot remove the waviness as effectively as the lapping operation because of elas-

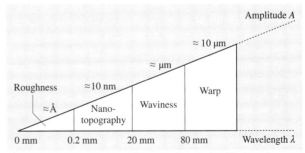

Fig. 52.6 Illustration of surface roughness classified as the *nanotopography* by SEMATECH. Nanotopography and waviness are typically found in large wafers

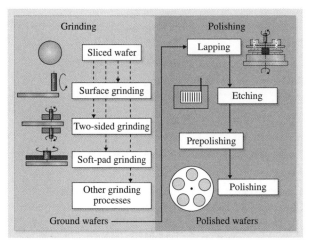

Fig. 52.7 Illustration of revised wafer manufacturing process with the insertion of the grinding operation, especially for very large silicon wafers (Fig. 52.1)

tic deformation of the wafer. When the same total force is applied on a wafer, the elastic deformation of the wafer in lapping is much smaller than that in grinding, resulting in a less pronounced spring-back effect in lapping. The amount of deformation in lapping is only 1/55 to 1/36 of that in grinding [52.77, 78, 97, 98]. In order to improve the removal of surface waviness in grinding operation, several methods have been proposed, including:

1. Wafer grinding followed by a lapping
2. Wax mounting to reduce the effect of elasticity
3. Reduced vacuum
4. Use of a soft pad to avoid elastic spring-back after the grinding operation

A revised wafer production flow is shown in Fig. 52.7 in which the grinding operation follows the slicing operation, followed by the lapping/polishing operations.

52.3 Modern Wiresaw in Wafer Slicing

In the following subsections, the technology of modern slurry wiresaw is introduced with discussions of various topics involved in the technology and research of modern slurry wiresaws.

52.3.1 Definition of Modern Wiresaw

The term *wiresaw* or *modern wiresaw* has been used throughout this chapter to refer to the equipment for slicing crystalline ingots using a bare wire (typically cold-drawn steel wire) pressing onto the ingot with the continuous supply of abrasive slurry at the cutting interface. This is to distinguish them from saws that cut into materials with diamond-impregnated wires under various arrangements such as rotary or prismatic reciprocating motions. In a typical arrangement of a slurry wiresaw, a wire web is formed with a single bare steel wire drawn from a supply spool by winding it along grooved cylindrical wire guides, as illustrated in Fig. 52.5. Although the alternative name *multiwire saw* is also used in the literature, this is potentially misleading because modern wiresaw equipment actually operates on a *single* wire wound on grooved cylindrical guides. In addition, multiple wires are often used in diamond-impregnated reciprocating saw, which could be more appropriately called a *multiwire saw*. Since separating materials with wire under a similar principle is a century-old technology when wire was used to slab stones in the 19th century, the term *modern* wiresaw has been adopted in this chapter and used interchange-

Fig. 52.8 Picture of a HCT wiresaw. The *circled region* indicates the wire web formed by the single wire winding over the grooved wire guides. To the *upper right* of this region is the ingot mounted on the jig, which is controlled to move downwards for slicing operation (source: HCT Inc., and GT Equipment Inc.)

Table 52.8 Operating parameters and their ranges and/or values for wiresawing

Parameter	Values and ranges
Speed of wire	5–15 m/s
Duty cycle of wire	Continuous feeding or reciprocating
Wire tension	25–35 N
Diameter of wire	150–175 μm
Wire bow angle	2–5 °C
Kerf loss	180–210 μm (typically)
Slurry, carrier fluid	Water soluble or oil based
Slurry, abrasive grits	Silicon carbide (usually)
Feed rate of ingot	100–300 μm/min
Consumables	Wire and slurry

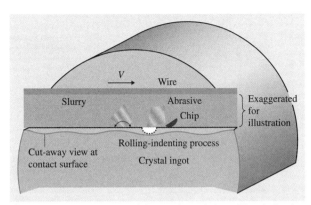

Fig. 52.9 Three-dimensional illustration of the rolling–indenting process in the wiresaw slicing process in which abrasive grits rolls and indents on the contact interface to remove materials from the substrate surface. This body abrasion process is also called *free abrasive machining* (FAM)

ably with *wiresaw*. The modern wiresawing is a much more rigorous process than the stone-slabbing operation using wire, and also is subject to more stringent requirements on surface finish for typical applications.

A typical industrial wiresaw is shown in Fig. 52.8, in which four wire guides are shown with a wire web face circled in the figure. A silicon ingot is shown to be fed top-down onto the top horizontal face of the wire web for slicing. To the left are the control console of the wiresaw and the wire management unit which controls the tension of the wire and manages the supply, feeding, tension, and speed of the wire.

52.3.2 Modern Wiresaw Technology

The *modern wiresaw* and associated technology discussed in this chapter utilizes a steel wire with abrasive slurry, consisting of abrasive grits suspended in carrier fluid, for the purpose of slicing crystals. Wire saws had been used in the 19th century in Europe to slab stones. Modern wiresaws, however, are employed to produce wafers with more stringent requirements of surface quality (Fig. 52.3). A typical modern wiresaw, as shown in Fig. 52.8, consists of three main components:

1. Wire management unit
2. Control and program console
3. The slicing compartment, where the ingot is sliced by feeding it onto the wire web

The slicing process of the modern slurry wiresaw belongs to the category of *free abrasive machining* (FAM). Modern wiresaws utilize a steel wire, under high tension, moving at high speed on the surface of the substrate submerged in abrasive slurry. The wire maintains a bow angle with the cutting surface, applying a normal load along the direction of cutting. Typical parameters of wiresaw and operating parameters are summarized in Table 52.8. A schematic drawing of the modern wiresaw is shown in Fig. 52.5 with the *rolling–indenting* process shown in Fig. 52.9. The FAM process of wiresaw slicing with slurry is theoretically less brutal than cutting with fixed abrasive grinding processes [52.90, 91], under the condition of the same energy input, because of the inherent cutting mechanism of rolling and indenting. The cutting interface in Fig. 52.9 is exaggerated for the illustration of free abrasive grits interacting with the substrate surface in rolling and indenting due to the speed differential within the film of fluid trapped between the wire and the substrate surface. Chips are removed from the surface of the work material as a result of repeated rolling–indenting with typical micro- and nanoindentation effects.

52.3.3 Modeling and Control of the Modern Wiresawing Process

Research and studies of the modern wiresaw slicing process started to appear in the 1990s, first in patent disclosures followed by academic and industrial research. Academic–industrial–governmental synergy is characteristic of such research due to the very nature of process modeling for the wiresaw process, which requires a high degree of collaboration in order to make research results

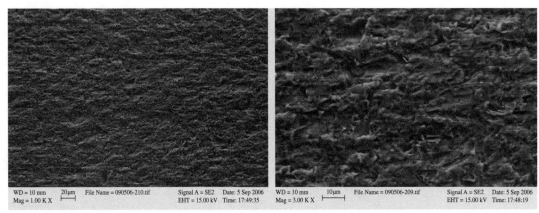

Fig. 52.10 SEM micrographs at scales of 10 μm and 20 μm illustrating the typical surface of an as-sliced wafer. The wafer surface shown here is from a polysilicon material sliced by an industrial wiresaw. The surface is full of random pits resulting from the rolling–indenting manufacturing process. Such a surface is very representative of wafers sliced by wiresaw under normal operating conditions

relevant. In the following subsections, a brief survey of various aspects of research and studies in wiresaw manufacturing processes is presented. Readers are referred to the bibliography at the end of this chapter for further reading.

SEM Study of Surface Characteristics of Wiresawn Wafers

Figure 52.10 shows a typical surface of an as-sliced wafer cut by an industrial wiresaw. The material of the wafer is polysilicon with rectangular surface area, to be used as a PV cell. As can be observed in the scanning electron microscopy (SEM) photo to the left with the scale bar of 20 μm, the surface is covered with a random pattern of pits caused by the rolling–indenting process. The abrasive grits used are silicon carbide 400 with an average grit size of 17 μm. A more detailed photo on the right with a scale bar of 10 μm shows magnified patterns of random pits, with very little or nearly no fracture, as opposed to typical surface of wafers sliced by ID saws which are populated with more visible fractures on the surface.

The Rolling–Indenting Process Model

Figure 52.10 illustrates the *rolling–indenting* manufacturing process model for slicing using a wiresaw. A three-dimensional (3-D) illustration of the rolling–indenting process associated with free abrasive machining is shown in Fig. 52.9. For more details on the rolling–indenting model, readers are referred to [52.1, 13, 23, 48].

Hydrodynamic and Elastohydrodynamic Interaction

It is known that even in a lubricated wire drawing manufacturing process, the hydrodynamic effect is present. In the wiresaw machining process, the wire moving with high speed (Table 52.8) in a slurry-rich environment will induce not only the hydrodynamic effect but also elastic interaction with the steel wire. The intertwined effects result in the elastohydrodynamic phenomenon common in the modern wiresaw process. In the initial study of such phenomena [52.23], hydrodynamic process modeling and computer simulation were implemented with realistic process parameters. Later, other work also addressed and advanced this topic [52.21, 24, 39, 40]. The hydrodynamic film between the wire and substrate surface supports and suspends the normal force applied to the ingot. The contact interface through the hydrodynamic film generally supports the free abrasive machining with abrasive grits rolling and indenting on the surface of substrate. In some areas, there may be intermittent direct contact made due to the thickness of the slurry film. In [52.50], the author suggested that direct contact may be a more regular occurrence.

Vibration of the Wire in Slicing Using a Wiresaw

Vibration analysis of a stationary wire is a century-old problem with well-known solution (for example, in a stringed musical instrument). The eigenvalue problem and solution of the analysis of vibration of a moving string, however, was only solved in the early 1990s, although the differential equation of motion was first proposed around 1970. Excessive vibration of the wire during slicing will increase the kerf – an undesirable effect. The application of eigenvalue problems of the moving wire in the wiresawing process has been presented in [52.7, 9, 20, 22, 31, 40].

Consumables in Wiresawing Process

The consumables of wiresawing process include the abrasive grits, carrier fluid, and wire. In recent years, recovery and regeneration of abrasive grits have received considerable attention due to environmental concerns. Several pieces of equipment that recycle abrasive grits have been constructed to recover a large percentage of grits. The carrier fluid has typically become water soluble instead of oil based. Wire used in wiresawing process is made of steel in a cold-drawn process, and is rather inexpensive with very consistent quality. Wires are typically discarded after each slicing operation. In the future, it is expected that more attention will be paid to the management of these consumables due to environmental concerns.

Material-Related Subjects and Thermal Issues

Wafer slicing also depends on the materials of the crystalline ingots. It may generally be fine to assume isotropy for consideration of slicing materials such as polysilicon; however, highly anisotropic materials such as lithium niobate will display specific behavior based on its anisotropy. Even single-crystalline silicon is anisotropic, although not to the extent seen in lithium niobate. Some recent work with consideration of materials properties can be found in, e.g., [52.38, 42, 43].

Thermal issues and correlation to the surface waviness of large wafers sliced by wiresaw have attracted much attention. Studies have been performed, resulting in disclosures such as [52.51, 52, 69].

52.4 Conclusions and Further Reading

The advent of the modern wiresaw as a generic wafer slicing tool has brought about a revolution in the wafer manufacturing industry, particularly in the silicon wafering industry. As a result of the versatility and capabilities of the modern wiresaw, it has entirely replaced the ID saw technology that from the 1970s to 1990s was the primary slicing tools for wafer production. Equipped with better control and monitoring of industrial processes and research knowledge base of wiresaw process modeling and control, today's industrial wiresaws are gradually becoming the primary tool of choice for slicing.

In this chapter, modern wafer manufacturing and wafering process was discussed and illustrated. Wafer production, based on the collection of operations shown in Fig. 52.1, produces prime wafers from crystalline ingots. Slicing is the first postgrowth wafer-forming process. Several technologies of slicing were discussed with historical perspectives. The modern slurry wiresaw and its technological development with research in process modeling to improve understanding and control of the process were also presented. Surveys of various topics in wiresaw processing and research include: modeling and control of modern wiresawing process, SEM study of surface characteristics of wiresawn wafers, the rolling–indenting process model, hydrodynamic and elastohydrodynamic interaction, vibration of wire in slicing using wiresaw, consumables in the wiresawing process, and material-related subjects and thermal issues.

Various references to research literature and patents are cited throughout the chapter. The list of references provides readers with suggestions for further reading on topics of their interests.

References

52.1   I. Kao: Technology and research of slurry wiresaw manufacturing systems in wafer slicing with free abrasive machining, Int. J. Adv. Manuf. Syst. **7**(2), 7–20 (2004)

52.2   I. Kao, V. Prasad, J. Li, M. Bhagavat, S. Wei, J. Talbott, K. Gupta: Modern wiresaw technology for large crystals, Proc. ACCGE/east-97 (Atlantic City 1997)

52.3   I. Kao, V. Prasad, J. Li, M. Bhagavat: Wafer slicing and wire saw manufacturing technology, NSF Grantees Conf. (Seattle 1997) pp. 239–240

52.4   S. Wu, S. Wei, I. Kao, F.P. Chiang: Wafer surface measurements using shadow moiré with Talbot effect, Proc. ASME IMECE'97 (ASME, Dallas 1997) pp. 369–376

52.5 J. Li, I. Kao, V. Prasad: Modeling stresses of contacts in wiresaw slicing of polycrystalline and crystalline ingots: application to silicon wafer production, Proc. ASME IMECE '97 (ASME, Dallas 1997) pp. 439–446

52.6 M. Bhagavat, F. Yang, I. Kao: Elasto-plastic finite element analysis of indentations in free abrasive machining, Proc. Manuf. Eng. Div. IMECE'98 (ASME, 1998) pp. 819–824

52.7 S. Wei, I. Kao: Analysis of stiffness control and vibration of wire in wiresaw manufacturing process, Proc. Manuf. Eng. Div. IMECE'98 (ASME, 1998) pp. 813–818

52.8 I. Kao, M. Bhagavat, V. Prasad: Integrated modeling of wiresaw in wafer slicing, NSF Des. Manuf. Grantees Conf. (Monterey, 1998) pp. 425–426

52.9 I. Kao, S. Wei, F.-P. Chiang: Vibration of wiresaw manufacturing processes and wafer surface measurement, NSF Des. Manuf. Grantees Conf. (Monterey, 1998) pp. 427–428

52.10 I. Kao, V. Prasad, F.P. Chiang, M. Bhagavat, S. Wei, M. Chandra, M. Costantini, P. Leyvraz, J. Talbott, K. Gupta: Modeling and experiments on wiresaw for large silicon wafer manufacturing, 8th Int. Symp. Silicon Mater. Sci. Technol. (San Diego 1998) p. 320

52.11 F.P. Chiang, M.L. Du, I. Kao: Some new applications on in-plane, shadow and reflection moiré methods, Int. Conf. Appl. Optic. Metrol. (Hungary 1998)

52.12 S. Wei, S. Wu, I. Kao, F.P. Chiang: Wafer surface measurements using shadow moiré with Talbot effect, J. Electron. Packag. **120**(2), 166–170 (1998)

52.13 J. Li, I. Kao, V. Prasad: Modeling stresses of contacts in wiresaw slicing of polycrystalline and crystalline ingots: Application to silicon wafer production, J. Electron. Packag. **120**(2), 123–128 (1998)

52.14 F. Yang, I. Kao: Free abrasive machining in slicing brittle materials with wiresaw, Tech. Rep. TR99-03 (SUNY at Stony Brook Department of Mechanical Engineering, New York 1999)

52.15 F. Yang, J.C.M. Li, I. Kao: Interaction between ingot and wire in wiresaw process, Proc. IMECE'99: Electron. Manuf. Iss., Vol. 104, ed. by C. Sahay, B. Sammakia, I. Kao, D. Baldwin (ASME, New York 1999) pp. 3–8

52.16 M. Bhagavat, I. Kao: Computational model for free abrasive machining of brittle silicon using a wiresaw, Proc. IMECE'99: Electron. Manuf. Iss., Vol. 104, ed. by C. Sahay, B. Sammakia, I. Kao, D. Baldwin (ASME, New York 1999) pp. 21–30

52.17 S. Wei, I. Kao: Hight-resolution wafer surface topology measurement using phase-shifting shadow moiré technique, Proc. IMECE'99: Electron. Manuf. Iss., Vol. 104, ed. by C. Sahay, B. Sammakia, I. Kao, D. Baldwin (ASME, New York 1999) pp. 15–20

52.18 F. Yang, I. Kao: Interior stress for axisymmetric abrasive indentation in the free abrasive machining process: slicing silicon wafers with modern wiresaw, J. Electron. Packag. **121**(3), 191–195 (1999)

52.19 M. Chandra, P. Leyvraz, J.A. Talbott, K. Gupta, I. Kao, V. Prasad: Challenges in slicing large diameter silicon wafers using slurry wiresaw, Proc. Manuf. Eng. Div. IMECE'98 (ASME, 1998) pp. 807–811

52.20 S. Wei, I. Kao: Free vibration analysis for thin wire of modern wiresaw between sliced wafers in wafer manufacturing processes, Proc. IMECE'00: Packag. Electron. Photon. Dev., Vol. 28 (ASME, Orlando 2000) pp. 213–219

52.21 L. Zhu, M. Bhagavat, I. Kao: Analysis of the interaction between thin-film fluid hydrodynamics and wire vibration in wafer manufacturing using wiresaw, Proc. IMECE'00: Packag. Electron. Photon. Dev., Vol. 28 (ASME, Orlando 2000) pp. 233–241

52.22 S. Wei, I. Kao: Vibration analysis of wire and frequency response in the modern wiresaw manufacturing process, J. Sound Vib. **231**(5), 1383–1395 (2000)

52.23 M. Bhagavat, V. Prasad, I. Kao: Elasto-hydrodynamic interaction in the free abrasive wafer slicing using a wiresaw: modeling and finite element analysis, J. Tribol. **122**(2), 394–404 (2000)

52.24 L. Zhu, I. Kao: Equilibrium elastohydrodynamic interaction analysis in wafer slicing process using wiresaw, Proc. IMECE'01: EEP, Vol. 1 (New York 2001), pp 123–128

52.25 F. Yang, I. Kao: Free abrasive machining in slicing brittle materials with wiresaw, J. Electron. Packag. **123**, 254–259 (2001)

52.26 I. Kao: The technology of modern wiresaw in silicon wafer slicing for solar cells, Invited talk at the 12th Workshop Cryst. Silicon Sol. Cell Mater. Process. (Brechenridge 2002)

52.27 I. Kao, F.-P. Chiang: Research on modern wiresaw for wafer slicing and on-line real-time metrology, Proc. NSF Des. Serv. Manuf. Grantees Res. Conf. (San Juan 2002)

52.28 I. Kao, L. Zhu: Computer simulation in backlapping of wiresaw sliced semiconductor wafers, Proc. Des. Serv. Manuf. Grantees Res. Conf. (Birmingham 2003)

52.29 S. Bhagavat, I. Kao: Nanoindentation studies on a non-centrosymmetric crystal: lithium niobate, Invited talk at the High Press. Phase Transf. Workshop NSF Focus. Res. Group (FRG) (North Carolina State University, Raleigh 2004)

52.30 S. Bhagavat, I. Kao: Nanoindentation studies on a non-centrosymmetric crystal: lithium niobate, High Press. Phase Transf. Workshop, NSF Focus. Res. Group (FRG) (North Carolina State University, Raleigh 2004)

52.31 S. Wei, I. Kao: Stiffness analysis in wiresaw manufacturing systems for applications in wafer slicing, Int. J. Adv. Manuf. Sys. **7**(2), 57–64 (2004)

52.32 M. Bhagavat, F. Yang, I. Kao: Elasto-plastic finite element analysis of indentations in free abra-

52.33 L. Zhu, I. Kao: On-line and real-time monitoring of moving wire with dynamic consideration in wiresaw slicing processes, Technical Report TR010-2005 (Department of Mechanical Engineering, Stony Brook University 2005)

52.34 S. Bhagavat, J. Liberato, I. Kao: Effects of mixed abrasive slurries on free abrasive machining processes, Proc. 2005 ASPE Conf. (ASPE, 2005)

52.35 S. Bhagavat, I. Kao: Ultra-low load indentation response of materials: in purview of free abrasive machining processes, High Press. Phase Transf. Workshop, NSF Focus. Res. Group (FRG) (2005)

52.36 I. Kao: Experiments of nanoindentation on non-centrosymmetric crystal and study of their implication in wafer manufacturing processes, Invited talk at the US-Africa Workshop Mater. Mech. (Cape Town 2005)

52.37 I. Kao: Experiments of nanoindentation on non-centrosymmetric crystal and study of their implication in wafer manufacturing processes, Invited talk at the US-Africa Workshop Mater. Mech. (Cape Town 2005)

52.38 S. Bhagavat, I. Kao: Nanoindentation of lithium niobate: hardness anisotropy and pop-in phenomenon, Mater. Sci. Eng. A **393**, 327–331 (2005)

52.39 L. Zhu, I. Kao: Computational model for the steady-state elasto-hydrodynamic interaction in wafer slicing process using wiresaw, Int. J. Manuf. Technol. Manag. **7**(5/6), 407–429 (2005)

52.40 L. Zhu, I. Kao: Galerkin-based modal analysis on the vibration of wire-slurry system in wafer slicing using wiresaw, J. Sound Vib. **283**(3–5), 589–620 (2005)

52.41 A. Gouldstone, I. Kao: Wafer slicing using slurry wiresaw and relevance of nanoindentation in its analysis, High Press. Phase Transf. (HPPT) Workshop (Kalamazoo 2006)

52.42 S. Bhagavat, I. Kao: Ultra-low load multiple indentation response of materials: in purview of wiresaw slicing and other free abrasive machining (FAM) processes, Int. J. Mach. Tools Manuf. **46**(5), 531–541 (2006)

52.43 S. Bhagavat, I. Kao: Theoretical analysis on the effects of crystal anisotropy on wiresawing process and application to wafer slicing, Int. J. Mach. Tools Manuf. **46**, 531–541 (2006)

52.44 S. Bhagavat, I. Kao: A finite element analysis of temperature variation in silicon wafers during wiresaw slicing, Int. J. Mach. Tools Manuf. **48**(1), 95–106 (2007)

52.45 I. Kao, S. Bhagavat: Single-crystalline silicon wafer production using wire saw for wafer slicing. In: *Semicondoductor Machining at the Micro-Nano Scale*, ed. by J. Yan, J. Patten (Transworld Research Network, Kerala 2007) pp. 243–270

52.46 C. Chung, I. Kao: Damped vibration response at different speeds of wire in slurry wiresaw manufacturing operations, Proc. Int. Manuf. Sci. Eng. Conf. (MSEC 2008) (ASME, 2008), paper number MSEC2008-72213

52.47 C. Chung, I. Kao: Comparison of free abrasive machining processes in wafer manufacturing, Proc. Int. Manuf. Sci. Eng. Conf. (MSEC 2008) (ASME, 2008), paper number MSEC2008-72213

52.48 H.J. Möller: Basic mechanisms and models of multi-wire sawing, Adv. Eng. Mater. **6**(7), 501–513 (2004)

52.49 H.J. Möller, C. Funke, M. Rinio, S. Scholz: Multicrystalline silicon for solar cells, Thin Solid Films **487**(1–2), 179–187 (2005)

52.50 H.J. Möller: Wafering of silicon crystals, Phys. Stat. Sol. A **203**(4), 659–669 (2006)

52.51 Y. Ariga: Wiresaw and cutting method, US Patent 6652356 (2003)

52.52 M. Bhagavat, D. Witte, S. Kimbel, D. Sager, J. Peyton: Method and apparatus for slicing semiconductor wafers, US Patent 2003170948 (2003)

52.53 U. Bismayer, E. Brinksmeier, B. Guttler, H. Seibt, C. Menz: Measurement of subsurface damage in silicon wafers, Prec. Eng. **16**(2), 139–144 (1994)

52.54 Y. Chang, M. Hashimura, D. Dornfeld: An investigation of material removal mechanisms in lapping with grain size transition, J. Manuf. Sci. Eng. Trans. ASME **122**(3), 413–419 (2000)

52.55 G.L. Contardi: Wire saw beads. Economic production, Indust. Diam. Rev. **53**(558), 256–260 (1993)

52.56 C. Hauser: Wire sawing device, US Patent 5910203 (1999)

52.57 R. Wells: Wire saw slicing of large diameter crystals, Solid State Technol. **30**(9), 63–65 (1987)

52.58 R.C. Wells: Wire saw, US Patent 4494523 (1985)

52.59 Corning Inc.: Corning to acquire world's largest multi-wire saw for manufacturing of image mask blanks (2005) http://www.corning.com/news_center/news_releases/2005/2005071201.aspx

52.60 A. Jindal, S. Hegde, S. Babu: Chemical mechanical polishing using mixed abrasive slurries, Electrochem. Solid State Lett. **5**(7), G48–G50 (2002)

52.61 O. Konunchuk, G. Preece: Apparatus and method for reducing bow and warp in silicon wafers sliced by a wire saw, US Patent 6352071 (2002)

52.62 A. le Scanff: New wire saw machine, Indust. Diam. Rev. **48**(527), 168 (1988)

52.63 H. Olkrug, H. Lundt, C. Andrae, J. Frumm: Process and device for producing a cylindrical single crystal and process for cutting semiconductor wafers, US Patent 6159284 (2000)

52.64 H. Shimizu: Wire-saw, US Patent 3942508 (1976)

52.65 G. E. Technologies: Advanced wiresaw for photovoltaic wafers, Technical Report, GT Equipment Technologies, Inc. (1998)

52.66 H.K. Tönshoff, B. Karpuschewski, M. Hartmann, C. Spengler: Grinding and slicing technique as

an advanced technology for silicon wafer slicing, Mach. Sci. Technol. **1**, 33–47 (1997)
52.67 K. Toyama, K. Hayakawa, E. Kiuchi: Method of slicing semiconductor single crystal ingot, EU Patent 0798092 (1997)
52.68 J. Verhey, U. Bismayer, B. Guttler, H. Lundt: The surface of machined silicon wafers, Semiconduct. Sci. Technol. **9**, 404–408 (1994)
52.69 T. Yamada, M. Fukunaga, T. Ichikawa, K. Furno, K. Makino, A. Yokoyama: Prediction of warping in silicon wafer slicing with wire saw, Theor. Appl. Mech. **51**, 251–258 (2002)
52.70 M. Yoshioka, J. Hagiwara: Simulation of time-dependent distribution of abrasive grain size in lapping, Jpn. Prec. Eng. **61**(9), 1270–1274 (1995)
52.71 S. Herbert: UK's biggest wire saw contract, Indust. Diam. Rev. **49**(534), 206–207 (1989)
52.72 M. Kojima, A. Tomizawa, J. Takase: Development of new wafer slicing equipment (unidirectional multi wire-saw), Sumitomo Met. Ind. **42**(4), 218–224 (1990)
52.73 Y. Xie, B. Bhushan: Effects of particle size, polishing pad and contact pressure in free abrasive polishing, Wear **200**(1–2), 281–295 (1996)
52.74 G. Wenski, T. Altmann, W. Winkler, G. Heier, G. Holker: Doubleside polishing – a technology mandatory for 300 mm wafer manufacturing, Mater. Sci. Semiconduct. Proc. **5**(4-5), 375–380 (2002)
52.75 S. Chonan, Z.W. Jiang, Y. Yuki: Stress analysis of a silicon-wafer slicer cutting the crystal ingot, J. Mech. Des. **115**, 711–717 (1993)
52.76 S. Chonan, Z.W. Jiang, Y. Yuki: Vibration and deflection of a silicon-wafer slicer cutting the crystal ingot, J. Vib. Acoust. **115**, 529–534 (1993)
52.77 W. Liu, Z. J. Pei, X. J. Xin: Finite element analysis for grinding and lapping of wire-sawn silicon wafers, J.Mater. Proc. Technol. **129**(1-3) 2–9 (2002)
52.78 X.J. Xin, Z.J. Pei, W. Liu: Finite element analysis on soft-pad grinding of wire-sawn silicon wafers, ASME J. Electron. Packag. **126**, 177–185 (2004)
52.79 P. Gise, R. Blanchard: *Modern Semiconductor Fabrication Technology* (Prentice Hall, New Jersey 1986)
52.80 Semiconductor Industry Association (SIA): Microelectronics: vision for the 21st century, Technical Report (1994)
52.81 Semiconductor Industry Association (SIA): National technology roadmap for semiconductor, Technical Report (1994)
52.82 Semiconductor Industry Association (SIA): Published data of Semiconductor Industry Association by P.S. Peercy, Technical Report (Semiconductor Industry Association/SEMATECH 1998)
52.83 SEMATECH: I. Project, assorted publications and web pages of sematech.org and i300i.org, Technical Report (SEMATECH and International 300 mm Initiative 1994-1998)
52.84 J.A. Slotwinski, N.N. Hsu, G.V. Blessing: Ultrasonic measurement of surface and subsurface structure in ceramics. In: *Machining of Advanced Materials*, ed. by National Institute of Science and Techology (NIST) (US Government Printing Office, Washington 1993), Special Pub. 847
52.85 I. Kao: Research towards the next-generation reconfigurable wiresaw for wafer slicing and on-line real-time metrology, Proc. 2001 NSF Des., Serv. Manuf. Grantees Res. Conf. (Tempa 2001)
52.86 H.W. Mech. Machine and method for cutting brittle materials using a reciprocating cutting wire, US Patent 3831576 (1974)
52.87 D. Hayes: Demolition – the modern method, Indust. Diam. Rev. **50**(537), 69 (1990)
52.88 W. Weiland: Railway bridge cut in two, Indust. Diam. Rev. **50**(537), 65–66 (1990)
52.89 HCT Shaping Systems SA: World leader in wire saw technology, Technical Report (1995)
52.90 W. Clark, A. Shih, C. Hardin, R. Lemaster, S. McSpadden: Fixed abrasive diamond wire machining – part i: process monitoring and wire tension force, Int. J. Mach. Tools Manuf. **43**(5), 523–532 (2003)
52.91 W. Clark, A. Shih, R. Lemaster, S. McSpadden: Fixed abrasive diamond wire machining – part ii: experiments design and results, Int. J. Mach. Tools Manuf. **43**(5), 533–542 (2003)
52.92 D. Hayes: Japanese granite trade embraces diamond wiresaw, Indust. Diam. Rev. **49**, 67–69 (1989)
52.93 T. Chang, T. Ueng, W. Lee: Study of mechanism of diamond wiresaw, Min. Metall. **37**, 73 (1993)
52.94 S. Chonan, T. Hayase: Stress analysis of a spinning annular disk to a stationary distributed, in-plane edge load, J. Vib. Acoust. Stress Reliab. Des. **107**, 277–282 (1987)
52.95 P. Moulik, H. Yang, S. Chandrasekar: Simulation of thermal stresses during grinding, Int. J. Mech. Sci. **43**, 831–851 (2001)
52.96 Z. Zhong, V.C. Venkatesh: Surface integrity studies on the grinding, lapping and polishing processes for optical products, J. Mater. Process. Technol. **44**, 179–186 (1994)
52.97 Z. Pei, X. Xin, W. Liu: Finite element analysis for grinding of wire-sawn silicon wafers: a designed experiment, Int. J. Mach. Tools Manuf. **43**(1), 7–16 (2003)
52.98 X. Sun, Z.J. Pei, X.J. Xin, M. Fouts: Waviness removal in grinding of wire-saw silicon wafers: 3d finite element analysis with designed experiments, Int. J. Mach. Tools Manuf. **44**, 11–19 (2004)

Subject Index

I_1 type step   1504
ξ anisotropy factor   67
α and β dislocation   1345
α-factor   67, 72
β-BaB$_2$O$_4$ (BBO)   746
β-BaB$_2$O$_4$ (β-BBO)   730
β-sitosterol   1626
γ-lithium aluminum oxide (LiAlO$_2$)   877
30° and 60°
– dislocation   178
60°
– dislocation   1344
I–III–VI$_2$ compound   917
– growth parameter   917
2,2-diphenyl-1-picrylhydrazyl (DPPH)   1533
II–VI compound semiconductor   912
2-D–3-D transformation
– in Stranski–Krastanov growth   44
– in Volmer–Weber growth   44
2-adamantylamino-5-nitropyridine (AANP)   398
2-methyl-4-nitroaniline (MNA)   569
III nitride   1244
300 mm large wafer   1731
III–V binary crystal growth technology   292
III–V compound   193
III–V material   328
4H-SiC   954
– p–n junction   960
– wafer   812

A

A defect   1283, 1292, 1322
ab initio method   1246
abrasive grit   1733
abrasive slurry   1730
absorption at laser Wavelength   499
absorption coefficient (α)   352
absorption edge   352
– spectroscopy (ABES)   1089
accelerated crucible rotation technique (ACRT)   175, 307, 316, 318, 334, 336, 734, 739
acceptance angle   714
acceptor activation   1541
acceptor passivation   1541

Acheson
– method   799
– process   798
acidic seeded growth   683
acoustooptic (AO)   162
activation energy   1273, 1352
active inhomogeneity   1471
additional absorption (AA)   495
additive decomposition of strain tensor   1342
adhesion energy   59
adhesive growth   1596
adhesive-type growth mechanism   136
Adornato–Brown pseudo-steady-state model (PSSM)   1383
ADP
– rapid growth   126
adsorbed species   1257
adsorption
– isotherm   80
– of impurity   580
– process   581
– site   565
adsorption isotherm
– Frumkin–Fowler adsorption   62
– Henry adsorption   62
– Langmuir adsorption   62
adsorption–desorption balance   62
advanced IR detector   1115
advanced photon source (APS)   1638, 1647
advanced protein crystallization facility (APCF)   594
advantage
– chemical vapor transport   904
– crystallization in gels   1613
AFM characterization   1597
afterheater   382, 396
agar   1608
agar-agar   1618
AgGaS$_2$ 731
AgGaSe$_2$ 731
– single crystal   923
agglomeration of point defects   1338
Al source instability   828
Al$_2$O$_3$ 442
Al$_2$O$_3$:Cr   395
Al$_2$O$_3$–ZrO$_2$ 418
Al$_2$O$_3$/GdAlO$_3$ 417

Al$_2$O$_3$-ZrO$_2$(Y$_2$O$_3$) (ZA)   416
Albon and Dunning model   79
Alexander–Haasen model   1351
alkaline halide flux   741
alkaline seeded growth   681
alkyl precursor   1134
alloy disorder   1109
alloy segregation in ternary semiconductor   302
alloying   665, 684
AlN
– bandgap   835
– bulk crystal   827
    etch pit density   834
    etching   827, 834
    growth   821
    prismatic glide   834
    seeded growth   829
– by HVPE   888
– cracking   830
– fundamental optical property   835
– growth habit   828
– growth rate   823
– powder sintering   838
– seed   831
AlN/GaN
– vapor deposition
    characteristic   1251
    mathematical model   1248
alternative model   1351
alternative wafer production process   1724
Aluminum garnet X$_3$Al$_5$O$_{12}$:Ce (X = Y, Lu, Y/Lu)-based scintillators   1677
aluminum nitride (AlN)   821
aluminum vapor pressure   823
ambipolar conduction   347
amelogenin   1622
ammonia (NH$_3$)   1138
ammonium chloride (NH$_4$Cl)   873
ammonium dihydrogen phosphate (ADP)   96
ammonothermal
– GaN   681
– growth system   666
– solvents   663
amorphous layer   1492
analog-to-digital converter (ADC)   1558
analysis

- dislocation 219
- gas flow 1251
- of defects on x-ray topographs 1445
analytical analysis of heat and mass transfer 1252
angle fixation 525
- boundary condition 523
angular vibration technique (AVT) 1237
anionic impurity 776
anisotropic material 1735
anisotropy 1735
- of properties 1464
annealing 190, 438, 466
- experiment 222
annihilation 1557
- characteristic 1553
annular capillary channel (ACC) 1391
annular dark field (ADF) 1490
anomalous birefringence 785
antimonide-based compounds 294
antimony (Sb) 328
antiphase boundary (APB) 944
antireflection (AR) 1715
antisite 167
aperiodic poled LN (APPLN) 258
apparatus 912
application 545, 713
- in crystal deposition diseases 1614
- of STEM, EELS, and EFTEM 1509
applied magnetic field 970
applying steady magnetic field 205
arbitrary Lagrangian Eulerian (ALE) 1397
arsenide
- activation 1106
- implantation 1105
- incorporation 1105
- indiffusion 1105
- monolayer 1087
- precursor 1102
- quantum dot 1133
arsenide-based compound 296
arthritis 1619
as-grown 711
- SiC crystal 806
- single crystal 922
aspect ratio trapping (ART) 1500
asymmetric reflection 341
atherosclerosis and gallstones 1625
atmosphere 406
atmospheric pressure (AP) 1157

atomic
- ordering 1485
- structure 762
atomic absorption spectroscopy (AAS) 1083
atomic energy 1195
atomic force microscopy (AFM) 15, 135, 140, 340, 765, 766, 827, 1081, 1083, 1595
- step velocity 779
atomic layer epitaxy (ALE) 1044
atomistic
- approach 55
- point of view 57
- theory of nucleation equilibrium size distribution of clusters 31
- view of equilibrium 62
attachment energy 56, 65
Auger electron spectroscopy (AES) 1083
Auger recombination 1107, 1108, 1112
autoclave-type cell 1647
automatic diameter control (ADC) 247, 399–402
- for Czochralski crystal growth 251
- of crystal 249
automation of VT 521
avalanche photodiode (APD) 1117
axial field growth 211
axial temperature gradient 266
axisymmetrical problem 523

B

B defect 1283, 1292
$Ba_{0.77}Ca_{0.23}TiO_3$ (BCT) 405
BaB_2O_4 (BBO) 691
back-contact cell 1717
background gas 1202
back-reflection 806
- SWBXT image 807
- topograph 806, 810
baffle design 670
balance of heat transfer 403
band anticrossing (BAC) 1140
- model 360
band structure 346, 1154
Bardeen–Herring mechanism 179
barium rare-earth fluoride (BaREF) 728
barium titanate 246
basal dislocation 14
basal plane 811

- dislocation (BPD) 15, 806, 810, 1466
- stacking fault (BSF) 1501
basal slip 488
basic dislocation analysis 1445
basic principle
- x-ray topography 1426
$BaTiO_3$ 410, 411
BCF (Burton–Cabrera–Frank) 6
- equation 74
benzil 100
benzophenone 99
$Bi_{12}GeO_{20}$ (BGO) 270, 442
$Bi_{12}Si_{12}O_{20}$ (BSO) 270
$Bi_{12}SiO_{20}$ (BSO) 404
$Bi_{12}TiO_{20}$ (BTO) 270, 416
$Bi_{20}SiO_{20}$ (BSO) 264
BiB_3O_6 (BIBO) 731
bidomain 705, 708
- structure 704
bile 1626
binary compound synthesis 293
binding energy 342, 563
biological macromolecules
- nucleic acid 1583
- polysacharide 1583
- protein 1583
biomacromolecular solution
- property 1584
biomimetic recognition 1627
bioseparation process 592
Biot 403
biotechnology 582
bipolar transistor (BPT) 162
bipyramidal 1625
birefringence 161, 692, 1092
bismuth sillenite $Bi_{12}MO_{20}$ (BMO) 264
Bi-Sr-Ca-Cu-O 411
bleaching impurity 268
blocker type additive 70
bootstrapping method 308
borate 730, 739, 743
Born approximation (BA) 1413
boron nitride (BN) 825
Borrmann effect 1438
bound exciton (BE) 887
boundary
- classification 705
- condition 515, 1218, 1362
- layer model 1137
boundary condition
- capillary 525
bow angle 1733
BOX (buried oxide) 1169
BPD (basal plane dislocation) 809

Bragg
- angle 14
- line 1488
- reflection 1486
- relation 1407
Bravais law 54
Bravais–Friedel 137
Bravais–Friedel–Donnay–Harker (BFDH) law 64
breaking strain 403
bridge layer 1031
Bridgman
- crystal growth technique 286
- method 263
- technique 10, 286, 335
Bridgman autoclave 618
Bridgman–Stockbarger (BS) 1379
- technique 437
bright field (BF) 1477, 1478
- image 344
brittle state 116
BSCCO 399, 408
bubble 191
- precipitation 99
bubbler 1135
buffer layer 1083–1085, 1170
buffered oxide etch (BOE) 960
building unit (BU) 64
built-in electric field 704, 705
bulk
- crystal growth of ternary III–V semiconductor 281
- GaN growth 875
- laser crystal 736
- species 1257
bulk growth 801
- InAs$_x$Sb$_{1-x}$ 335
- InBi$_x$Sb$_{1-x}$ 337
- of InSb 334
bunched growth step 141
Burgers vector 109, 114, 219, 766, 806, 1446, 1480
- determination of 114
- direction 1445
- magnitude 1446
- sense 1446
buried-contact cell 1716
bursitis 1619
Burton–Cabrera–Frank (BCF) 54, 577
- theory 119
Burton–Prim–Slichter (BPS) 134, 303
- relation 175
- theory 416

- uniform-diffusion-layer model (UDLM) 1380

C

C/Si ratio 959
Ca$_{1-x}$Sr$_x$MoO$_3$ (CSMO) 406
Ca$_2$FeMoO$_6$ (CFMO) 406
cadmium (Cd) 333, 1617
CaF$_2$ 162, 163, 412
CaF$_2$–MgO 418
calcification 1629
calcite 628
calcium
- carbonate 1629
- gallium germanate, Ca$_3$Ga$_2$Ge$_4$O$_{14}$ (CGG) 1055
- hydrogen phosphate pentahydrate (octacalcium phosphate, OCP) 1622
- phosphate 1619
- phosphate dihydrate 1615
- sulfate 1623
- tartrate 1610
calcium oxalate 1617
- crystallization 1617
- dihydrate (COD) 1615, 1617
- monohydrate (COM) 1615, 1617
calculation of phase diagram (CALPHAD) 447
calibration of etching 1467
calorimetry curve 343
CaMoO$_4$ (CMO) 400, 405, 412
CaO–ZrO$_2$ 418
capacitance–voltage (C–V) 960, 1541
capillary
- boundary condition 525
- boundary problem for TPS 522
- boundary problem solution 527
- problem 517
- problem – common approach 514
- shaping technique (CST) 509
capillary shaping technique (CST) 510
capping 96
carrier gas 872, 1260
carrier lifetime 357
CaSrCu$_2$O$_4$ 401
catching boundary condition 523, 525, 526
cathode luminescence (CL) tomography 149
cathode-ray luminescence (CL) 135
cathodoluminescence (CL) 680, 835, 884, 1453, 1467

CCD (charge-coupled device) 1642
CdS single crystal 915
CdTe 162, 168, 171, 184–186, 190, 191
- growth 1101
- growth nucleation 1101
CdTe/GaAs substrate 1077
CdTe/Si substrate 1076, 1077
CdZnTe substrate 1076, 1077
- characterization 1082
- screening 1082
Ce-doped rare-earth halide single-crystal 1684
Ce-doped silicate single crystal 1681
cell
- formation 183
- pattern 181
- patterning 182, 188
- size 184
- structure 182
cellular
- growth 376
- interface 176
- structure 162, 456, 457
central capillary channel (CCC) 1391
cesium ion exchange 1654, 1656
chalcopyrite 731
challenges 684
Chang–Brown quasi-steady-state model (QSSM) 1381
change of the face character 82
Chapman–Enskog formula 902
character of the face 55
characteristic configuration of growth dislocation 110
characteristics of CVD process 1248
characterization 239, 832, 1405
- method 1668
- of crystals 377
- tool 1090
charge dislocation 351
charge state 1560, 1562
charge-coupled device (CCD) 236, 1449, 1643
chemical
- characterization 1057
- etching 340, 806, 881, 1454, 1459
- inhomogeneity 1339
- potential 1136
- reaction 1254
- reduction 1610
- transport reaction 902

chemical vapor deposition (CVD) 800, 899, 900, 939, 946, 1044, 1157, 1206
chemical vapor transport
– advantage 904
chemical vapor transport (CVT) 135, 897, 899–901
– technique 1573
– transport kinetics 901
chemomechanical polishing (CMP) 827, 1167, 1724
– method 1167
Chernov mechanism
– direct integration 74
chloride VPE (Cl-VPE) 926
choice and change of solvent 76
cholesterol 1626, 1627
cholesteryl acetate 1627
cholic acid 1627
ciprofloxacin 1625
circular cone shaper wall 523
circular or polygonal spiral 140
circumferential stress component 1367
citric acid 1618
classical nucleation theory 1291
cleaning 1725
cleaning procedures for growth chamber, crucible, and charge 299
climb 1350
close-core screw dislocation 806
cluster 562
– balance 1301
clustering in heteroepitaxy 43
Cl-VPE gallium trichloride ($GaCl_3$) 926
CO_2 laser 395
coalescence front 1014
cobalt 1621
cobble texture of quartz (0001) face 104
coefficient of thermal expansion (CTE) 500
coercive field 709
coherent
– x-ray source 1417
coincidence lattice 1501
cold crucible (CC 437, 441, 442, 466
cold crucible (CC) 434, 436
cold-cone seal autoclave 617
cold-drawn steel wire 1730
collagen gel 1620
colony 462
color center 494

colored quartz 632
combining electric and magnetic fields 1593
commercial solar cell concept 1716
common crucible material 249
comparison of ammonia and water as solvents 657
compensating center
– dominant 1568
compensating defect 1542
compensation mechanism 221
compensation ratio 226
complex dilution 1609
complexity 1413
composition 698, 910, 1488
– amplitude 175
– profile 340
– sensitivity 1414
– variation 1420
compound 165
– semiconductor 8
computational fluid dynamics (CFD) 1361
computational issue 1258
computer tomography (CT) 1687
concave interface 267
concentration sensor 774
concentric ring 1611
conductivity 349
congruent
– composition 269
– lithium niobate 254
– lithium niobate crystal 255
– melting 373, 409
– melting fibers – the search for stoichiometry 409
– melting point 253
connected net analysis 69
conoscopic pattern 12
conservation
– equation 1217
– law of Burgers vectors 109
– of mass 402
constant-strain-rate compression test 1346
constitutional supercooling 176, 188, 375, 457
construction ceramics 444
consumables in wiresawing process 1735
contact plane 121
contact sintering 1715
contactless chemical vapor transport technique (CCVT) 1573
continuous feeding during growth 745

continuous filtration 781, 788
– system (CFS) 771
continuous wave (CW) 260
continuum model 977
continuum-scale quantitative defect dynamics in growing Czochralski silicon crystal 1281
contrast 1426, 1448
– associated with cracks 1448
– from inclusions 1446
– transfer function (CTF) 1482
contrast on x-ray topographs 1440
controlling the growth 1090
convection 1215
– diffusion (CD) 1398
– flow 144
– in the melt 1362
– pattern 407
convectional stirring 439
convective frequency 175
conventional method 771
convergent-beam electron diffraction (CBED) 1478, 1486
cool-down period 1371, 1372
cooled sting assembly 584
cooled sting technique 583
cooperating spiral 119
coordinate measuring machine (CMM) 781
copper gallium diselenide ($CuGaSe_2$) 898
copper indium diselenide ($CuInSe_2$) 898
copper indium disulfide ($CuInS_2$) 898
core 268
– effect 267
– energy of dislocation 113
correlation with electrically detected trapping centers and defect levels 1541
corrosion 567
corundum 630
Cottrell atmosphere 1462
counterdiffusion method 1590
counterelectrode (CE) 1454, 1455
covalent state 332
CP analysis 1097
crack 1448
crack formation 304
crack formation in ternary crystal 304
cracker cell 1090
creep curve 1349
criterion for characteristic defect formation 237

critical condition   1324
critical condition for radiation
   dominance   1248
critical layer thickness of dot
   nucleation   1143
critical point (CP) model   1095
critical supersaturation   75
critical thickness   1161, 1497
critical-point (CP) energy   1091
critical-resolved shear stress (CRSS)
   177, 179, 1340
$CrO_3/HF/H_2O$ in the dark   1469
cross-slip   178, 1350
CRSS-based elastic model   1340
crucible free   393
crucible material   249, 825, 826
crystal   691
– characterization   11
– chemical aspect of Bi substitution
   in InSb   333
– cooling   377
– defect   11, 73, 541, 1478
   observation   12
– density   251
– edge   767, 768
– grown under unconstrained
   condition   133
– habit   565, 578
– originated particle (COP)   188
– potential   347
– rotation and pulling arrangement
   249
– seed holder   572
– shaping measure   207
– structure   445
– structure and bonding   331
– structure and bonding of InSb
   331
– surface   761
– truncation rod (CTR)   762
– twin   213
crystal deposition
– disease   1614
– related disease   1616
crystal face
– atomically rough   1002
– imperfect singular   1002
– perfect singular   1002
crystal growth   4, 574, 854, 1335,
   1724
– and ion exchange in titanium
   silicates   1637
– classic method   1587
– control of crystal defects   237
– from low-viscosity solutions   736
– Haasen model   1350

– high-viscous solution   739
– history   394
– hydrodynamic effect   781
– in bile   1625
– in space   583
– in the presence of a magnetic field
   1610
– limitation   375
– new trend   1591
– nucleation   847
– of laser fluorides and oxides from
   melt   479
– of lithium niobate   252
– process   1336
   classification   7
– rate for crack-free ternary crystal
   308
– SiC   798
– system   1380
– technique   6
   traveling solvent   369
– termination   377
– theory   4
crystal orientation
– nonstandard   951
crystal quality   783, 1592
– enhancement   1591
– spectroscopic study   783
crystalline
– defect   1079
– fiber   407
– imperfection   805
– layer structure   1405
– quality   13
– SiC   799
– silico titanate (CST)   1650
– silicon   240
crystallization   452, 561, 1587
– capillary tube   1590
– electric field   1591
– energy   65
– front   516
– front instability   376
– gel   1589
– high-throughput   1594
– in gels   1613
– magnetic field   1592
– microgravity environment   1588
– of calcium oxalate   1617
– of hormones   1628
– of hormones: progesterone and
   testosterone   1628
– of the constituents of crystal
   deposits   1616
crystallizer   569
– reciprocating motion   575

crystallographic
– orientation   454
– plane   694
– shape   768
crystallography
– law   54
crystal–melt system   517
$CsTiOAsO_4$ (CTA)   692
$CuAlSe_2$ crystal   919
cubic solid solution   444
cubic zirconia   435
$CuGaS_2$ 922
– based single crystal   921
$CuInTe_2$ crystal   920
Curie temperature   697–699, 701,
   702, 705
Curie–Weiss law   697
cusped field growth   212
CVD (chemical vapor deposition)
   798
– epitaxial film   801
– reactor configuration   947
CVT
– reaction   903
– $ZnSe-I_2$ system   905
CVT growth
– chemical parameter   903
CVT growth of crystals   904
– geometrical parameter   904
CVT method
– advantage   904
– limitation   904
cystinosis   1624
cystinuria   1624
cytochrome c   1598, 1599
CZ defect dynamics   1291, 1316,
   1318, 1330
– absence of impurity   1329
– lumped model   1297
– the quantification   1299
Czochralski (CZ)   312, 501, 1215,
   1281, 1706
– crystal growth system   249
– defect dynamics   1304, 1313
Czochralski growth   192, 335, 713
– of organic crystal   99
Czochralski growth system
– design   247
– development   247
Czochralski method   9
– of crystal growth   246
Czochralski silicon   232
– conventional   232
– crystal
   vacancy   1321
Czochralski technique   1336

Czochralski technique (CZT) 441, 480, 509

D

D defect 1283, 1322
dark field (DF) 1477, 1478
Dash seeding 216, 218
Davey and Mullin model 79
dead zone 76, 80, 81, 780
Debye–Waller factor 1487
deceleration of plasma 1202
decomplexation 1613
decomposition of nitrogen precursor 1138
decorated dislocation 15
deep positron state 1554
deep-level transient spectroscopy (DLTS) 950, 1539, 1541, 1542
defect 1565
– at an interface 1535
– characterization 1551
– control 205
– density 15, 883
– formation energy 165
– impurity effect 776
– in AlN/GaN films originating from SiC substrate steps 1503
– in crystal 380
– in SiC 942
– level 1541, 1543
– mapping 225
– of crystal 776
– passivation 1087
– selectivity 1459
– site 35
defect dynamics 1284, 1310, 1324
– general 1322
– in one-dimensional crystal growth 1308
defects
– in a thin film on a substrate 1534
defects at surfaces 1536
defect-selective etching (DSE) 1453, 1461
deformation
– behavior of semiconductor 1346
– plastic 156, 183
– potential (b) 356
– stage 1347
degenerate 346
degeneration 453
– crystal orientation 453
– crystal size 453
– geometric selection 453
deionized (DI) water 299

deleterious effects of dislocation 1337
density 733
dental calculus 1619
depolarization 1091, 1093, 1099, 1100
deposition rate 1269
– expression 1272
– prediction 1270
deposition uniformity 1269
desolvation at surface site 76
detectivity (D^*) 357, 590
detector 1642
– noise 590
– response 1116
determination of Burgers vector
– direction 1445
– sense and magnitude 1446
determination of line direction 1445
detrapping 1557
deuterated KDP crystal 786
deuterated potassium dihydrogen phosphate (DKDP) 96
deuterated triglycine sulfate (DTGS) 575, 579
developing new material 1044
developments in liquid-phase ELO growth 1027
deviation from calculated direction 113
deviation from stoichiometry 165
device-grade ternary substrate 284
dewetting 265
diameter control 401
diamond 149
– abrasive grit 1728
– cubic 1485
– growth (DIA) 1647
– impregnated wire 1731
– structure of the crystal 1343
dicalcium phosphate
– (DCP) monetite 1619
– dihydrate (DCPD) 1615, 1620
dichlorosilane (SiH_2Cl_2) 882
dielectric function library 1092–1095
differential interference contrast
– (DIC) 1464
– microscopy (DICM) 135
differential scanning calorimetry (DSC) 261
differential thermal analysis (DTA) 331, 370, 373
diffraction
– contrast 1478
– contrast imaging 1479

– efficiency 259
– image 589
– technique 1484
– theory 1638
diffuse scattering 1416
diffuse scattering of x-ray 333
diffuser plate 1088, 1089
diffusion 166, 800, 1003
– coefficient 583, 902
– in solids 1552
– layer 375
– theory 5
diffusion-controlled
– crystallization apparatus for microgravity (DCAM) 594
– method 1588
– process 591
diglycine sulfate (DGS) 578
dilute nitride quantum well laser 1141
diluted nitrides 1133
diluted Sirtl with light (DSL) 175, 1460
dimensionless group 1251
dimethylhydrazine $(CH_3)_2NNH_2$ 1138
dipole of dislocations 178
direct current (DC) 408
direct dislocation image 1442
directional crystallization 437, 440, 450, 454
directional solidification (DS) 1707, 1709
– by normal freezing 309
– by solute diffusion and precipitation 310
discrete lattice structure 113
discrete rate equation 1299
disilane (Si_2H_6) 882
dislocation 107, 160, 161, 176, 340, 497, 543, 566, 588, 676, 883, 1086, 1110, 1409, 1469
– 30° and 60° 178
– 60° 1344
– analysis 177, 1488
– bunching 185, 186
– bundle 181, 187
– core observation 1483
– density 12, 179, 216, 344, 404, 785, 1085, 1365
– different types 1351
– dipole 109
– dynamics 178
– dynamics (DD) 177
– edge dislocation 109
– engineering 187, 1170

– free mechanism 767
– generation 152, 155, 1335
– geometrical position 1465
– glide 184
– glide velocity 1352, 1355
– in ELO layers 1011
– in semiconductor material 1337
– in SiC 1466
– jungle 182
– lineage 181
– loop 168, 180
– loop–hole configuration 813
– mechanism 765
– misfit 1417
– multiplication 180, 1338, 1355, 1356
– nucleation 1338
– pivot 812
– redirection in AlN/sapphire epilayer driven by growth mode modification 1507
– reduction during seeding 217
– screw dislocation 109
– slip 220
– threading 1417
– type 177
– wall 187
disorientation angle 181
dispersion surface 1437
displacement
– rate 516
– reaction 1623
– reaction method 1609
dissipative structuring 184
dissolution 151
distinction between natural and synthetic gemstones 155
distorted-wave Born approximation (DWBA) 1413
distributed Bragg reflector (DBR) 1176
distribution coefficient 170, 333, 456, 460, 738
distribution of dislocation density 1366
distribution of impurity 778
domain 704
– boundary 705, 706
– formation 704
   kinetics 704
– polarization 257
– switching 252
domain structure 258, 462, 463, 691
– artificial 708
– ferroelectric 691
Donnay–Harker 137

donor concentration 348
donor defect EL2 173
dopant
– activation 1104
– concentration 1059
– distribution 406
– recharging 173
– solubility 171
doped crystal 918
– of $CuAlS_2$ 918
– of $CuAlSe_2$ 919
doped LGT LPE film 1057
doped lithium niobate crystal 260
doped TGS 580
doping 665, 678, 684, 1355, 1565
– extrinsic 1104
– incorporation technology 949
– n-type 1104
– of sillenite 270
– p-type 1104, 1105
– technique for GaN in HVPE 882
Doppler broadening 1553
– spectroscopy 1559
DOS (density of states) 789
dot formation 1177
dot-in-a-well (DWELL) 1147
double crucible in the CZ (DCCZ) 254
double diffusion 1612, 1613
double layer ELO (2S-ELO) 878
double tungstate 736, 737
double-crucible Czochralski (DCCZ) 255
double-crucible technique 443
driving force 136
drug design 592
DSE of InP 1464
DSL system 1470
Dupré's formula 58
dyeing of crystal 103
dynamic polygonization 183, 185
dynamic reflectance spectroscopy (DRS) 1089
dynamic stability of crystallization (DSC) 509
dynamical image 1443
dynamical theory 1413
– of x-ray diffraction 1436
dynamical x-ray theory 215

E

Eagle–Picher (EP) 1573
early theoretical and modeling study 971
eccentricity of spiral steps 154

EDAX spectrum 344
edge defined film fed growth (EFG) 538
edge dislocation 6, 119, 178, 219, 342
– growth-promoting 119
edge facet 213
edge ring 1322
edge-defined film fed growth (EFG) 394, 1379, 1389, 1706
edge-ring 1327
edge-supported pulling (ESP) 1706
EELS 1509
– application 1509
– application in microanalysis 1509
– elemental analysis 1491
– fine edge to study interface material 1509
– spectrum 1491
– study of Mn diffusion 1509
E-etch 1463
effect of additives on crystallization of calcium phosphates 1620
effect of convection in solution growth 563
effect of decoration and composition 1465
effect of dislocation 785
effect of flow rate on substrate temperature 1266
effect of impurities 564
effect of impurities on TGS crystal growth 579
effect of impurity concentration and supersaturation 80
effect of magnetic field on crystal twinning 210
effect of magnetic field strength 984
effect of seed 142, 145
– crystal 577
effect of tartaric and citric acids 1618
effective diffusion length 1274
effective distribution coefficient 149, 407, 409, 712
effective mass 346
effective medium approximation (EMA) 1099
effective nonlinear coefficient 709
effective segregation coefficient 172
effective stress 1357
effusion cell 1089
EFTEM 1509
– to enhance contrast 1512
– to map elemental distributions 1512

– to reduce diffraction contrast 1512
Ehrlich–Schwoebel barrier 39
eight-membered ring (8MR) 1650
EL2 defect 167
elastic energy of dislocations 1466
(elastic) stress response 1342
elastohydrodynamic interaction 1734
electric field 407
– poling 708
electric level 990
electrical assisted laser floating zone technique (EALFZ) 408
electrical compensation 1565
electrical conductivity 261
electrically active inhomogeneity 1471
electrically detected magnetic resonance (EDMR) 1546
electrically detected trapping 1541
electrochemical etching 1454, 1459
electrochemical/electrolysis 1610
electroless etching 1456, 1460
electroless etching for revealing defects 1469
electroless etching in the dark 1457, 1460
electroless photoetching 1457, 1460
electromagnetic (EM) 207
electromagnetic casting (EMC) 1707
electromagnetic Czochralski (EMCZ) 1238
electromagnetic mobility 992
electromigration 971, 1028
electron beam (EB) 1181
– interaction 1199
– plasma generator 1199
electron cyclotron resonance (ECR) 1087
electron energy-loss spectroscopy (EELS) 380, 1478, 1486, 1489
electron ionization 271
electron microprobe analysis (EPMA) 380
electron nuclear double resonance (ENDOR) 1546, 1669
electron paramagnetic resonance (EPR) 15, 495, 680, 695, 1521
electron spin resonance (ESR) 1521, 1669
electron trap center 716
electron-beam induced current (EBIC) 549, 944, 1453
electronegativity 359

electronic balance 249
electron-nuclear double resonance (ENDOR) 715
electrooptic (EO) 262, 691
– effect 271
element partitioning 134, 147, 149
– in different growth sectors 150
elemental spiral 141, 153
elimination of crack 306
ellipsoidal mirror 369
ellipsometer design 1091
ELO
– choice of growth technique 1004
– filtration of dislocation 1001
– filtration of dislocations 1011–1014
– growth anisotropy 1005
– growth enhanced by dislocation 1006, 1008, 1025
– growth retarded by doping 1009, 1010
– mask-induced strain 1017–1024
– perfection of coalescence front 1014–1016
– surface supersaturation in LPE 1007
– thermal strain 1024
ELO growth
– new concept 1030
emerald 141, 631
emissivity of the liquid surface 272
emitter formation 1714
encapsulation height 1371
end chain energy (ECE) 67
energetic condensation 1193
energy factor of dislocation 111
energy gap 329
energy minimization 405
energy-dispersive detectors 1643
energy-dispersive x-ray analysis (EDAX) 340, 380
energy-dispersive x-ray spectroscopy (EDS) 1478
energy-dispersive x-ray spectroscopy (TEM-EDS) 1486
energy-filtered transmission electron microscopy (EFTEM) 1478, 1490, 1515
energy-loss near-edge structure (ELNES) 1492
enthalpy 164
entropy 164
– of fusion 192
environmental concern 1735
environmental effect 664
epilayer

– $InAs_xSb_{1-x}$ 338, 355
– $InBi_xSb_{1-x}$ 339, 355
– InSb 337, 355
epilayer uniformity 1108, 1109
EPIR Technologies 1102
epitaxial film 746, 802
– of laser material 746
epitaxial lateral overgrowth (ELO) 113, 877, 999, 1000, 1002, 1042
– of semiconductors 999
epitaxial relationship 1049
epitaxy 593
– GaN/AlN/SiC 1506
epitaxy of nitride
– substrate 1061
epitaxy within the structural field of KTP 748
EPR analysis 1524
EPR technique 1534
equation
– conservation 1217
equilibrium
– concentration of clusters 29
– crystal 57
– crystal–ambient phase 18
– curve 1585
– distribution 170
– form 137
– of infinitely large phase 18
– of small crystal with the ambient phase 20
– phase diagram 330, 331
– shape (ES) 55, 60, 66
– shape crystal 58
– shape of crystals 22
– surface profile 66
– thermodynamics 1075
– vapor pressure 1260
equipment for time-resolved experiments 1642
error function (erfc) 1714
estimation of the electromagnetic mobility value 993
etch pit density (EPD) 177, 188, 334, 1077, 1083, 1341, 1461
etch pit pattern 340
etchant composition 1464
etching anisotropy 215
etching in the dark 1457
etching of multilayer laser structures 1468
etching of semiconductors 1453
ethyl alcohol 575
ethyl vinyl acetate (EVA) 1705
ethylene dithiotetrathiafulvalene $(CH_2NH_2)_2C_2H_4O_6$ (EDT) 569

EuAlO$_3$ 412
eutectic 411
– fiber 416
– solidification 383
evaporating fiber 416
evolution of crystal growth under applied magnetic field 206
evolution of crystal habit 85
evolution of interfaces 989
evolution of internal variables and plastic strain 1342
evolution with growth 1099
Ewald sphere 1486
excess shear stress 1341
exchange mechanisms 1659
expansion of plasma 1199
experimental details of ceramics preparation for OFZT 372
experimental simulation study 733
experimental tests of the capillary shaping statement 530
expression system
– membrane protein 1594
extended atomic distance mismatch (EADM) 1502
extended imperfection 676
external force 1228
extinction contrast 1441
extract
– cereal 1618
– fruit 1618
– plant 1618
extrinsic atom 172
extrinsic point defect 170

F

F (flat) face 138
face character 66
faces of TGS 578
facet 160, 191, 192, 490, 544
– formation 193
– interface 191
– sector 104
faceting of rounded surfaces 96
factor affecting growth form 143
factors influencing morphology of pits 1462
factors influencing the crystal habit 71
fast Fourier transform (FFT) 1506
fault
– Shockley 813
feed rod 369
femtosecond laser irradiation (FSLI) 1591

fermentation process 592
Fermi level 353, 1355
– effect 171
ferritin 1599
ferroelastic switching 125
ferroelectric
– domain 700, 703, 707
– material 273
– phase transition 697
FFT (fast Fourier transform) 1506
fiber growth 409
fiber pulling 394
field effect transistor (FET) 860, 863, 968, 1153
figure of merit (FOM) 483
film growth 1204
filtration 780
– of substrate dislocations in ELO 1011
fingerprinting of cut stones 155
finite crystal 57
finite element 1361
fitting to a library 1097
flame fusion technique 9
flat 562
– bottomed etch pit (F-type) 139
– face (F-face) 55
– interface 263
flight crystal growth cell 586
flight hardware 584
flight optical system 586
floating zone (FZ) 367, 394, 509, 583, 1281, 1304, 1706
– advantage 370
– limitation 370
– temperature gradient 370
flow
– control 1228
– pattern 669, 670, 672
– simulation 872
– structure 1219
flow and heat transfer 666
fluctuation of growth conditions 143
fluctuation of growth conditions (growth accidents) 96
fluid experiments system (FES) 584
fluid field 585
fluids experiment system (FES) 586
fluorescence quenching 496
fluoride 728, 738
flux 375
– growth 9, 725
– technique (FT) 487, 503
focal-plane array (FPA) 1069, 1118
focused ion beam (FIB) 1467, 1493

foggy inclusion 776
Fokker–Planck equation (FPE) 1290, 1299, 1301
forced convection 266, 564
foreign adsorption 61
foreign particle 95, 1339
foreign substrate 1275
forest dislocation 1358
formation of 3-D nuclei on unlike substrate 25
formation of quantum wells, superlattices, and quantum wire 1173
fourfold symmetry 1341, 1366
Fourier-transform infrared spectroscopy (FTIR) 261, 1079, 1083, 1089, 1108
Frank fault 813
Frank partial dislocation 807
Frank's conservation law 109
Frank–Read mechanism 178
Frank–Read source 812
Frank–van der Merwe growth 20
Frank–van der Merwe growth mode 1073
free abrasive machining (FAM) 1724, 1730, 1733
free carrier absorption (FCA) 354
free convection 266
free energy 562
Frenkel
– defect 166
– reaction 1304
– reaction dynamics 1284
frequency conversion 760
frequency doubling 692, 710
friction coefficient 464
from crystal ingot to prime wafer 1723
front stability 375
full width at half maximum (FWHM) 12, 830, 886, 1077, 1079, 1082, 1110, 1409, 1558, 1639
fully overgrown ELO structure 1014
fundamental dislocation theory 816
fundamentals
– LPE 1042
furnace construction 247

G

GaAs 162, 166, 169–171, 175, 179, 182, 183, 186, 190, 231, 350, 355, 876
– system 994

– wafer 181
gallbladder 1626
gallium
– berlinite 625
– evaporation 1262
– iodine vapor growth 1245
– monochloride (GaCl) 926
galvanomagnetic application 358
galvanomagnetic device 358
GaN 655, 663, 1410
– by HVPE 927
– by VPE 925
– deposition rate 1271
– film growth 925
GaN film 928
– characterization 928
GaN IVPE growth
– modeling 1258
GaN/AlN
– film growth 1274
– vapor-growth system 1244
GaN/AlN/SiC epitaxy 1506
GaP 170
gas convection 1368, 1373
gas mixing process 1267
gas phase 1256
gas-phase reaction 1256
– analysis 1259
gas-source MBE (GSMBE) 946, 1141, 1160, 1177
Gaussian reflector 397
Gd_2SiO_5:Ce (GSO) 1681
Ge condensation method 1169
Ge substrate 1088
$Ge_{1-x}Si_x$ 163
gel
– acupuncture method (GAME) 1590
– method 560
– system
   pattern formation 1610
– technique 1611
gel growth 566
– of crystals 1608
gelling 1609
gemstone 629
general
– defect dynamics 1322
general formulation 28
general purpose autoclave 618
generation
– of defects 93
– of micropipes 804
– of point defect 1345
genomics 1584
geometric factor 343

geometric partial misfit dislocation (GPMD) 1503
geometrical partial misfit dislocation (GPMD) 1506
geometrical position of dislocations with respect to the surface 1465
geometrically necessary boundary (GNB) 182
geometry optimization 1268
Ge-on-insulator (GOI) 1170
GeTe 168
g-factor 329
Gibbs free energy equation 1586
Gibbs–Thomson effect 1004, 1008, 1030
Gibbs–Thomson equation 5
g-jitter 584
glacial acetic acid (CH_3COOH) 299
glancing incidence 345
glass synthesis by skull melting 465
glass-forming melt 438
glass-forming region 467
glide dislocation 1344
global modeling 184
globular cell morphology 183
glow discharge mass spectrometry (GDMS) 211, 839, 1522
glow discharge mass spectroscopy (GDMS) 222
governing equation 1305, 1314
graded buffer 1165, 1498
– and insertion of strained layers 1498
graded double layer heterojunction (DLHJ) 1116
graded layer 1414
gradient freezing technique 288
grain boundary 160, 162, 184, 464
grain expansion 831
grain-free growth 185
graphite component 803
graphitization of SiC 805
Grashof number 564
gravitational force 588
gray track 691, 700, 715
– center 715
– formation 716
grazing incidence
– imaging 807
– small-angle scattering 1418
– SWBXT 808
– XRT 815
green-radiation-induced infrared absorption (GRIIRA) 716
grinding 1724, 1731

grooved cylindrical wire guide 1732
ground-based cooled sting apparatus 585
group I dopant 1106
– diffusion 1106
– incorporation 1106
group III nitride 821
group V dopant incorporation 1106
growing CZ crystal
– reactions 1304
growth 239, 1027, 1711
– angle 515
– angle certainty 530
– axis 1372
– band 151
– banding 146
– chemistry 869
– condition 152, 776, 1569
– control 1101
– defect 269
– dislocation 107, 805, 1358
– facet 460
– from melt 9
– habit 577
– hardware 1088
– hillock 103, 114
– history 147
– interface (G) 306
– interruption (GRI) 1145
– law 74
– mechanism 582, 970, 1052
– parameter 1052
– period 1365
– pit 943
– polarity 1572
– procedure 1101
– process 1135
– spiral 806
– stoichiometry 1570
– striae 456
– striation 213, 455
– surface evolution 1274
– technology for silicon photovoltaics 1706
– temperature 578, 910
– twin 121
– under controlled atmosphere 405
– unit 164
– using compositionally graded feed 315
growth form 137, 143
– of polyhedral crystals 143
growth from
– crystal edge 767
– large-volume melt 311

– supercooled melt   99
growth history and internal morphology   148
growth kinetics   73, 561, 577, 664, 1249
– of CVD process   1250
growth of
– $AgGaS_2$ 923
– bulk crystals   1215
– bulk sillenite crystal   264
– compounds   8
– GaN films   1206
– III-nitrides with halide vapor-phase epitaxy (HVPE)   869
– lattice-mismatched ternary on binary using quaternary grading   311
– lithium niobate crystal   252
– organic semiconductor   846
– photorefractive bismuth silicon oxide crystal   272
– silicon crystals of semiconductor grade by Czochralski (CZ) technique   232
– sillenite crystals and its characteristic   264
– single crystals based on zirconium dioxide   443
– thick layer   337
growth of CZ crystal   449
– equipment   449
– flowsheet   450
– impurity   451
– melt formation   450
– melt propagation   451
– melting   449
– mode of melting   451
– raw material   449
growth on
– spheres   96
– templates   877
– two-dimensional nucleus   767
growth rate   5, 306, 457, 805, 970, 994
– anisotropy   376
– determination method   290
– dispersion   106
– in LPEE   992
– linear   712
– parameter   769
growth sector (GS)   101, 102, 146, 147, 677, 700, 701, 706
– boundary   101, 102, 105
growth shape   66
– structural and bond-energy approach   64

growth system   1062
– and optimization   1062
growth technique   134, 769
– for single crystals   769
growth unit (GU)   64
gypsum   1623

H

habit change   69
– with supersaturation   75
habit modification   576
habitus   144
Hagen–Strunk mechanism   1494
hair inclusion   98
hairlike inclusion   780
hairpin dislocation   119
half-crystal position   19
half-loop array (HLA)   811
halide vapor-phase epitaxy (HVPE)   869, 1245
Hall effect   1107, 1108
Hall factor   348
hanging drop method   592
Hartman–Perdok theory   64
heat and mass transfer   1251, 1252
heat shield   439
heat transfer   563
heat treatment   1543
heat-exchanger method (HEM)   480, 483, 502
heating method   247
helical Liesegang ring   1611
hematite   141
heteroepitaxial semiconductor system   1494
heterogeneous nucleation   58, 910
heteropolar crystal   6
heterostructure bipolar transistor (HBT)   162, 1153
hexagonal dislocation loop   1351
hexamethyldisilane (HMDS)   947
Hg absorption   1098
Hg adsorption   1098
(Hg,Cd)Te   175
HgCdTe   1069, 1071, 1072
HgCdTe (MCT)   328
HgCdTe growth   1078, 1079, 1102, 1103
HgTe/CdTe superlattice (SL)   1112
– Auger recombination   1113
– energy gap   1113, 1114
– experimentally observed property   1114
– growth   1114
– growth quality   1115

– interdiffusion   1113
– interfacial roughness   1115
– inverted band   1113
– optical absorption   1113
– theoretical property   1113
high nitrogen pressure (HNP)   1564
high resolution x-ray diffraction (HRXRD)   929
high speed IR detector   1117
high-angle annular dark field (HAADF)   1490
high-angle annular dark field in scanning transmission electron microscope (HAADF-STEM)   380
high-angle annular dark-field scanning TEM (HAADF-STEM)   1484
high-density protein crystal growth system (HDPCG)   594
high-electron-mobility transistor (HEMT)   968, 1059, 1172
higher-order Laue zone (HOLZ)   1486, 1488, 1496
high-frequency device   798
high-frequency heating   248
high-index surface   1077
high-level waste (HLW)   1649
highly n-type Si   1566
highly oriented pyrolytic graphite electrode (HOPG)   1597
highly vacancy-rich condition   1323
high-operating-temperature (HOT)   1118
high-potassium KTP   710
high-power device   798
high-power electron beam   1199
high-pressure ammonothermal technique (HPAT)   684
high-pressure cells   1647
high-quality bulk crystal
– characterization   832
– structural property   832
high-quality crystal   1584
high-resolution multiple-crystal diffractometer   1412
high-resolution transmission electron microscopy (HRTEM)   177, 380, 1478, 1482
high-resolution x-ray diffraction   1406
high-resolution x-ray diffraction (HRXRD)   340, 833
high-temperature
– CVD (HTCVD)   801
– glass   465

- growth 560
- materials compatibility 825
- solution (HTS) 9, 731
- solution growth 725, 799
  bulk growth 801
- superconductor (HTSC) 373, 1042, 1046
hillock 704
hippuric acid 1625
historical development of LPE 1042
HIT cell 1717
hollow core (micropipe) 806
hollow morphology 1625
hologram thermal fixing 261
holographic
- image 260
- interferometry 586
- optical element (HOE) 587, 589
- tomography 584
HOLZ line 1488
homoepitaxial ELO layer 1017
homoepitaxial layer 811
homoepitaxial LGT LPE film growth 1056
homoepitaxy 1056
homogeneity 133, 784, 1421
homogeneous nucleation 909
homogenization 466
homojunction device 1713
homopolar crystal 6
horizontal Bridgman (HB) 169, 1216
- crystal 287
- technique (HBT) 485, 502
horizontal gradient freezing (HGF) 288
horizontal ZM (HZM) 1216, 1223
hot-wall CVD 801
hot-wall Czochralski (HWC) 170
hourglass inclusion 97
hydrazine (H_2NNH_2) 1138
hydride vapor-phase epitaxy (HVPE) 656, 681, 899, 925, 926, 1001, 1564, 1568, 1572
hydrochloric (HCl) 299
hydrodynamic
- condition 782
- effect 781
- film and hydrodynamic interaction 1734
- principle 209
hydrodynamics 85, 98
- of the solution 733
hydrofluoric acid (HF) 299
hydrogen ion concentration (pH) 1609

hydrogen passivation of defects 1087
hydrogen passivation of Si surface 1101
hydrothermal (HT) 1573
- condition 610
- method 8
- steel autoclave-type cell 1647
hydrothermal growth 599
- apparatus 615
- growth kinetics 674
- hydrodynamic principle 606
- morphology 674
- of fine crystals 634
- thermodynamic basis 606
- thermodynamic modeling 608
- ZnO crystals 674
Hydrothermal ZnO 674
hydroxyapatite (HAP) 608, 1615, 1620

I

ID saw 1730
identification flat (IF) 1463
idler 714
image force 110
image plates (IP) 1642
imaging 1478
immersion-seeded KTP 694
immobilization 468
impact ionization 1117
imperfect layer 1408
imperfect structure 1413
imperfection 566, 590
impurity 160, 458, 577, 580, 678, 699, 838, 1284
- concentration 11, 770, 777
- decoration 1560
- distribution 534, 1380, 1389
- effect 144, 776
- effectiveness 80
- getter region 268
- incorporation 711, 839
- segregation 220
impurity adsorption 78
- theoretical growth 69
In bump connector 1119
in situ
- cell 1645
- control 169, 170, 1727
- ion exchange 1656
- studies of titanium silicates 1649
- study 1658
- synthesis of Na-NbTS 1655

- x-ray experiment 1419
InAs/GaAs
- quantum dot 1142
$InAs_xSb_{1-x}$ 342, 345, 349, 350, 352, 353
- transmission spectra 353
$InAs_xSb_{1-x}$ 330
$InBi_xSb_{1-x}$ 331, 344, 345, 350, 352, 355
InBiSb 347
incandescent heating 367
incidental dislocation boundary (IDB) 182
inclusion 95, 160, 163, 186, 255, 456, 458, 491
- incorporation 190
- primary 95
- secondary 95
- trapping 191
- zonal 97
incongruent melting 370, 416, 726
incongruently melting 373
incorporation coefficient 172
indirect laser-heated pedestal growth (ILHPG) 397
indium (In) 328
- bismuth (InBi) 328
- bismuth arsenic antimonide ($InBi_xAs_ySb_{1-x-y}$) 329
- phosphide (InP) 205, 231
- tin oxide electrodes (ITO) 1597
induction
- furnace 272, 803
- heater 253
- heating 247
- heating system 249
- period 764
industrial
- bulk growth 802
- crystallization 77
- production 435
inertial confinement fusion (ICF) 759
infinite crystal 57
infrared (IR) 15, 162, 283, 328
- absorption 224, 488, 716
- active lattice mode 355
- detector 1118
- laser scattering tomography 189
- photodetector 357
InGaAsN
- electronic property 1139
- nitrogen precursor 1139
- quantum wells 1137
- valence band offset 1140
inhibitor 1465, 1616

inhomogeneity of impurity 544
inhomogeneous 337
initial dislocation density 1358
initial incorporation 1285
InN by HVPE 890
inner diameter (ID) saw 1728
inner-diameter (ID) 1722
InP 180, 185, 194
in-plane scattering 1418
InSb 191, 192, 330, 340, 350, 352
– wafer 342, 352
InSb substrate 1088
inspection 1726
integrated circuit (IC) 1706
interaction between dislocations 1356
interaction coefficient 1358
interface 359, 1501
– defect 1543
– diffusion 167
– growth kinetics 268
– kinetics 193
– of epitaxial systems 1483
– processing 1193
– roughness 1415
interface type
– castellated 1416
– fractal 1416
– staircase 1416
interference-contrast microscope (DICM) 140
interferogram 591
intermediary image 1444
internal detector noise 1116
internal morphology 146
internal stress 543
International Microgravity Laboratory (IML-1) 582, 583
intersecting stacking fault 1506
interstitial 167
interstitial atom 160
intrasectorial sector 146
intrinsic
– carrier concentration 349
– defect 260
– point defect 172, 1284
– point defect balance 1285
– point defect property 1327
inversion domain (ID) 1464, 1472
– boundary (IDB) 1501
inverted temperature gradient method 832, 839
in vitro crystallization 1614
iodine
– gallium reaction 1261
– vapor pressure 1260

– vapor-phase epitaxy (IVPE) 1243, 1245
ion acceleration 1200
ion beam etching (IBE) 625
ion chamber (IC) 1645
ion current 1205
– of plasma propagating in ambient gas 1205
ion energy 1200
– spectrum 1201
ion exchange 708, 1652
– of Cs^+ into Na-TS 1652
ion implantation 1105
– method 1168
ion-beam-assisted deposition (IBAD) 1204
ionic conductivity 694, 710
ionized impurity scattering 351
ion-scattering spectroscopy (ISS) 1087
island
– formation 1074, 1075, 1103, 1104
– growth 1596
– morphology 338
– structure 461
isopropyl alcohol (IPA) 1714
isothermal evaporation 575
isotropic thermal strain response 1343
iterative target transform factor analysis (ITTFA) 1644

J

Jackson factor 167, 192
jewelery 4
jog 1356
joint density of states (JDS) 1094, 1096
Jones matrix 1091
junction FET (JFET) 940
junction isolation 1715

K

$K_6P_4O_{13}$ 751
$K(D_xH_{1-x})_2PO_4$ (DKDP) 759
$K(Gd_{0.5}Nd_{0.5})(PO_3)_4$ 732
$K(Ta_xNb_{1-x})O_3$ (KTN) 162
$K_2W_2O_7$ 737
$K_6P_4O_{13}$ 749
KDP
– rapid growth 126
kerf loss 1730
$KGd(PO_3)_4$ (KGdP) 742
$KGd(WO_4)_2$ 738

KGdW 737
KH_2PO_4 (KDP) 730
KHoW 738
Kikuchi line 1488
Kim model 1095–1098
kinematic viscosity 564
kinematical theory 1413
– of x-ray diffraction 1436
kinetic
– deposition model 1271
– Monte Carlo method 1246
– of crystallization 610, 614, 761
– related conditions 1461
– roughening 72, 593
– step coefficient 168
– theory 1075
– trapping model 1556
kinetic model
– Bliznakow mechanism 78
– Cabrera–Vermilyea (CV) mechanism 79
kinetic modeling 1256
– of surface reaction 1257
kinetically limited growth 1136
kink 55, 562, 581, 1354
kinked face (K-face) 57
$KLiYF_5$ (KLYF) 738
KLuW 737
$KNbO_3$ (KN) 730
$KNd(PO_3)_4$ (KNP) 742
knife-edge 587
Knoop microhardness 1617
Knudsen cell (K-cell) 1156
Kossel crystal 55
KREW 732, 737, 746
KTA crystal 702
$KTi_{1-x}Sn_xOPO_4$ 749
$KTi_{1-x}Ge_xOPO_4$ 750
$KTi_{1-x}Ge_xOPO_4$ 751
$KTiOAs_xP_{1-x}O_4$ 749
$KTiOAsO_4$ (KTA) 692
$KTiOPO_4$ (KTP) 740, 746
$KTiOPO_4$ (KTP) 691, 692
$KTiOPO_4$ crystal 697
$KTiOPO_4$ (KTP) 730, 746
KTP 714, 739, 742, 751
KTP crystal 702
KTP crystal growth 694
KTP crystal structure 692
KTP hydrothermal growth 694
KTP isomorph 702, 710
KTP-type 691
Kubota and Mullin model 79
KYbW 737
KYF_4 (KYF) 738

L

$La_{0.67}Ca_{0.33}MnO_3$ 410
$La_3Ga_{5.5}Nb_{0.5}O_{14}$ (LGN)   1055
$La_3Ga_{5.5}Ta_{0.5}O_{14}$ (LGT)   1055
$La_3Ga_5SiO_{14}$ (LGS)   1055
$LaAlO_3$ 402
labile zone   562
laboratory instrument   1419
$LaGaO_3$ (LGO)   1049
L-alanine doped triglycine sulfo-phosphate (ATGSP)   579
Landau level spacing   351
Lang projection technique   1430
Lang technique   13
Langmuir isotherm   79
lanthanide   746
lapping   1724, 1725, 1731
large crystal   659
large-angle convergent-beam electron diffraction (LACBED)   1478, 1488
large-angle grain boundary   185
large-eddy simulation (LES)   1216
large-mismatch heteroepitaxial system   1500
large-mismatch interface   1502
L-arginine phosphate (LAP)   569
L-arginine phosphate monohydrate (LAP)   568
L-arginine tetrafluoroborate (LAFB)   569
laser   4, 161
– and nonlinear optical material   727
– beam scanning (LBS)   135
– beam scanning microscope (LBSM)   140
– beam tomography (LBT)   135
– conditioning   789
– crystal defect   487
– crystal growth   480
– damage threshold   787
– diffraction   1728
– diode (LD)   162, 879, 898, 1059, 1244
– emission microanalysis (LEM)   535
– gas breakdown   1198
– heated   393
– heated pedestal growth (LHPG)   174, 393, 395, 399, 486, 503
– heated pedestal growth method (LHPG)   480
– host fluoride   479
– induced damage (LID)   789
– induced damage threshold (LDT)   787
– ion source   1197
– lift-off process (LLO)   879
– material   727, 746
– plasma ion source   1200
– plasma range   1203
– scattering tomography (LST)   177, 181, 1453
lateral epitaxial overgrowth (LEO)   953, 1086
lateral incorporation of vacancies   1331
lateral incorporation of vacancy   1321
lateral overgrowth   1026
lattice
– constant   693
– distortion   493
– matched substrate   877
– mismatch   1110
– near-coincidence   1502
lattice parameter   344, 446, 1411, 1488
– InAsSb   332
– InBiSb   332
Laue pattern   340
Laue photograph   379
Lawrence Livermore National Laboratory (LLNL)   760, 775
layer-by-layer growth   167
LBO   743, 744
L-cystine   1624
lead tungstate   1670
lead zirconium titanate (PZT)   608, 634
ledge   562
Lely method   798, 799
– modified   800
Lely platelet   800
L-histidine tetrafluoroborate   573
L-histidine tetrafluoroborate (LHFB)   569
LHPG system   397
Li_2O 416
$Li(Nb,Ta)O_3$ 410
$LiAlO_2$ (LAO)   1060
LiB_3O_5 (LBO)   691
$LiBO_3$ (LBO)   730
Liesegang ring   1608
$LiGaO_2$ (LGO)   1060
light- and heavy-hole effective mass   350
light scattering   458
light-beam induced current (LBIC)   549
light-emitting diode (LED)   162, 328, 798, 898, 1059, 1244
– performance   802
$LiIO_3$ 731
limitation of chemical vapor transport   904
limitation of kinematical theory   1436
$LiNbO_3$ (LN)   162, 168, 192, 252, 401, 404, 406, 413, 415, 416, 708, 729
line defect   11
line direction   1445
lineage   187
liquid and solid phase   561
liquid encapsulated Czochralski (LEC)   163, 188, 206, 289, 1465
liquid inclusion   96
liquid phase   402, 1002
– diffusion (LPD)   979
– electroepitaxy (LPEE)   338, 967, 968, 1028
– electroepitaxy of semiconductors   967
– ELO   1027
– epitaxy (LPE)   9, 283, 328, 337, 725, 732, 734, 735, 746, 748, 751, 802, 946, 975, 1001, 1041, 1072, 1679
   requirement   1044
– epitaxy (LPE) of nitride   1059
– lateral overgrowth   1007
liquid-crystal display (LCD)   1723
liquid–solid interface   337
$LiTaO_3$ 413
lithium
– gallate ($LiGaO_2$)   877
– niobate ($LiNbO_3$)   273
– strontium aluminum fluoride (LiSAF)   728
lithium niobate
– near-stoichiometric   252, 255
lithium niobate ($LiNbO_3$)   246
– crystal   253
$LiYF_4$ (YLF)   738
load cell   249
local electronic properties of shaped silicon   549
local lattice distortion   270
local shaping technique (LST)   540
local vibrational mode (LVM)   222
locking stress   1342, 1354
Lomer–Cotrell mechanism   187
long-range stress   114
long-wavelength infrared (LWIR)   358, 1105

Lothe theorem 111
low pressure (LP) 1157
low to high complexity 1413
low-angle grain boundary (LAGB) 181, 185, 543, 1060
low-defect crystal 661
low-energy electron-beam irradiation (LEEBI) 882, 1060
lower yield stress 1348
low-level waste (LLW) 1649
low-temperature growth 560
low-temperature method 1166
low-temperature poling 710
low-temperature-grown GaN (LT-GaN) 876
low-thermal gradient 1374
low-viscosity melt 695
L-pyroglutamic acid crystal 573
L-tyrosine 1625
Lu_2SiO_5:Ce (LSO) 1681
LY 1666
Lyapunov equation 513
lysozyme 593

M

macroscopic motion of the fluid 563
macrosegregation 172
magnesium 1618
magnesium ammonium phosphate (MAP) 1615, 1617, 1622
magnetic circular dichroism (MCD) 1670
magnetic Czochralski (MCZ) silicon 235
magnetic field 85, 175, 194, 970, 1374
– effect 220
– interaction with the melt 209
– level 990
magnetic liquid encapsulated Czochralski growth (MLEC) 205
magnetic liquid encapsulated Kyropoulos growth (MLEK) 205, 208
magnetic liquid-encapsulated Czochralski growth (MLEC) 207
magnetite (Fe_3O_4) single crystal 441
magnetoresistive random-access memory (MRAM) 1509
majority-carrier reaction 1454, 1459
malformed form 144
malic acid 1621

Marangoni convection 370
Marangoni number (Ma) 1391
mask width 958
mask width-to-window width ratio 958
mask-induced strain 1017
mass thickness contrast 1478
master equation for equilibrium 57
material synthesis and purification 333
materials compatibility 825
Maxwell–Jeffries-formula 1386
MBE growth
– technique 1072
– theory 1073
mean escape depth 343
mean lattice site 55
mean separation work 22
mean size of crystals 454
mechanical characteristics 468
mechanical polishing 881
mechanical stability 404
mechanical stirring of the solution 745
melt 437, 480, 1219
– based compound 1335
– convection 1369
– density 251
– epitaxy (ME) 328, 339
– growth 9, 855
– replenishment (MR) 1390
– replenishment model (MRM) 1390
melt meniscus 515
– shaping condition 514
melting point (mp) 370, 393
melt–solid (M–S) 319
– equilibrium 436
– interface 452
– interface shapes on radial uniformity of ternary crystal 318
membrane protein 1594
meniscus instability 192
meniscus surface equation 514
meniscus wetting 273
merohedral twin 121
metabolic stone 1620
metal impurity 1728
metal ion complex 580
metal wire 394
metallization 1715
metalorganic chemical vapor deposition (MOCVD) 829, 899, 901, 1044, 1072, 1133, 1245, 1541, 1569

metalorganic MBE (MOMBE) 1072, 1141
metalorganic vapor-phase epitaxy (MOVPE) 10, 113, 283, 328, 869, 890, 901, 925, 1001, 1072, 1133
metal–oxide–semiconductor (MOS) 162, 1154
metal–oxide–semiconductor field-effect transistor (MOSFET) 940, 1165, 1541
metal-semiconductor field effect transistor (MESFET) 163
metamorphic rock 152
metaphosphate concentration 777
metastable condition 331
metastable phase boundary 330
metastable zone 561, 764
methyl-(2,4-dintropheny)-aminopropanoate (MAP) 569
methyltrichlorosilane (MTS) 949
metrology 1726
microanalysis 1489, 1509
micro-area x-ray fluorescence (MXRF) 135
microdefect 1283, 1286
microelectronics (ME) 162, 1721
microfaceting 376
microgravity 265, 335, 582
– condition 8
– diffusion-controlled crystallization 594
– environment crystallization 1588
– grown TGS crystal 589
– handheld protein crystallization apparatus 594
– protein crystallization apparatus (PCAM) 594
microinhomogeneity 174
Microphysics Laboratory (MPL) 1089, 1095, 1102, 1103
micropipe (MP) 98, 802, 806, 942
– density 805
micro-pulling-down method (μ-PD) 480
microsegregation 172
microstructure 461
microstructured material 1687
microtwin 1110
microtwinning 1079
microvoid 168
mid-wave IR (MWIR) 1105
mineral 133
minimum-energy theorem 110
minority-carrier

– lifetime 1083
– reaction 1456, 1460
– recombination lifetime 1107, 1108
miscut 1077
misfit dislocation 186, 1494
mismatch between the substrate and the film 751
mismatch heteroepitaxy 359
mismatched epitaxy 355
Mn diffusion 1509
mobility 1087
– ratio 348
model for dislocation generation 1339
model-based prediction 1380, 1389
modeling 434
modeling of AlN/GaN vapor deposition 1246
modern wiresaw 1729, 1730, 1732, 1735
modifications of TPS 540
modified non-stationary model (MNSM) 1386
modified quasi-steady-state model (MQSSM) 1386
modulation-doped field-effect transistor (MODFET) 1165
moiré fringe 1502
moiré interferometry 1727
mold-pushing melt-supplying (MPMS) 255
molecular dynamics (MD) 76
molecular-beam epitaxy (MBE) 10, 18, 283, 328, 605, 869, 890, 946, 1004, 1044, 1069, 1070, 1133, 1156, 1162, 1206, 1245, 1567, 1569
mollusk 1614
molten zone profile 405
monitoring 1090
monoatomic crystal 86
monochromatic beam 1434
monochromator 1645
monoclinic (m) 445, 574
– phase 764
monodomain crystal 257
monohalide (GaCl) 926
monolayer (ML) 1172
monolithic microwave integrated circuit (MMIC) 162, 216
monosilicic acid 1609
monosodium urate monohydrate (MSUM) 1615, 1623
Monte Carlo simulation 69, 73
Morey autoclave 617

morphodrome 70, 82
morphological evolution 136, 1062
morphological habit 701
morphological importance (MI) 64
morphological instability 136, 173, 175, 781
morphological shape 706
morphology 133, 136, 578, 848, 1459
– and faceting 268
– of growth spiral 140
– of pits 1462, 1465
– of sillenite crystal 268
mosaic block 1409
Moss–Burstein effect 353
mother phase 57
mounting of the substrate 1088
MOVPE of InAs quantum dot 1144
MOVPE precursor 1133
Mueller matrix 1091
multicarrier conduction 350
multicrystalline (MC) 162
– ingot growth 1709
multidomain
– crystal 698
multilayer model 1092
multiple quantum well (MQW) 1469
multiple reflection 353
multiple-beam interferometry (MBI) 135, 140
multiple-exposure holography 585
multiplication rate 1348
multiwire saw 1732

N

Na-NbTS
– cesium ion exchange 1656
– in situ synthesis 1655
nanocrystal 637
nanomaterial 916
nanostructure 463
nanotopography 1731
National Aeronautics and Space Administration (NASA) 583
National Ignition Facility (NIF) 761, 774
National Institute of Standards and Technology (NIST) 588
National Physical Laboratory (NPL) 301
National Renewable Energy Laboratory (NREL) 898
National Synchrotron Light Source (NSLS) 1431, 1638

native point defect concentration 165
natural and synthetic diamond 150
natural and synthetic quartz 142, 145
natural convection 563, 970
natural crystallization 135
natural diamond 151
Navier–Stokes (NS) 1397
Nb_2O_5 413
$Nd_3Ga_5O_{12}$ 443
Nd:YAG 399
$NdBa_2Cu_3O_{7-x}$ (NdBCO) 1046
Nd-doped congruent LN (Nd:CLN) 252
Nd-doped laser crystal 489
$NdGaO_3$ (NGO) 1049
near-band-edge (NBE) 838
near-coincidence
– lattice 1502
near-stoichiometric
– lithium niobate (nSLN) 252, 255
necking 382, 1359
– of seed crystal 256
needle defect 1082
negative ion 1564
neighboring confinement structure (NCS) 1175
neutron diffraction 380
– cell 1648
– theory 1640
newberyite 1622
Newton–Raphson 1361
niobate 729
nitric acid (HNO_3) 299
nitride 1059
nitrogen (N_2) 822
nitrogen precursor 1138
NLO single crystal 736
Nomarski image 218
noncentrosymmetric 271
noncongruent melt 176
nonconservative system 173
noncritical phase matching (NCPM) 692, 714
nondilute system 1537
nonflat interface 269
nonintrusive wafer inspection 1727
nonlinear coefficient 708, 713
nonlinear optical (NLO) 162, 691, 1625
– crystal 691
– material 726, 728
nonmerohedral twin 121
nonmetabolic stone 1620
nonpolar layer 1572

nonstationary model (NSM)   1384
nonstoichiometry   171, 189, 699
nonuniform composition   1080
normal growth rate   65
n-type doping of GaN   882
nuclear hyperfine interaction   1526
nuclear magnetic resonance (NMR)
   693, 1522, 1592
nucleation   45, 192, 339, 452, 565,
   735, 763, 808, 1101
– at surface   17
– calculation   908
– control   1610
– exclusion zone   35
– of intrinsic point defects   1304
– phenomenon   1586
– study   763
nucleus (N)   1462
numerical
– implementation   1360, 1361
– method   983
– model for vapor growth systems
   1247
– modeling of GaN IVPE growth
   1258
– modelling of CVD process   1246
– result   1362
– scheme   667
– simulation   179
– solution   1251

O

observation of dislocations   1500
observation of growth rate   970
O-cluster   1306, 1309
octacalcium phosphate   1618
off-centered Czochralski system
   258
one-dimensional
– crystal growth   1316
– initial incorporation   1288
– model   973
– nucleation   27
one-step ELO structure (1S-ELO)
   877
operational condition   1264
OPO interaction   714
opposite domain LN (ODLN)   258
optical
– absorptivity   1108
– afterheater   397
– anomaly of growth sectors   105
– breakdown of gases   1198
– ceramic   1687
– characterization   224

– cutoff   1112
– dielectric function   1091
– glass   465
– material   726
– nonuniformity   713
– parametric oscillation (OPO)   691
– phonon occupancy   354
– plasmatron   1198
– property   679
– pyrometer   249
– pyrometry   1089
– transmittance   1108
– uniformity   458
optical absorption   1112
– spectrum   784
optical floating zone (OFZ)   368
– application for oxides   368
– composition evolution   374
– crack   382
– furnace   371
– high pressure   371
– inclusion   382
– modelling   370
– overheating   370
– self-flux   375
– technique   368, 369
optically detected magnetic
   resonance (ODMR)   1546, 1669
optimization   1007, 1062, 1264
optimization of growth of GaN films
   – a materials example   1206
optimization of liquid-phase lateral
   overgrowth procedure   1007
optimization of plasma flux for film
   growth   1204
optimum growth parameter   905
optoelectronic devices and integrated
   circuit (OEIC)   968
optoelectronic integrated circuit
   (OEIC)   216, 1153
optoelectronics   1336
ordinary differential equation (ODE)
   1361
organic additive   778, 1622
organic light-emitting diode (OLED)
   865
organic semiconducting single crystal
   862
organic semiconductor
– Bridgman technique   856
– Czochralski technique   857
– gas phase growth   857
– single-crystal   845
organometallic crystal   1612
organometallic vapor-phase epitaxy
   (OMVPE)   901

orientation
– contrast   1440
– dependence   956
– determination   340
– flat (OF)   1463
– state   121
oriented film   338
origin of dislocation   1338
origin of screw dislocation   808
Orowan relation   180, 1347
orthodox etching   1461, 1468
orthophosphate   692
Ostwald ripening   189, 1143
Ostwald's step rule   86
overheating   194, 436
oxidation-induced stacking fault
   (OSF)   168, 1284
oxide   393, 479, 728
– crystal   433, 434
– glass   433
– photorefractive crystal   262
oxygen
– contamination   824
– redistribution   463
– stoichiometry   441
– vacancy   715

P

packaging   1725
pancreatic stone protein (PSP)   1629
parabolic band   349
partial differential equation (PDE)
   1397
partial dislocation   1344, 1497
partial pressure   1135
particle acceleration   1199
particle diagnostics   586
particle imaging   586
partly stabilized zirconium dioxide
   (PSZ)   444
passivation   1086
pathological biomineralization
   1614
pattern formation in gel systems
   1610
patterned domain   257
patterned substrate   952, 1086
$PbMoO_4$ 170
PbTe   168, 185
pearl   1614
PED technique   1194
Peierls
– barrier   1343
– energy   113
– potential   178, 1346

Peltier interface demarcation (PD) 291
Peltier-effect 1028
Peltier-induced growth kinetic 971
Pendellösung fringes (PF) 1444
pendeo-epitaxy (PE) 1031
– of GaN 1032
penetration twin 122
perfection 133, 152
– limit 164
– of crystals grown rapidly from solution 125
– of single crystal 152
periodic bond chain (PBC) 54, 64, 137
periodic domain 259
– structure (PDS) 708, 711, 713
periodic modulation 259
periodic poled LN (PPLN) 258
periodic poling lithium niobate (PPLN) 398
periodic solute feeding process 315
periodically poled KTP (PPKTP) 691, 708
periodically poled lithium niobate structure 258
peripheral ring 1322, 1327
peritectic decomposition 374
peritectic transformation 373
permeability 670
perturbation frequency 175
pH of solution 577, 579
phase
– composition 460
– contrast microscopy (PCM) 135, 140
– extent 165, 169
– modulator 273
– relation 1046
– stability 661
– transformation 448
– transition 125, 446
phase conjugated optical waveguide 273
phase diagram 330
– of the ZrO_2–Y_2O_3 system 445
phase equilibrium 663
– for binary compound 292
– for ternary compound 300
– of ZnO 662
phases with different composition 270
phase-shifting interferometry (PSI) 135, 140

phase-shifting microscopy (PSM) 135, 140
phlogopite 141
phosphate 730, 739
– flux 740
– solution 748
phosphide-based compound 296
phospholipid lecithin 1626
phosphorus glass removal 1715
photo-assisted MBE (PAMBE) 1072, 1107
photochromic property 271
photoconductive-decay lifetime 1082
photoconductivity 260
photoconductor (PC) 358, 1116
photodiode (PD) 162, 357
photo-EPR 1539
photoetching 1470
photogalvanic etching 1458, 1461, 1471
photoionization cross section 343
photoluminescence (PL) 15, 683, 835, 918, 1082, 1158, 1207
– mapping 1083
photorefractive (PR) 252
– crystal 264
– damage 258, 261
– gain 221
– oxide material 246
photovoltaic (PV) 898, 1703, 1722, 1723
– efficiency (PVE) 162
– module 1705
– value chain 1705
physical laws for transport processes 1217
physical property 346
physical vapor deposition (PVD) 135, 900
physical vapor transport (PVT) 135, 800, 821, 899, 900, 946, 953
physicochemical properties of the solution 733
piezoelectric 262, 264, 703
planar defect 11
planar doping 1107
plasma 1202
– acceleration 1197
– energetics 1193
– enhanced chemical vapor deposition (PECVD) 1715
– etching 881
– expansion 1197
– flux 1204
– formation 1198

– formation in PED 1198
– formation of vaporized material 1196
– processing 1193
– propagation in gas 1203
plastic
– deformation 156, 183
plastic relaxation 178, 180, 184
plastic state 116
plate shaped crystal 521
PLD technique 1194
ploughing 1730
plume range 1203
Pockels cell 760
point defect 11, 160, 161, 163, 1556, 1569
– characterization 1521
– concentration 164
– engineering 168
– generation 163
– identification 1560
– kinetics 167
point group symmetry 697
point seed 773
point-bottomed (P-type) etch pit 139
polar growth 1572
polar surface 72
polarity 1463, 1479, 1484, 1486
– of III–V material 1464
– of twinning 214, 215
polarizer 4
poling of crystal 252
poling of lithium niobate 257
polishing 1725
polyacrylamide 1629
polycrystal 192
polycrystalline SiC 804
polyelectrolyte 82
polyethylene oxide (PEO) 1609
polygonal or circular spiral 141
polyhedral crystal 138, 143
polyhedral seed 572
polyimide environmental cell (PEC) 1647
polymorphic transition 726
polypyrrole (ppy) 1597
polyscale crystal 599
– growth 4
polytype formation 805
polyvinyl alcohol (PVA) 1609
porous bed 670
– height 672
position-sensitive detector (PSD) 1643
positron

– annihilation spectroscopy 1551, 1552
– density at a vacancy 1555
– emission tomography (PET) 1682
– implantation 1552
– lifetime 1553
– lifetime spectroscopy 1557
– state 1553
– trapping 1556
– trapping rate 1556
– wavefunction 1553
postgrowth dislocation 107, 118
postgrowth movement of dislocations 116
postgrowth treatment 788
potassium amide (KNH_2) 664
potassium azide (KN_3) 664
potassium dihydrogen phosphate, KH_2PO_4 (KDP) 96, 560, 568, 569, 759
potassium double tungstate (KREW) 728
potassium iodide (KI) 664
potassium niobium tantalate (KTN) 398
potassium stoichiometry 716
potassium titanyl phosphate (KTP) 691
potassium vacancy 700
power rectifier 798
practical results of the theoretic analysis 519
Prandtl number 734
precipitate 160, 163, 189, 1339
precipitation 376
precursor
– decomposition 1135
– for SiC CVD epitaxial growth 946
– ligand 1134
– vapor pressure 1134
predicting the growth morphology 65
prediction 1254
pregrowth purification 850
primary agglomeration 1616
primary crystallization field (PCF) 1046
primary nucleation 763
prime wafer 1723
prismatic stacking fault (PSF) 1501, 1504, 1508
prismatic zone 577
probabilistic model of second-layer nucleation 42
process simulation 669

processed seed 572
profile of growth spirals 150
progesterone 1628
propagation
– of growth dislocation 110
– of twin boundary 123
propagation of defects 93
properties of CZ crystal 455
proportional–integral–derivative (PID) 251
proportional–integral–differential (PID) 248, 286
proportional–integral–differential (PID) controller 913
protein crystal growth 592
– AFM 1596
– electrochemistry 1596
– facility (PCF) 594
– in microgravity 593
– mechanism 593
– method 592
protein crystallization 1588
– high throughput 1593
protein immobilization 1597
proteomics 1584
pseudobinary phase diagram 300
pseudodielectric function 1091
pseudo-equilibrium theory 1074
pseudohexagonal twin 124
pseudosemicoherent interface 1501
PSZ crystal 459
– composition 459
– cooling rate 459
– phase transformation 459
Pt wire 400
p-type doping of GaN 882
pulling 407, 441
– on a seed 443
– rate and temperature gradient 238
– rate of crystal 247
pulsed electron beam source (PEBS) 1196
pulsed electron deposition (PED) 1193, 1194
pulsed laser deposition (PLD) 1193, 1194, 1206
pure and doped lithium niobate crystal 256
purification 913
purity 163
PV technology 1706
PVT crystal growth 822
pyroelectric 703
– effect 262
pyrolytic boron nitride (pBN) 291, 296

Q

Q-switch 273
qualify control 1726
quality variation 1420
quantification of the microdefect formation 1290
quantitative estimation of dopant concentration 1059
quantum dielectric theory (QDT) 355
quantum dot (QD) 1142
– arsenide 1133
– buried 1417
– growth interruption 1145
– InAs/GaAs 1142
– laser 1147
– multimodal size distribution 1146
– ripening 1146
– strain energy 1143
– structural property 1147
– subensemble 1146
– surface 1417
quantum efficiency (QE) 1116
quantum well (QW) 1160, 1173
quantum wire 1418
quartz 149, 620
quasi-equilibrium deposition model 1270
quasi-equilibrium model 1270
quasi-phase-matched (QPM) 691
quasi-phase-matching (QPM) 708
quaternary phase diagram 301
quenching 336

R

R_2O_3–Al_2O_3–SiO_2 system 467
radial morphology 263
radial temperature gradient 1364
radiation 1368
– detector 4
– model selection criterium 1249
radiative recombination 1107, 1108
radioactive waste 468
radiofrequency (RF) 248, 251, 367, 435, 831, 1044
– generator 248
Raman 693
– peak 355
– spectrum 783
ramping mode 256
random alloy scattering 350
rapid analysis 1419
rapid growth 773
– method 73

– of ADP   126
– of KDP   126
rapid-thermal annealing (RTA)   1567
rapid-thermal chemical vapor deposition (RTCVD)   1162
rare earth (RE)   259, 262, 368, 1681
– lithium fluoride (RELF)   728
– vanadate (REVO$_4$)   633, 728
rate equations approach   36
rate of nucleation   28, 32
– on single-crystal surface   30
rate-dependent   1342
ray-tracing simulation   807
Rb$_x$K$_{1-x}$TiOPO$_4$ 751
RbTiOAsO$_4$ (RTA)   692
RbTiOPO$_4$ (RTP)   740
RbTiOPO$_4$ (RTP)   691, 692
RbTiOPO$_4$ crystal   700
RbTiOPO$_4$ (RTP)   741
RCA passivation of Si surface   1101
reaction free energy   1263
reaction in growing CZ crystals   1314
reaction involving no aggregation   1304
reaction of dislocations   116
reactive ion etching (RIE)   881
reactor geometry   872
readout integrated circuit (ROIC)   1118
real-time inspection   1728
reciprocal space   1407
– map (RSM)   833, 1082, 1411
reciprocating motion   572
recombination   1082, 1107
– enhanced dislocation glide (REDG)   810
– of electrons and holes on dislocations   1470, 1471
recording geometry   1435
– back reflection   1435
– transmission   1435
recovery process   150
reduced pressure (RP)   1157
reference electrode (RE)   1454, 1455
reflaxicon   398
reflection high-energy electron diffraction (RHEED)   1072, 1083, 1089, 1099, 1100, 1102
reflection topograph   342
reflectometry   1413
refraction of dislocation lines   110
refractive index (RI)   587, 698, 713, 714, 1715

refractory   434
– material   433
– melt   440
regeneration   768, 774
relation of Dupré   20
relative growth rate   105
relaxed SiGe layer   1165
replenishment model (RM)   1397
reported model   1502
residual impurity   883
residual strain   356
residual stress   1728
– measurement   1728
resistance   358
resistivity   711
responsivity   591
retardance   1091
retrofitting the MBE chamber   1092
reverse current (RC)   408
reverse diffusion   1628
Reynolds number   564, 734
RF (radiofrequency)   1709
RGS (ribbon growth on substrate)   1712
ribbon-to-ribbon (RTR)   522
RMS (root-mean-square)   1167
robotics   1593
rocking curve   341, 379, 1058
– measurement   1058
rod preparation for OFZT   372
rolling–indenting
– model   1730
– process model   1734
room temperature (RT)   350, 1664
– photoluminescence (RTPL)   1207
rotating analyzer ellipsometer (RAE)   1091, 1092
rotating compensator ellipsometer (RCE)   1091, 1092
rotating disc technique   574
rotating magnetic field   191
rotating thermal field   745
rotation   1233
– of crucible   1370
– of crystal   1370
– Reynolds numbers   1369
rotatory Bridgman method (RBM)   328, 336
rough interface   136
roughening transition   72
round cylindrical crystal   519
RTA crystal   702
RTP   742
– crystal   702
ruby   630
Rutherford scattering   1490

S

salol   102
sapphire (Al$_2$O$_3$)   630, 656, 875, 899
– fiber   396
saturation
– nucleus density   35
– temperature   575, 576
scaling analysis   1258
– of a CVD reactor   1258
scaling exponent in diffusion   37
scanning Auger microprobe (SAM)   829
scanning electron microscopy (SEM)   15, 331, 340, 346, 380, 929, 1006, 1079, 1162
scanning force microscopy (SFM)   259
scanning photoconductivity   225
scanning photocurrent (sPC)   224
scanning transmission electron microscopy (STEM)   1163, 1489
scanning tunneling microscopy (STM)   135, 140, 1163
scattering   491, 1412
– amplitude   1407
Scheil equation   173
Scherzer point resolution   1483
schlieren system   587
Schmidt contour   186
Schottky defect   166
Schwuttke technique   1429
Scientific Production Company (SPC)   1574
scintillation material   1663, 1670
scintillation parameter   1665
scintillator
– aluminum perovskite   1673
– device   264
screw dislocation (SD)   6, 15, 178, 180, 219, 806, 942, 1352, 1466
– mechanism   769
– theory   6
screw-oriented BPD   813
SE data analysis   1092
SECeRTS cell   1646
secondary agglomeration   1616
secondary ion mass spectrometry (SIMS)   1564
secondary nucleation   765
secondary Peierls potential   1354
secondary-ion mass spectrometry (SIMS)   259, 681, 682, 785, 787, 829, 1059, 1083, 1157, 1415, 1522, 1572

second-harmonic generation (SHG) 691, 697, 729, 775
second-layer nucleation in homoepitaxy 38
second-phase particle 189
sector boundary 786
Seebeck 408
seed 146, 152, 382, 441
– crystal 573, 804
– generation for new material 289
– length 251
– orientation 570
– rod 369
– rotation mechanism 571
– sublimation 802
– temperature 800
seeded
– growth 822
– sublimation growth 800
seeding 400
– process 263, 290
segregation 172, 270
– coefficient 171, 172, 337, 375
selected area diffraction (SAD) 829, 1478, 1484
– pattern 345
selective epitaxial growth (SEG) 953, 1181
selective-area growth 1086
self-assembly of islands 1074
self-flux 695, 750
self-seeded growth 822, 827
self-separation 880
self-trapped exciton (STE) 1671
Sellmeier equations 715
semiconductor 967, 1459, 1551
– alloy 1485
– grade 232
– grade silicon 802
– single crystal 897
– single–crystalline wafer 1723
– structure 1026
– substrate 661
– wet etching 1459
– zincblende structure 1485
sense of screw dislocation 807
separation work 55
sequential etching 1468
sex hormone 1628
shaded area 209
shallow positron
– state at negative ions 1555
– trap 1564
Shanghai Institute of Optics and Fine Mechanics (SIOM) 482

shaped
– crystal 517, 519
– crystal by FZT 522
– crystal growth (SCG) 509
– silicon structure 546
shear modulus 806
shear stress 178, 184
shield gas flow rate 1266
Shockley
– fault 813
– partial 178, 1497
– partial dislocation 809
Shockley–Read–Hall (SRH) 1085, 1107
– recombination 1108
shoulder facet 214
shoulder zone 266
shuffle dislocation 1344
Si substrate 1076
Si/Ge
– growth technology 1156
– heterostructure 1154
Si_3N_4 powder 802
Si-based compliant substrate 1086
Si-based substrate 1084
SiC 141
– Acheson method 153
– based device 798
– based device technology 816
– boule 805, 816
– bulk growth 802
– epitaxial growth technique 946
– growth system 803
– homoepitaxy 945
– Lely method 153
– polytype 805
– seed 829
– selective growth 956, 959
Si-face epilayer 811
SiGe heterostructure 1153
SiGeC structure 1415
SiGe-on-insulator (SGOI) 1169
signal-to-noise ratio (SNR) 1118
silane (SiH_4) 800, 882
silica gel 1608
silica nozzle angle 1268
silicon (Si) 166, 231, 358
– multicrystal 239
– on cloth (SOC) 551
– on insulator (SOI) 944, 1086, 1169
– photovoltaics 1703, 1706
– ribbon growth 1711
– solar cell 1703
– strip detector 1644
– wafer 1722

silicon carbide (SiC) 656, 797, 876, 939, 1734
– growth 944
– polytype 941
– substrate 944
silicon crystal 232
– semiconductor grade 232
sillenite 266
simulation 1412
– model 789, 982
– of LID 789
single crystal 4, 152, 334, 769, 923
– by OFZ 383
– fiber (SCF) 393
– growth 1707
– growth process 297
– organic semiconducting 862
– perfection 152
– ternary seed generation process 308
single diffusion 1612
single slip 1346
single-domain
– crystal 698
– growth 707
single-ended pinning point 813
Sirtl-type etchant 1460
site stability and adsorption 63
sitinakite topology 1649
sitting drop method 593
sixfold defect distribution 1341
sixfold symmetry 1372
size of pits 1467
skull melting (SM) 433, 434
slice energy 56, 65
slicing 1725, 1728
– technology 1730
slip
– band 834
– dislocation 1359
– system 341, 1340
slow cooling (SC) 9
– bottom growth (SCBG) 9
small environmental cell for real-time studies (SECeRTS) 1646
smooth interface 136
SnTe 168
sodium
– azide (NaN_3) 664
– metasilicate 1609
– niobium titanosilicate (Nb-TS) 1655
– nitrate ($NaNO_3$) 1219
– nonatitanate (SNT) 1651
– titanium oxide silicate (STOS) 1651

– titanium silicate (Na-TS) 1649
– urate monohydrate 1624
software 1644
solar cell 240, 1703
– application 239
– binning 1716
– fabrication 1713
– performance 1716
– testing 1716
solid solution 330
solid–liquid interface (SLI) 136, 253, 265–267, 735, 762
solid-state
– electrochemistry 1597
– laser (SSL) 726, 727
solidus 331
solubility 567, 610, 613, 661, 696, 1585
– curve 561, 1046, 1047
– determination 568
– gradient 567
– of ammonothermal GaN 663
– of berlinite 613
– of gallium orthophosphate 614
– of hydrothermal ZnO 662
– of quartz 613
– parameter 568
– plot 1585
– reduction 1610, 1613
soluble system
– reverse-grade 672
solute boundary layer 172
solute feeding
– crucible oscillation 314
– homogeneous alloy growth 311
– in vertical Bridgman method 313
– using double-crucible 312
solute precipitate 99
solute trail 99
solution
– circulating method 771
– flow 154
– grown KTP 704
– inclusion 782
– preparation 573
– stirring (SS) 1591
– temperature 737
solution growth 1042
– chemical/gel method 567
– crystallizer 570
– high temperature 696
– high-temperature 731
– low temperature 566
– method 566
– of triglycine sulfate 582
– slow cooling method 566

– slow evaporation 566
– temperature gradient method 567
– temperature lowering 567
– top-seeded 696, 732
solvent 568
– adsorption on crystal surface 77
– based growth method 862
– effect on the crystal habit 75
– inclusion 96
– selection 567
– system 1046, 1047, 1055
solvent–solute interaction 77
Soret effect 1388
source preparation 400
source-current-controlled (SCC) 975
space group 332
space grown TGS crystal 591
space-charge electric field 254
space-charge grating 221
Spacelab-3 (SL-3) 583
spatial instability 437
spatially resolved x-ray diffraction (SRXRD) 1019
spectroscopic ellipsometry (SE) 1072, 1089, 1090
– calibration 1093
– composition determination 1094
– temperature determination 1094, 1098
spectroscopic library 398
sphalerite structure 1487
spherulite 1621
spinning disc growth 576
spin–orbit splitting 347
spintronics 4
spiral growth 136, 1596
– linear and parabolic growth rate 74
spiral morphology 141
splitting energy 356
spontaneous nucleation 770
spontaneous polarization 704, 709
spontaneous self-separation 881
spreading resistance (SR) 549
spring plate 1089
sputtering (SP) 1206
Sr_2RuO_4 414, 416
$SrMoO_4$ 400
stability
– criterion for nuclei 847
– of a crystal site 61
– of crystallization 512
– of solution 770
stability analysis (SA) 509, 516, 529

– and crystal growth 519
stable growth 373
stacking fault (SF) 178, 810, 814, 885, 944, 1466, 1479, 1489, 1497
– and partial dislocations 1497
– energy 193
stacking mismatch boundary (SMB) 1501, 1506
stacking sequence rule 813
staining of crystal 102
standard testing conditions (STC) 1716
start melting 435
starting compound 914
starting material 913
state at negative ions
– shallow positron 1555
static stability 515
stationary magnetic field 978
stationary state 1349
stationary temperature profile (STP) 900, 918
steady magnetic field 1229
STEM 1509
– imaging 1490
step flow 813
– growth 1074, 1075, 1110
step separation 140
step source
– longitudinal 120
– transverse 120
stepped 562
– face (S-face) 57
steroid 1610, 1626
– hormone 1628
sticking probability 1273
stimulated Raman scattering (SRS) 728
Stockbarger method 9
stoichiometric crystal 254
stoichiometric LN (SLN) 252
stoichiometry 170, 188, 409, 691, 697
– potassium 691
strain 785, 787, 1488
– in ELO layers 1016
– measurement 1489
– reducing layer (SRL) 1147
– relaxation 1163
strained layer 1498
Stranski–Kaischew criterion of the mean separation work 60
Stranski–Krastanov
– growth 20, 1104
– growth mode 1073, 1074
Stranski–Krastanow

– 2-D–3-D transition 1142
strength property 463
striation 101, 174, 383, 489, 711
strontium
– barium niobate (SBN) 396
– carbonate 1611
– tartrate 1610
structural
– characterization 1057
– coherence 345
– defect 848, 1110
– dynamics 341
– form 137
– instability 70
structural perfection 676, 1014
– of coalescence front in fully overgrown ELO structures 1014
structural property 340, 1139
– HgCdTe 1110
– $InAs_xSb_{1-x}$ 332
structure of small clusters 32
structure perfection 664
structure–properties relationship 713
struvite 1622
subgrain 1440
sublattice 1562
sublimation 852
– growth 802, 1274
– sandwich method (SSM) 800
submerged heater method (SHM) 320
substrate 875, 1083, 1087
– dislocation in ELO 1011
– epitaxial relationship 1049
– for epitaxy of nitride 1061
– for homoepitaxial LGT LPE film growth 1056
– impurity 1080
– manufacturing with system-oriented approach 1725
– material 1076
– orientation 1077
– preparation 1101, 1102
– removal technique 879
– rotation 1269
– rotation effect 1269
– roughness 1081
– surface curvature 1082
– temperature 1264
substrate–nozzle distance 1269
succinonitrile (SCN) 1223
sulfate-containing flux 741
supercooling 4, 192, 194, 338
supercritical fluid technology (SCF) 619

superheated aqueous solution 8
superheating 185, 186
superlattice 1410
supersaturated solution 763
supersaturation 5, 6, 21, 73, 339, 561, 578, 735, 779, 804, 908, 1273, 1586
– and driving force for LPE 1048
– ratio 57, 561, 908
– relative 561
surface
– acoustic wave (SAW) 657, 1059
– anisotropy 1091
– attachment 735
– characterization 1100
– damage 1339
– damage removal 1714
– energy 5, 137, 194, 405
– energy theory 5
– microtopography 134, 139
– modification 1193
– morphology 339, 461
– nonstoichiometry and contamination 1082
– reaction 1255–1257, 1272
– reaction analysis 1259, 1262, 1263
– reconstruction 69
– relief of the growth face 114
– roughness 67, 1093, 1099, 1725
– segregation 1157
– site 55
– structure 72
– tension 58, 733
surface defect 1110
– crater 1111
– cross-hatching 1112
– flake 1111, 1112
– hillock 1112
– microvoid 1111
– needle defect 1112
– roughness 1112
– triangle defect 1111
– void 1111
surface-mount technology (SMT) 1705
surfactant 45
– efficiency 46
– mediated growth (SMG) 1160
SWBXT back-reflection image 806
symmetry 848, 1373, 1462
– of growth spiral 141
synchrotron radiation 13
– source 1642
– topographic study 341
synchrotron topography (ST) 181

synchrotron white beam x-ray topography (SWBXT) 209, 677, 806, 827, 1359
synchrotron x-radiation diffraction 588
synchrotron x-ray topography (SXRT) 1019, 1021
synovial fluid 1623
synthesis
– of Na-TS and Na-NbTS 1658
synthesis problem 1650

T

TaC mask 954
tailor-made additive 70, 83
tangled dislocation 181, 186
tantalum carbide (TaC) 826
tartaric acid 1618, 1622
Tatarchenko steady-state model (TSSM) 1389
Te monolayer 1087
Te precipitate 1080, 1082
technique
– surface characterization 1595
technique of pulling from shaper (TPS) 509
TEM
– application 1493
– CL observation of prismatic stacking faults 1508
– EDS (energy-dispersive x-ray spectroscopy) 1486
– study of dislocations 1467
temperature 1362
– control technique 248
– dependent Hall measurement (TDH) 1541
– distribution 438, 517
– effect on adsorption 84
– effect on crystal habit 84
– field 1363
– gradient technique (TGT) 403, 404, 439, 480, 482, 501
– gradient zone melting (TGZM) 335
– oscillation 712
– pattern 670
– ramping 1089
– reduction method (TRM) 771, 773
template 875
ternary phase diagram 300
ternary substrate 283
terrace 563
– of growing faces 581

tertiarybutylhydrazine
   $(C_4H_9)(H)NNH_2$ 1138
testosterone   1628
tetragonal (t)   445
– DKDP crystal   772
– phase (TZP)   444
tetragonal–monoclinic phase
   transition   772
tetramethoxysilane (TMOS)   1609
texturing   1714
theorem of Herring   96
theoretical calculations of hyperfine   1528
theories of urinary stone formation   1616
theory of preferred direction   110
thermal
– annealing   788
– condition   440
– conductivity   733, 1078
– cycling   1085
– diffusion model (TDM)   1387
– expansion coefficient   467
– fluctuation in slicing   1735
– gradient   396
– lensing   711
– mismatch   1110
– property of InSb   356
– shock   1364, 1373
– stimulated conductivity (TSC)   1669
– stimulated luminescence (TSL)   1668
– strain   1350
– strain in ELO layers   1024
– stress   499, 813, 1339
– treatment   699
thermocouple   248, 1089
thermocouple calibration   1103
thermodynamic   163, 869
– analysis of gas-phase   1254
– approach   61
– consideration   847, 905
– factor   1466
– kinetic analysis   1254
– potential   164
– prediction   1255
– prediction of surface reactions   1255
– property   1265
– supersaturation   57
thermoelastic stress   179
thermoelectric effect   969
thermophysical   583
– characteristics   468
thermoplastic relaxation   177

thickness determination   1486
thin film   1534
– deposition   1193
thiophene (C_4H_4S) 854
third harmonic-generation (THG)   775
Thomson–Gibbs
– equation   21, 22
– formula   58
Thomson–Gibbs–Wulff equation (TGW)   59
threading dislocation (TD)   875, 899, 1060, 1110
threading edge dislocation (TED)   811
three-dimensional (3-D)   18, 20, 232, 1172, 1222
– characterization   1598
– generalization   1352
– simulation   981
threefold symmetry   1372
three-phase boundary (TPB)   193, 210, 214
three-vessel solution circulating method (TVM)   772
$Ti:Al_2O_3$ 406
Ti:sapphire   487
Tiller criterion   176
tilt variation   1420
time of flight   1201
time varying temperature profile (TVTP)   918
time-resolved experiment   1642
time-varying temperature profile (TVTP)   900
TiO_2 414
titanium oxide   703
titanium silicate   1649
Tm-doped epitacial layer   748
Tm-doped KLuW   747
Tokyo Denpa (TD)   1574
top-seeded solution growth (TSSG)   9, 487, 691, 725, 732, 734, 1042
total gas flow rate   1264
total number of spins   1533
total thickness variation (TTV)   1727, 1729
TPS (technique of pulling from shaper)
– brief history   537
– capillary shaping   522
– definition   537
– metal growth   551
– peculiarity   552
– sapphire growth   539
– silicon growth   546, 551

trace element   1617, 1618
tracht   144
transducer   4
transformation
– hardening   444
– of atomic energy in PLD and PED   1195
– twin   125
transition
– metal   252
– stress   1353
transmission   352
– electron microscopy (TEM)   15, 189, 340, 380, 462, 829, 1006, 1079, 1162, 1426, 1453, 1477
– spectrum   354, 783
– topograph   218, 810, 813
transmitted wavefront (TWF)   786
transparent spectrum   783
transport
– agent   904
– equation   1248
– growth model   659
– kinetics   901
– limited growth   1136
– model   666, 901
– phenomena   1587
– property   347
– rate   906
transverse magnetic field   231
– applied Czochralski (TMCZ) method   235
transverse magnetoresistance   346
transverse optic (TO)   1158
trapiche ruby   151
trapping   1557
– center   1541
– coefficient   1556, 1567
traveling heater method (THM)   174, 315, 335
traveling solvent floating zone (TSFZ)   367, 368
– self-flux   373, 374
traveling solvent zone (TSZ)   9, 373
triamterene   1625
triangular inclusion   943
tribotechnical property   464
tricalcium phosphate (TCP) whitlockite   1619
trichloroethylene (TCE)   927
triglycine sulfate
– crystal growth   574
– single crystal   574
triglycine sulfate
   $(NH_2CH_2COOH)_3H_2SO_4$ (TGS)   569, 574

TSSG (top-seeded solution growth) 708
tube shaped crystal 520
tungstate
– flux 740, 741, 750
– melt 741
– solution 748, 750
tunneling current 1112
twin 160, 544, 1440, 1485
twin boundary 121
– growth-promoting effect 124
– propagation 123
twin formation
– after growth 125
– by inclusions 123
– by nucleation 122
– during growth 122
twin law 120
twinning 120, 193, 194, 334, 341
– dislocation 121
twin-plane reentrant-edge effect (TPRE) 124
two shaping elements technique (TSET) 551
two-dimensional (2-D) 18
– characterization 1595
– crystal growth 1310, 1318
– epitaxial adsorption layer 71
– epitaxy 82
– equilibrium shape 61
– mechanism 74
– nucleation 167, 767
– nucleation growth (2DNG) 136
– nucleus 767
type I diamond 156
type II diamond 156
types of gels 1608
types of in situ cells 1645
TZM vessel 618

U

ultrahigh pressure high temperature (UHPHT) 155
– metamorphic rock 155
ultrahigh vacuum (UHV) 398, 1156
ultralarge-scale integrated circuit (ULSI) 231
ultraviolet (UV) 162, 728, 888
undercooling 712
undoped crystal of $CuAlS_2$ 918
undoped crystal of $CuAlSe_2$ 919
UNi_2Al_3 398
universal compliant (UC) 944
upper yield stress 1348

Urbach tail energy 1109
uric acid 1623
urinary stone disease 1616
urinary stone formation 1616, 1617
– theory 1616
use of a defective seed 1339

V

V/III ratio 1267
vacancy 160, 167, 694
– concentration 1567
– condensation 190
– in Si 1560
– in ZnO 1562
– potassium 699
– related complex 1575
vacancy defect 1554
vacancy–donor complexes 1566
valence band (VB) 1457, 1676
valved cell 1090
van der Pauw (vdP) 358
van der Waals force 6
vanadate 739
vapor composition 805
vapor condensation 798
vapor diffusion apparatus (VDA) 1589
vapor growth 799
– classification 899
– of III nitride 1244
vapor phase (VP) 897, 898, 1573
– epitaxy (VPE) 11, 901, 925, 926, 954, 1001, 1041, 1046, 1206
– growth 10
vapor pressure 5
– controlled Czochralski (VCz) 169, 170, 188, 190
vapor–liquid–solid (VLS) 138, 193
– mechanism 138, 146
Vegard's law 332
Verdet constant 468
Verneuil
– method 9
– technique (VT) 509
vertical and horizontal Bridgman method 334
vertical Bridgman (VB) 169, 1216, 1227
– crystal 286
– technique (VBT) 480, 484, 502
vertical gradient freeze (VGF) 169, 188, 288
vertical magnetic field applied Czochralski method (VMCZ) 235

vertical-cavity surface-emitting laser (VCSEL) 1414
very large-scale integrated circuit (VLSI) 939, 1153
vibration 1237
– of wire 1734
vicinal
– facet 103
– plane 703
– pyramid 103, 114
– sector 103
– sectorality 786
Vicker's microhardness 1617
virtual interface approximation 1092
virtual-crystal approximation (VCA) 350
viscoplastic model 1342
viscosity 568, 733
– melt 695
void-assisted separation (VAS) 881
voids 676
volatility 567
volatilization 333
Volmer–Weber growth 20
– mode 1073
volume and surface diffusion 76
volume defect 12
von Mises contour 180
von Mises invariant 179
VPE system 926

W

wafer
– annealing 190
– bowing 880
– characteristics 1734
– forming 1724, 1725
– micromapping 1420
– polishing 1724
– preparing 1724
– slicing 1722, 1732
wafer manufacturing 1721, 1722
– and slicing using wiresaw 1721
warp 1727, 1731
warpage 1728
waste recycling 469
waviness 1731, 1735
weak beam dark field (WBDF) 1479
welded closure 618
wet-etching of semiconductor 1459
– mechanism 1454
wetting 59
– angle hysteresis 525

– condition   523
– function   24
– to-catching condition transition   523
whisker   146
white-beam x-ray topography   1431
wide-bandgap semiconductor   821
Wilson plot   1589
window retardation   1092
window width (W)   958
wire web   1730
wiresaw   1721, 1722
– operation   1728
– process parameter   1733
work for nucleus formation
– general definition   24
work hardening   1347, 1357
work of formation of 2-D crystalline nuclei on unlike and like substrate   27
working electrode (WE)   1454, 1455
Wulff
– plot   137
– theorem   96
Wulff–Kaischew theorem   23
wurtzite   832
– structure   1487

X

xanthine   1615
x-ray
– and diffraction theory   1638
– anomalous scattering   215
– crystallography   1599
– diffractometry   1405, 1406
– method   12, 1637
– photoelectron spectroscopy (XPS)   342, 1083, 1087, 1158
– powder diffraction   379
– powder diffraction pattern (XRPD)   1639

– refractive index   1408
x-ray diffraction (XRD)   12, 341, 829, 885, 889, 1017, 1057, 1084, 1110, 1405, 1436, 1598
– analysis   1407
– dynamical theory   1436
– kinematical theory   1436
– limitation   1436
– pattern
  instrument   1412
x-ray scattering
– established method   1413
– new method   1416
– rapid analysis   1419
x-ray topograph   766, 810, 1440, 1445
– analysis of defects   1445
– image   781
x-ray topography (XRT)   13, 14, 135, 805
– basic principle   1426
– Berg–Barrett topography   1430
– conventional   1430
– Lang   1430
– monochromatic-beam   1434
– synchrotron-radiation-based   1431
– technique   1430, 1431
– theoretical background   1435
– white-beam   1431

Y

$Y_3Al_5O_{12}$ (YAG)   162
(Y_2)-$Lu_2Si_2O_7$:Ce (YPS and LPS)   1681
Y_2SiO_5:Ce (YSO)   1681
YAG   414
$YBa_2Cu_3O_{7-x}$ (YBCO)   1044
Y-Ba-Cu-O   415
Yb-doped KLuW   747
Yb-doped laser crystal   492
yellow luminescence (YL)   887
Young's relation   59

yttrium aluminum garnet (YAG)   728
yttrium aluminum perovskite (YAP)   728
yttrium iron garnet (YIG)   1042

Z

Z-contrast STEM   1509
Zeeman effect   1524
Zeldovich factor   29
zeolite   8
zero-force theorem   110
zero-loss peak (ZLP)   1491
zinc   1617
– oxide (ZnO)   655, 1446
zirconia   433
zirconia crystal   448
– cubic   448
– cubic lattice   447
– ion radius   447
– mechanism of stabilization   448
– stabilization   447
– tetragonal   448
zirconium dioxide   445
ZM configuration   1219
$ZnGeP_2$ 731
ZnO
– bulk growth   1573
– electrical properties   678
ZnS_xSe_{1-x} single crystal   908
ZnS_xSe_{1-x} system   908
ZnSe single crystal   914
ZnTe buffer   1102
ZnTe growth nucleation   1101
ZnTe on Si   1101
zoisite   633
zonal inclusion   97
zone
– melting (ZM)   173, 334, 1216
– refinement   851
– refining   333
ZrO_2 417
ZrO_2–Y_2O_3 phase diagram   446